Organizational Innovations and Economic Growth

To my parents
and
to Xenophon

Organizational Innovations and Economic Growth

Organosis and Growth of Firms, Sectors and Countries

Elias Sanidas

University of Wollongong, Australia

Edward Elgar

Cheltenham, UK • Northampton, MA, USA

Published by
Edward Elgar Publishing Limited
Glensanda House
Montpellier Parade
Cheltenham
Glos GL50 1UA
UK

Edward Elgar Publishing, Inc.
136 West Street
Suite 202
Northampton
Massachusetts 01060
USA

A catalogue record for ths book
is available from the British Library

Library of Congress Cataloguing in Publication Data

Sanidas, Elias.
Organizational innovations and economic growth: organosis and growth of firms, sectors and countries/Elias Sanidas.
p. cm.
Includes bibliographical references and index.
1. Technological innovations—Economic aspects. 2. Technology transfer—Economic aspects. 3. Organizational change. 4. Corporations—Growth. 5. Economic development. 6. Technological innovations—Economic aspects—United States. 7. Technological innovations—Economic aspects—Japan. I. Title.
HD45.S273 2005
338'.064—dc22

2004057462

ISBN 1 84376 721 X

Typeset by Manton Typesetters, Louth, Lincolnshire, UK.
Printed and bound in Great Britain by MPG Books Ltd, Bodmin, Cornwall.

Contents

Figures

Tables

Acknowledgements

My special appreciation to Professors Paul Robertson, Franco Malerba, Leif Hommen, and Charles Harvie for their helpful comments on my initial manuscripts.

I would also like to thank the following who have kindly given permission for the use of copyright material:

Elsevier Ltd: for two figures and abstracts reprinted from Elias Sanidas (2004), 'Technology, technical and organizational innovations, economic and societal growth', *Technology in Society*, **26** (1), 67–84. Copyright 2004, with permission from Elsevier.

Serial Publications, New Delhi, for material from: Elias Sanidas (2004), 'Impact of the lean production system on economic growth: evidence from US manufacturing industries', *International Journal of Applied Business and Economic Research*, **2** (1), 21–45.

Every effort has been made to trace all the copyright holders but if any have been inadvertently overlooked the publishers will be pleased to make the necessary arrangements at the first opportunity.

Abbreviations

ADF	augmented Dickey-Fuller
AIC	Akaike information criterion
ASI	American semiconductor industry
ASQ	American Society for Quality
CIs	Contract innovations
Dell CC	Dell Computers Corporation
DIs	Decision innovations
DVD	digital versatile disc
ECM	error correction model
FDI	foreign direct investment
FFS	focal factory system
FIML	full information maximum likelihood
FMS	flexible manufacturing system
GDP	gross domestic product
ICs	integrated circuits
IMPV	International Motor Vehicle Programme
IRY	Inventories to shipments ratio
ISIC	International Standard Industrial Classification
IT	Information technology
JIC	just-in-case
JIT	just-in-time
JIT/QC	just-in-time/quality control
JNR	Japanese National Railways
JSI	Japanese semi-conductor industry
JUSE	Japanese Union of Scientists and Engineers
KCs	kinetic costs
LCD	liquid crystal display
LEs	Large enterprises
LP	Linear programming
LPS (or LPs)	lean production system
LSI	large-scale ICs
NBER	National Bureau of Economic Research
OE	organizational entity
OIs	organizational innovations

OLS	ordinary least squares
PC	personal computer
POC	process of contracts
POM	process of movements
POS	process of strategies
POW	process of wisdom
PPF	production possibility frontier
PPP	purchasing power parity
PROSIBB	process(es) of the system of inputs of the black box
PS	Production system
QC	quality control
R-A	resource-advantage
R&D	research and development
ROI	Return on investment
SBC	Schwartz Bayesian criterion
SCs	strategic costs
SEE	standard error estimate
SI	system of innovations
SIC	Standard Industrial Classification
SIS	sectoral innovation system
SSIP	sectoral system of innovation and production
SM	scientific management
SMEs	small and medium enterprises
SS	sectoral system
SURE	seemingly unrelated regressions
TCs	transaction costs
TCT	transaction costs theory
TFP	total factor productivity
TFS	Taylor's cum Ford's system
TI	technical innovation
TIOP	tangible inputs of production
TNCs	transnational corporations
TQC	total quality control
TQM	total quality management
TR	technological regime
TVAL	Toyota Verification of Assembly Line
TVE	township and village enterprise
UNCTC	United Nations Centre on Transnational Corporations
UNIDO	United Nations Industrial Development Organization
VAR	vector autoregressive
VECM	vector error correction model
VI	Vertical integration

WCs	wisdom costs
WIs	Wisdom innovations
WIP	work-in-process
2SLS	two-stage least squares

1. Introduction

1.1 BACKGROUND

The objective of this research is to answer the following questions: have organizational innovations (OIs) significantly and positively impacted on economic growth, and can this be shown both quantitatively and qualitatively? As adjuncts to this main question, some other issues need to be answered. First, can OIs be successfully transferred from one country to another? In particular, is it possible to provide quantitative evidence that the so-called lean production system of Japan, with its corollary of just-in-time/quality control (JIT/QC), has been successfully operating in the USA with a significant effect on this country's sectoral and hence macro growth? Second, what are the links between OIs and economic growth from a theoretical point of view?

Table 1.1 summarizes the major OIs that have taken place over the last 150 years. This non-exhaustive list is based upon the major findings presented in Chapter 2. This study is devoted to analysing and exploring these OIs in relation to economic growth, with emphasis on the internal OIs of firms (taking place inside the firm, for example *kanban*). Economic activities have been variously organized since ancient times. Farms, estates, mines, shops, craft production and governments are organizational entities that have existed in many areas of the world for at least 3000 years. A large estate was the object of the first known economic or management manual ever written, that of Xenophon (*circa* 370 BC [1990]). Closer to our era, the Industrial Revolution that began in Britain about 200 years ago generated new forms of economic organization, such as factories and the appearance of managers. Later, in other parts of Europe, in the USA and in Japan, new forms of industrial organization were developed, for example, the big integrated business, the M-form of firms, 'scientific management', and lean production. Many of these new OIs took place in the USA from the 1870s to the 1920s, and many of them in Japan, in particular from the 1950s to the 1970s.

These examples raise the following questions: what is the definition of an organizational entity (OE) and when does this OE become an OI? An OE can be a production form, or an industry structure, or a firm structure, or a production process, or a network of OEs, or an institution in general.[1] This

Table 1.1 Organizational entities and innovations historically

Object of OE	Type of OI	Approximate date of introduction and country of first appearance
Production or distribution	Craft production	From ancient times
	Putting-out production	18th century (Britain), 19th century (USA)
	Factory production	End of 18th century (Britain)
	Mass production	End of 19th century (USA)
	Mass distribution	End of 19th century (USA)
	Integration of mass production and distribution	End of 19th century (USA) and beginning of 20th century
	Lean production	1960s and 1970s in Japan
Industry structure	Small, medium and large enterprises	End of 19th century (USA)
	Oligopolies, monopolies, etc.	End of 19th century (USA)
	Clusters and industrial districts (also as networks)	1950s, 1960s, 1970s in the USA, Japan, Germany, Italy
Networks	*Zaibatsus*	End of 19th century (Japan)
	Keiretsus (with subcontractors)	Since 1950s (Japan)
Firm structure	Diversified	1920s (USA)
	Focal	1930s (Japan)
	Vertically integrated	End of 19th century (USA)
	Conglomerate	End of 19th century (Japan)
	Functionally managed	1870s and 1880s (USA)
	Divisionally managed	1920s (USA, Japan)
	Teamwork based	1950s and 1960s (Japan)
Production process	U-shaped machine layout	1960s (Japan)
	Scientifically managed	1910s (USA)
	Quality controlled	1950s (Japan)
	Just-in-time oriented	1960s (Japan)
Institutions	Government agencies	After World War II (Japan)
	Anti-trust laws	1890s (USA)
	Stock exchanges	End of 19th century (USA)

Source: Based on the findings presented in Chapter 2.

list of OEs depends on the object of analysis; for instance, if we consider the production form as an OE, then the object of analysis could be the craft production mode or the mass production mode and so on (see Table 1.1). If a factory producing furniture is taken as a concrete example of a firm structure, it is an OE, which involves the following concepts: ownership, people working in it to produce output, tools and machinery used by these people, some targets to be achieved by everybody; relations between various employees, between employees and employers, between employees and tools and machinery; relations between this factory and other factories or other OEs, and a learning process for everybody involved in this factory.

All these concepts encompass dynamic organizational interactions for the OE to survive. Any one of these interactions continually change and, eventually, new OEs appear. These new OEs become OIs when they are very different from previous OEs. For example, the putting-out OE was eventually transformed to the factory OE, hence the latter became an OI; the mass production OE became a lean production OE and so on. In these cases, the factory OE and the lean production OE became OIs in relation to their previous states.

In this introduction, Table 1.1 summarizes the OIs that will be the object of further exploration in this study.[2] A more detailed and economics oriented analysis of these OIs will be undertaken in subsequent chapters, thus seeking to justify their existence and value. Note that all these OEs are interrelated, either through their object or their type. For example, a lean production OE includes focal factories, the JIT/QC philosophy and so on. Also, all the OEs are a consequence of the type of underlying ownership, core competences, strategies and other features that will be analysed in subsequent chapters. Furthermore, since OEs and OIs are virtually the same concept, with the difference that OIs mean new and very important OEs, reference will only be made to OIs in the remainder of this book.

All these major OIs, as recognized by numerous researchers, took place in the two largest economies of the world today, the USA and Japan. These two countries' gross domestic products (GDPs) together constituted approximately 30 per cent of the world's total GDP (purchasing power parity – PPP) in 1999, or 43 per cent of the world's total GDP (non-PPP) in 1999 (World Bank Group, 2001). Also, these two countries experienced, at different times in their economic history, a very high industrial growth. Table 1.2 summarizes the growth rates in the two countries for some key periods (the period 1925 to 1950 was excluded, as it was exceptional in terms of the great depression, and the preparation or outbreak of wars. However, this period is included in this study of OIs).[3]

Before we proceed, another question must be addressed. What is the relationship between OIs and technological innovations or technology, at least

Table 1.2 Average annual real growth rates of industrial production for some key periods in the USA and Japan, 1866–1996

Period	USA	Japan
1866–1891 (for Japan: 1876–1901)	6.0	4.4
1895–1920 (for Japan: 1900–1925)	5.6	5.1
1951–1976	4.3	12.6
1971–1996	3.3	3.2

Source: Based on Mitchell (1998) and World Bank Group (2001).

from a definitional point of view? The definition of technology provided by an international institution, the United Nations Centre on Transnational Corporations (UNCTC), in 1985 is revealing: 'Technology may be embodied in the form of capital goods, such as machinery, equipment and physical structures; or it may be disembodied in such forms as industrial property rights, unpatented know-how, *management and organisation* (my emphasis) and design and operating instructions for production systems' (UNCTC, 1985, p. 119).[4] A key objective in this study is to separate embodied from disembodied in the definition of technology. Hence, whenever we refer to OIs we mean disembodied technology (according to the above definition of the UNCTC). The term technical innovations (TIs) is adopted to refer to embodied technology (for example, new machines, new products and so on). As argued in Chapter 3, TIs have been the main target of the immense literature on economic growth. The aim of this study is to introduce a systematic awareness that OIs may play a distinct and indeed preponderant role in the process of such growth (as distinct from the role of TIs or technology as a whole).

The following example is used here to illustrate the importance of OIs. The semiconductor industry has been very important for both the Japanese and the American economies. A recent detailed account of this industry by Langlois and Steinmueller (1999) reached some notable conclusions:

> One of the factors driving the success of Japanese firms in memory products in the early 1980s was the higher quality of the chips they produced. For Japanese chips, defect rates – the fraction of chips that prove to be defective – were probably half to one-tenth the rates for American products. By the second half of that decade, however, American firms had dramatically increased expenditures for quality control, imitating Japanese practices such as total quality management (TQM), greater attention to preventive maintenance, and automated process control and monitoring. By the early 1990s, American manufacturers had probably begun to match the defect levels of their Japanese counterparts. Intel reportedly reduced its

> defect rate by a factor of 10. There is also evidence that American firms have improved manufacturing yield rates and direct labour productivity since the early 1990s … (ibid., p. 49)

A brief comparison between TIs and OIs is necessary at this point. Overall, technology in the ordinary sense is responsible for changes in economic growth. Certainly TIs have played a primary role in promoting such growth. Effectively, the theorists of economic growth have been concentrating their research on the impact of four main categories of factors affecting such growth: capital, labour, technology (in the sense of TIs) and human capital. However, despite the writings of such eminent economists like Marshall (1890) and Schumpeter (1934; 1942a/b) that OIs significantly generate economic growth, and despite the writings of management scholars that OIs do actually matter significantly in the growth of firms, the vast majority of modern economists have consistently ignored OIs in their analysis of economic growth models. This is true for both exogenous and endogenous types of models, and yet, on a priori grounds, OIs are no less important than factors such as TIs.

What are the implications of not explicitly including OIs in the theories and empirics of economic growth? The famous residual of Solow's analysis (1957), the so-called total factor productivity (TFP) was found to be substantial in many studies. This residual has *usually* been explained by technological factors (or TIs in the present terminology). However, how do we know that OIs have not contributed equally to TFP, or even more? Unless we carefully explore the contribution of OIs we will not be able properly to refute or accept their importance. Consequently, the issue is there to be explored. This is exactly what this study is about: do the major OIs significantly affect economic growth and how do they do this? Is it also possible to provide quantitative evidence to answer this question?

A further issue must be identified before we proceed to address the core points of the study. Are OIs related to entrepreneurship? The immediate answer is yes, as it is the entrepreneurs and managers (in a broad sense) who instigate, introduce and implement OIs. However, in this book, the specific contribution of these people is not explored, except on an ad hoc basis, as this would involve much more space than is available. Future research should attempt to investigate such contributions in a more systematic way (a preliminary analysis is provided in Sanidas, 2004c). After all we do not a priori know whether or not OIs are exogenously, or endogenously, determined by entrepreneurs and managers, and whether, in general, OIs, entrepreneurship and management are part of a larger and more complicated mechanism of determination involving other parameters as well such as law, institutions, culture and so on.[5]

1.2 MAIN POINTS

1.2.1 Evidence on the Impact of OIs on Economic Growth

There is a need to explore the significance of OIs in promoting economic growth on a firm, sectoral and macro basis. Furthermore, this significance ought to be both qualitative and quantitative in kind. Management research has already provided ample evidence to suggest that some OIs matter considerably for the firms' expansion (for example, Davidson and Griffin, 2000; Thomson and Strickland, 2001). This study, effectively, extends the theoretical literature by introducing new concepts integrated into existing ones in order to show that OIs, or disembodied technology, have a significant positive impact on economic growth. This study also provides some empirical evidence on the significance of OIs in promoting industrial growth in the context of manufacturing sectoral growth.

1.2.2 Exploration of OIs Ignored by Other Scholars

Chandler (1977; 1990) and to a lesser extent Lazonick (1990) amongst others have provided considerable evidence (mainly qualitative) that 'organizational capabilities', as these authors call them, played a key role in shaping the growth of firms (mainly big business) and industrial sectors in an evolutionary manner. These capabilities were linked with some of the OIs listed in Table 1.1, but have not included some others, such as the production process of JIT/QC. Consequently there is a need to explore the ignored OIs and in particular JIT/QC as well more systematically.

1.2.3 A Comparison between the USA and Japan Regarding the Impact of OIs

An extensive comparison between the two countries, in terms of industrial growth and OIs, is needed in order to further support the overall evidence. By comparing the two largest economies in this study, a gap in the literature is filled. Chandler has briefly compared the two countries, although he concentrated primarily on the USA. Lazonick has made a more conscious effort to bridge the gap, but he too concentrates mainly on the USA. Fruin (1992), a Chandlerian scholar, has concentrated his work on exploring Japan. Best (1990) has made a good qualitative comparison between the two economies in terms of some OIs, but again without quantitatively linking them to their respective economic growth.

1.2.4 Essence of Contributions

The contributions of this study are linked to providing quantitative and theoretical evidence as to the relationship between economic growth and OIs. Thus, an 'atomic/systemic' theoretical economic growth model, which incorporates the OIs and is based on microeconomic foundations, is suggested (see Figures 5.2 and 5.3 in Chapter 5, for a brief description). To construct that model, it is necessary to identify OIs through history, to provide evidence of their link with economic growth and to group OIs into three main axes. It also becomes indispensable that OIs are further scrutinized in order to explore them in terms of fundamental firm operations and activities (such as the 'process of movements') that pinpoint the exact substance of OIs and their relationship with economic growth. In this way, the links between OIs and economic theory are thoroughly established. The atomic/systemic model also incorporates 'systems of innovations', 'sectoral systems' and 'evolutionary learning capabilities'.

This theoretical model is also used to determine proxies for measuring OIs. One of these proxies is in turn used to econometrically measure the impact of the main axis of OIs, namely the JIT/QC system, on American manufacturing sectors during the last 40 years. At the same time, it is shown that it is possible to successfully transfer the main axis of OIs from Japan to the USA. Furthermore, an explanation of the recent American continuous economic growth, and of the recent continuous Japanese recession through a set of propositions related to OIs and leading sectors is provided, according to the overall findings of this study.

In summary, the main innovations introduced in this study are:

- A synthetic approach of the identification of OIs and their contribution to economic growth. This approach includes the 'three axes of OIs' and is comparative in terms of the two countries, the USA and Japan.
- An extension to the theory of the firm with the introduction of the four fundamental firm operations that include the process of movements, the process of contracts, and so on.
- The proposition of an atomic/systemic model of economic growth that incorporates OIs through the four fundamental firm operations, leading firms and leading sectors within a more general framework of sectoral systems, systems of innovations and an evolutionary learning process. The concept of an evolutionary negentropic open system is crucial to this analysis, for instance, in order to appreciate the role of 'waste' in the growth of firms.
- A thorough quantitative analysis that provides evidence for the role of OIs in promoting economic growth. This analysis is primarily sectoral

and applies to both countries, the USA and Japan. For the USA, the econometric analysis is carried out for the American manufacturing sectors which successfully transferred the JIT/QC system (a major OI) from Japan to the USA.

1.3 ORGANIZATION OF THE BOOK

In Chapter 2, entitled 'Organizational innovations from the 1860s in the USA and Japan', the major OIs are identified historically in the USA and Japan from about the 1860s to the present by mainly referring to some major works of eminent economic historians. This chapter pinpoints the main forces that shaped the appearance of OIs, their chronology and their evolution, together with examples of leading firms, leading sectors and sometimes links between OIs and TIs. The importance of this chapter for subsequent chapters is paramount. Directly or indirectly, elements of this chapter are used in various ways in the whole study to bring evidence to the study.

The next three chapters lay the theoretical foundations of why OIs should be included separately in explaining economic growth. In particular, in Chapter 3, entitled 'The role of OIs in the theory of economic growth', a general theoretical background to economic growth is proposed in order to take into account the impact of OIs. This background first reviews the relevant literature and introduces some important concepts and theories of the significance of OIs in explaining growth. Then, it suggests that OIs are not necessarily a 'handmaiden' of TIs, but rather they should be separately examined. It also proposes (in a non-mathematical way) a general form of a production function that has explicitly taken into consideration various types of organizational inputs. The latter are linked with some fundamental concepts such as division of labour, transaction costs, institutions and others. However, a more thorough and comprehensive explanation as to why OIs should be included in theories of economic growth will take place in Chapters 4 and 5.

Chapter 4, entitled 'Contribution of specific OIs to economic growth and the three axes of OIs', is a continuation of the themes of Chapter 3. However, in Chapter 4 the links between industrial growth and the OIs are established in a concrete manner. Thus, a theoretical explanation is offered regarding the replacement of the putting-out system by the factory one; the role of scientific management and its 'successors'; the appearance and evolution of mass production in the USA; the appearance and evolution of lean production in Japan, and so on. The links between these OIs and industrial growth are made according to various economic concepts and propositions, like those mentioned in the previous chapter, but in a more concrete and analytical manner.

In Chapter 5, entitled 'The process of firm operations: core of the atomic/systemic model of economic growth', the proposition is made that in order properly to explain the close link between OIs and economic growth, we must explore the heart of this growth which is the firm. Thus, a taxonomy of operations and activities of running a firm (hence including costs and benefits) is proposed, according to which OIs, as analysed up to Chapter 4, are a mixture of these operations. It becomes apparent in this chapter, that OIs are a consequence of what is called here the 'process of movements', as opposed to the 'process of contracts' which is related to the transaction costs theory, and so on. This chapter also analyses the role of leading firms, leading sectors and networks in the context of OIs and economic growth, so that finally a basis for an atomic/systemic model of economic growth incorporating the role of OIs and firms is suggested. This model also incorporates systems of innovation, entropy, sectoral systems and evolutionary learning process. This chapter extends the existing theory of the firm by introducing the negentropic system of four processes of firm activities, their corresponding opportunity costs and a set of related propositions. Within this extended theory the role of OIs and other types of innovations in economic growth is explicitly and intrinsically included.

In Chapter 6, entitled 'OIs and manufacturing sectors' growth in the USA and Japan', a quantitative sub-sectoral analysis is carried out for the USA and Japan, first from the 1880s up to World War II in a brief manner (due to lack of data) and with emphasis on the American economy, and second from 1960 to 1996 in a more detailed manner. Some general trends and tendencies are discerned concerning the leading sectors in terms of OIs. Real output and TFP per industrial sector are scrutinized for different sub-periods so that more appropriate conclusions can be drawn both for the USA and Japan. The statistical methods used in this chapter are relatively simple, from graphs and tables to cointegration analysis. One of the sub-sections explores some of the main reasons for the recent prolonged recession in Japan. In order to support the findings as to the role of OIs, some of the TIs are also briefly explored for both countries by referring to some leading sectors.

In Chapter 7, entitled 'Empirical evidence on the links between industrial sectoral growth and the lean production system (JIT/QC) in the USA', the focus is on the American manufacturing sub-sectors for the period between 1958 and 1996, and in terms of OIs the JIT/QC system becomes the object of analysis. First, the chronology of implementation of the lean production system (LPS) is established in the USA in relation to Japan, both in terms of the just-in-time (JIT) and the quality control (QC) processes. Then, a thorough econometric analysis is conducted by exploring models with various lengths of cross-sectional or pooled time-series data. These models are either simple regression models or more complex ones such as the vector error

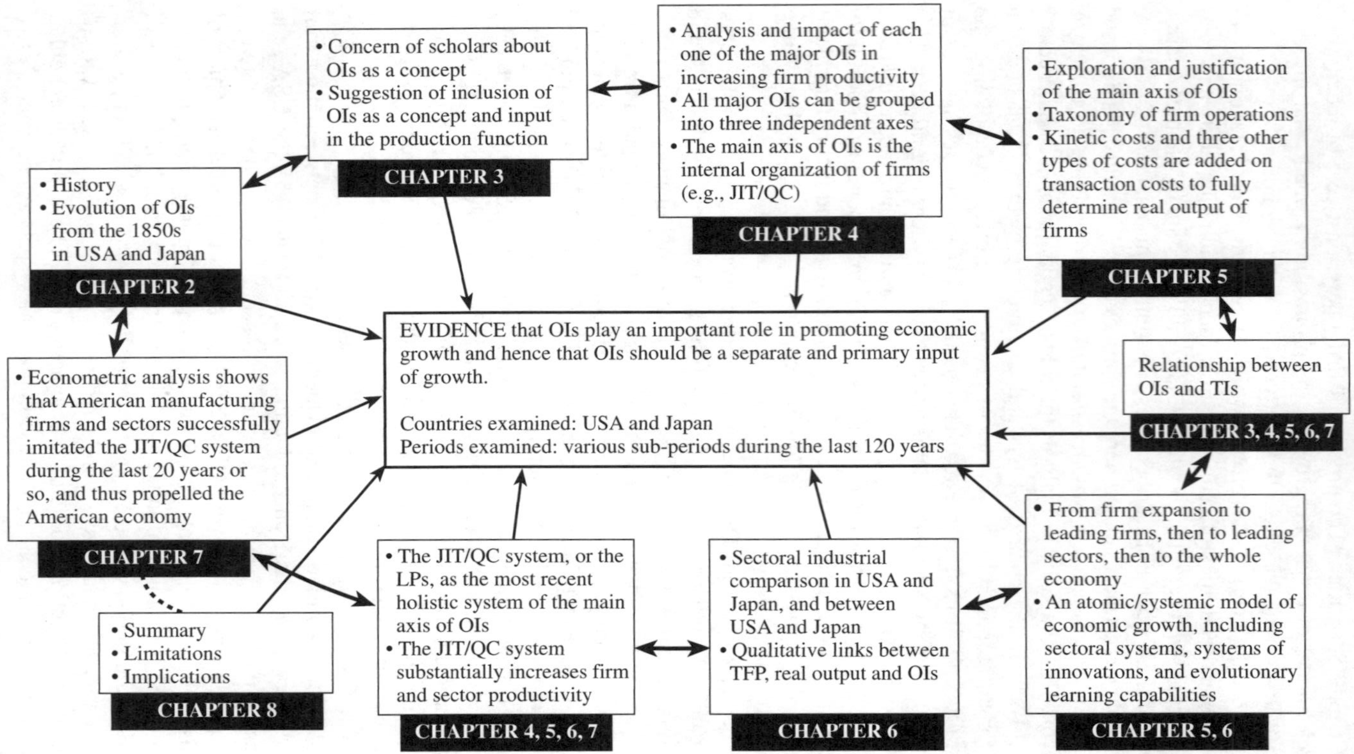

Figure 1.1 The circular flow of the main points for each chapter

correction models (VECM), and incorporate or include the TFP or labour productivity variables as the main variables to be quantitatively explained. The main purpose of this chapter is to show empirically that the LPS or its equivalent JIT/QC has been successfully transplanted in the USA from Japan, and that this imitation has been the main reason for the recent American economic revival (from about the mid-1980s up to the end of the 20th century).

Finally, Chapter 8 draws out the conclusions, limitations and implications of this study.

In short, Chapter 2 reveals the nature and evolution of OIs, Chapters 3, 4 and 5 establish the necessary theoretical background for the inclusion and impact of OIs in explaining economic growth, whereas Chapter 6 discusses the importance of sectoral growth in the two countries regarding OIs, and Chapter 7 presents empirical evidence for the impact of OIs using various statistical techniques. Figure 1.1 summarizes the main points that each chapter contains in a circular flow of interdependent nodes.

NOTES

1. Van de Ven and Poole (1995, p. 512) provide the following similar definition: 'The entity may be an individual's job, a work group, an organisational strategy, a program, a product, or the overall organisation … '.
2. According to this research, there were no discernible or emerging new OIs in the 1980s and 1990s, except perhaps an Internet system that started to take shape in the second half of the 1990s.
3. World Wars I and II are mostly excluded in this study for being unusual historical periods.
4. Also quoted in Dicken (1998, p. 248).
5. All these issues, briefly mentioned in this introduction, will be fully explored in this book.

2. Organizational innovations from the 1860s in the USA and Japan

2.1 INTRODUCTION

The primary aim of this chapter is to pinpoint the most significant and essential OIs in the USA and Japan. The period examined is approximately from the 1860s to the present. However, two sub-periods will be more closely scrutinized, the 1880s to 1920s and the 1950s to 1990s, of approximately 40 years each. During these two sub-periods the so-called Second Industrial and Third Industrial Revolutions took place. The first two sections will examine the historical evolution of OIs up to World War II (the first section for the USA and the second for Japan). The next two sections will examine the historical evolution of OIs from World War II onwards (the third section for the USA and the fourth for Japan). Finally, in the fifth section, a brief comparison between the two countries in terms of OIs will be undertaken. In this historical survey the analysis will concentrate on the detection of the main OIs and their evolution, thus avoiding as much as possible any links with economic theories. These links will be made in the following three chapters. Finally, note that this chapter does not provide a comprehensive history of all OIs in the two countries; rather, its aim is selectively to pinpoint the aspects and nature of the major OIs that will be useful for the present study.

2.2 OIs IN THE USA UP TO WORLD WAR II

This section is primarily based on Chandler's pioneering work on the evolution of big business. However, it is important to note that exploration of big business is not the only type of business that has shaped the American economy. Scranton (1997) emphasizes that 'specialty production' paralleled, complemented and sometimes conflicted with the mass production corporations; this other side of the Second Industrial Revolution certainly played an important role in shaping the American economy, but major OIs that took place during the period examined were more related to mass production and other features (which will be explored in the following pages). Thus, Boyce and Ville (2002, p. 150) remarked: 'Specialist manufacturers relied extensively on skilled labour, while mass producers de-skilled workers. The methods employed by the former relied on "smart workers", while the latter depended on "smart (production) systems" … '.

2.2.1 The Appearance of Big Business, Full Integration, Hierarchies and Oligopolies

The US westward movement of population[1] was the moving force for the rapid development of the railway system in the 1840s and 1850s, which eventually created a national market. However, its growth and restructuring was not complete until after the recession years of the 1870s. The development of the railway system coincided with the USA's economic industrial take-off period in the mid-19th century, which later led to the Second Industrial Revolution. Many famous entrepreneurs, who played a marked innovation role in their own business, were originally trained in the railway firms. Another precondition for the Second Industrial Revolution was the superabundance of land, combined with a favourable inclination for entrepreneurship (Americans were always fascinated by the success of their business people) and a shortage of labour (Habakkuk, 1962; McGraw, 1997).

Major and minor technological breakthroughs (TIs according to the terminology followed in this study), although continuous in trend, were particularly noticeable in the 1870s and 1880s. At the same time big business, through mass production, began to gather steam in the 1870s and 1880s. Out of the Fortune 500 largest American firms in the mid-1990s, 53 were founded in the 1880s (for example, Kodak, Johnson & Johnson, Coca-Cola, Westinghouse Electric), 39 in the 1890s (for example, General Electric, Pepsico, Goodyear), and 52 during the 1900s (for example, Ford Motor, Gillette, General Motors) (source: Harris Corporation, 1996). The first industries to grow very large were consumer goods industries, such as fresh meat, harvesters, sewing machines, sugar, salt and leather, which all built large marketing departments and expanded horizontally. The same story was repeated in other industries processing agricultural products, for which a number of machines were invented in the late 1870s and 1880s to permit large-batch and continuous process production (cigarettes, matches, flour, breakfast cereals, canned foods and soap).

In the 1880s and early 1890s, many small manufacturers in the sugar, leather, salt, distilling, linseed and cotton oil, biscuit, petroleum, fertilizer and rubber boot and glove industries, jointly formed large horizontal combinations. The latter resulted mainly as a response to overproduction by numerous small firms in an expanding national market during the 1860s and 1870s that threatened prices falling below production costs (Chandler, 1959, p. 10). In most of these industries, consolidation and vertical integration followed horizontal combinations. Examples in this category are the whisky enterprises and the rubber boot, shoe and glove industries. In both the petroleum and fertilizer industries, control over raw materials came after the combination and consolidation of groups of small manufacturing firms. For instance, the Standard Oil Trust, formed in 1882, consolidated its production activities and created a domestic marketing system first, and only in the late 1880s did it start to adventure into the production of crude oil. The producer goods industries developed later. Until the depression of the 1890s, most of the combinations and consolidations had been in the consumer goods industries. By the first decade of the 20th century the leading firms in many consumer goods sectors had become departmentalized and centralized as a consequence of vertical integration, thus creating oligopolies and sometimes monopolies for their markets. For instance, by 1898 the Rubber Company controlled 75 per cent of the nation's rubber boot, shoe and glove output.

Like the companies making consumer goods, those making producer goods also set up nation-wide and worldwide marketing and distribution organizations, consolidated production into a few large plants and established purchasing departments. Also, except in steel, integration usually followed combination in the producer goods industries. The growth of such large enterprises often led to diversification of the types of products these manufacturing companies made and sold. The 'full line' strategy, pioneered by General Electric and Westinghouse, was soon adopted by many other consolidated concerns. Continuing diversification came largely in industries where science could be most easily applied. Thus,

> The chemical, automotive, power machinery, rubber and petroleum industries led the way to the development of new processes and products, new ways of internal organisation and new techniques of external competition as the new century unfolded. The metals industries and those processing agricultural goods have, on the other hand, changed little since the beginning of the century … For them the greatest period of change came in the last decade of the 19th century … The middle of the first decade of the new century might be said to mark the end of an era. (Chandler, 1959, p. 26)

These large industrial firms became by the turn of the 20th century vertically integrated, forward into distribution and backward into supplies of inputs, as well as centralized and functionally departmentalized organizations. They became increasingly bureaucratic internally and oligopolistic externally, despite some tendencies for monopolies. Integration and combination by one manufacturer forced others to follow. The oligopolistic structure of most of the leading industries meant that, by and after 1900, costs rather than interfirm competition determined prices, and that strategies such as price leadership and price umbrellas became common practices. Therefore the top management of these industries paid more attention to issues such as advertising, product differentiation, brand names, improvements in further integrating manufacturing, marketing and buying processes, and developing more diversified lines of products.

In short, the American economy was consolidated and strengthened during the approximate period 1880 to 1920.[2] This consolidation took place through the process of appearance of big national and international corporations, which usually were backward and forward vertically integrated (full integration of mass production and mass marketing for both the consumer goods and producer goods industries). In addition, these corporations were multi-functional hierarchical business organizations (to some extent similar to that of the railways firms), and they were markedly changed in the organization of the shop floor, especially through various applications of scientific management (see next sub-section). Their salaried managers increasingly took over the running of business from the founding entrepreneurs; in particular, middle management and especially top management for the first time replaced market forces in an oligopolistic environment (the learned skills and knowledge within each oligopolistic market were company-specific and industry-specific).

2.2.2 The Rise of Scientific Management and Fordism

The coordination and control of the flow of materials at a high volume and speed through many departments in which many workers were employed in each of the production processes created challenging administrative and managerial problems in the second half of the 19th century. Litterer (1963) said that the growth of internal disorder fell into two major categories: first, a breakdown of coordination among subordinates, and second, a breakdown in relationships between top and lower levels of management. These problems became acute in the late 1870s and 1880s but their solutions only started to take place in the 1890s and 1900s. The movement of 'Systematic Management' attempted to suggest such solutions from the 1870s (cf. Litterer, 1961; 1963); Nelson (1974, p. 479) said in this regard: 'The Taylor system was a comprehensive answer to the problems of factory coordination, a refinement and extension of the earlier ideas known as systematic management … '

During the expansion years of the 1870s and 1880s, industrialists relied on skilled foremen to recruit, train and manage the workers. That was the 'inside contracting' system of labour organization, according to which these departmental foremen were contracted with the owners to produce a specified number of parts or mechanisms in a specified time (usually a year) for a specified price. In this way, the owners were able to pass on to the contractors all management-related problems. With this system, the owners lost control over costs and the coordination of the flow of goods through the many departments. The contractors had obvious reasons to conceal valuable information from the factory owners, and did not have the need to coordinate shop floor operations in an efficient way.

H.C. Metcalfe, in 1885, prescribed a solution to the 'inside contracting' system in his book *The Cost of Manufactures and the Administration of Workshops, Public and Private* (the first book ever published in the USA on factory management). His solution was an adaptation of the voucher system of accounts developed in railroad repair shops to meet the needs of interchangeable-parts manufacturing (Chandler, 1977). However, Metcalfe's solution had major disadvantages: first, it required an excessive amount of paperwork from the foremen and, second, the latter were still not willing to provide all necessary information. H.A. Towne of Yale and Towne, F.A. Halsey of the Rand Drill Co. and other metalworking manufacturers, subsequently developed the 'gain-sharing' system, according to which any reduction in unit costs achieved by improved machinery and plant design, better scheduling, or more productive labour would be equally shared by the company, the workers and the foremen (the latter would get 60 per cent to 80 per cent of the non-company share) (ibid.). A further suggestion came from F.W. Taylor, who, in 1895, pronounced his 'scientific management', according to which the standards of premiums and bonuses paid should be determined 'scientifically' and not historically. He also urged that a number of 'functional foremen' and a 'planning room' should replace the contracting foremen. However, American manufacturers rarely adopted Taylor's full system, which was developed during the period 1895 to 1915 (Taylor's detailed principles will be elaborated in subsequent chapters).

There is another way of looking into the relationship between workers on the *shop floor* and management of the production processes. Skilled workers who played a central role in shop-floor operations were transformed into lower managers, thus a valuable cooperation was gained from these transformed salaried employees. As Lazonick said: 'making skilled workers members of the firm helped management to divide and conquer the labour force' (Lazonick, 1990, p. 229). Lazonick regards the shift in control over shop-floor organization from craftsmen and inside-contracting on the shop floor to line and staff managers as a key feature of managerial capitalism in the USA during the period from the 1880s to the 1920s. The main reason for this shift, deliberate and deterministic in nature as Lazonick describes it (1990, ch. 7), was the fact that the shop was really run by the craftsmen or inside contractors and hence the upper management was obstructed from introducing new technologies and methods of production.

By now, it is perhaps becoming apparent that the gradual disappearance of the inside-contracting system coincided with the parallel advent of various methods of scientific management. As Lazonick (1990, p. 223) said: 'Inside contracting, and with it craft control, was particularly prevalent in the metal-working industries of the northeastern United States. It was also these machine shops that gave birth to the scientific management movement … ' One of the consequences of this transformation towards a management-controlled mass production system was a sharp increase in managerial, engineering, scientific and, generally, professional personnel during the period 1890 to the 1920s. For example, 'the total number of salaried employees in manufacturing in the United States rose from 348 000 – one for every 12.9 wage earners in manufacturing – in 1899 to 1 496 000 – one for every 5.6 wage earners – in 1929' (Lazonick, 1990, p. 229). During the same period, the higher education system developed in a similarly spectacular way (ibid.). All this important organizational transformation, which accelerated at the end of the 19th century and became a common policy at the beginning of the new century, was observed by other researchers as well. For instance, Montgomery (1987, p. 215) has argued that 'proliferation of white shirts and conversion of craftsmen into supervisors were universal and continuing phenomena' in the first decades of the 20th century.

Furthermore, this organizational transformation was also linked to new technologies (TIs). For instance, the period 1880 to 1920 saw an immense increase in the development of machine tools for highly specialized purposes. This development started earlier with the production of firearms, continued with the production of sewing machines from the 1850s to the 1870s, then culminated in the 1890s through to the 1910s with the emergence of the bicycle and automobile industries (Rosenberg, 1976, pp. 18–31).

An application of scientific management was Fordism. Effectively, Fordism was a not so fortuitous continuation of the Taylorist principles. As Littler (1982, p. 56) said:

> By the time that Henry Ford came into business Taylorism had begun to affect the US engineering industries. Consequently, Ford took over some of the essential aspects of Taylorism (the divorce of planning and doing, the fragmentation of jobs, each task allotted a specific time, etc), but he also went further by introducing two further principles. These were the flow-line principle and a new method of labour control. The new method of labour control used by Ford revolved around the 'Five Dollar Day' …

The contributions of Taylor and Ford to OIs will be more extensively explored in sub-section 2.4.2 and subsequent chapters.

2.3 OIs IN JAPAN UP TO WORLD WAR II

Although this section draws extensively from Fruin (1992), several other sources were consulted that confirm some aspects of Fruin's pioneering work; these sources include Yui and Nakagawa (1989), Lockwood (1968), Morikawa (1992) and Hirschmeier and Yui (1981).

2.3.1 Conglomerates (*Zaibatsus*)

Fruin (1992) distinguishes three types of enterprises, which made an impact during the initial period of industrialization (from around 1870, beginning of the Meiji era until World War II), namely the *zaibatsu*, the independent urban firms, and the independent rural firms. Most studies have concentrated on the first type, which, admittedly, was the most prominent in shaping Japanese industries. These groups (the *zaibatsus*) were, from that starting point, a collection of mostly unrelated commercial, industrial and service enterprises and were family owned and for a while family managed. To use a modern term, they became conglomerates and they found a way to create interdependence between enterprises of the same *zaibatsu* conglomerate.

For this type of enterprise, from this early period, there was a situation of joint production and distribution generating economies of scope, thus firms were becoming interdependent. As Fruin (1992, p. 90) remarked 'In Japan, *zaibatsu* grew for the most part by unrelated diversification, which is to say that economies of scale in production and distribution were not the forces behind the development of national or *zaibatsu* business groupings … '. Unrelated diversification was a salient feature of these groupings before the 1930s. The initial clusters were transportation, mainly in shipping, energy production and finance. Manufacturing was crafted on to these initial activities. These early Japanese *zaibatsu* gradually became more efficient and lowered their costs by economizing in scope much more than in scale.

Furthermore economies of scale did not take place until the boom associated with World War I, and in a more extensive way in the late 1950s: 'In Japan before WWII, indeed before the late 1950s, the domestic and proximate East Asian markets for volume goods were not large, and it was extremely risky to compete on the basis of economies of scale in most instances' (Fruin, 1992, p. 113). Some sustainable scale economies were present before World War I only in some industries, such as textiles, paper, some metals production, food (sugar, beer, milled grains), beverages, some chemicals and cement. However, only textiles were exported in bulk, thus realizing more economies of scale. A significant corollary of the importance of economies of scope present in this early Japanese industrial development, is the 'nearly universal separation of production and distribution' (Fruin, 1992, p. 109).

The parallel existence of these two phenomena, first, the divorce between distribution and production and, second, the economies of scope, had another consequence, namely the non-existence of organizational centralization; hence Japanese firms of this period had a non-complex U-form organizational chart. The Japanese holding companies concentrated managerial resources in factories and at the lower level of the firm and, consequently, they did not act as a capital market to allocate funds to subsidiary enterprises. Even the most managerially advanced companies, the textile firms, kept a relatively simple form. Nonetheless, and despite a low proportion of middle managers, textile firms were the first to introduce managerial hierarchy, coordination of production and distribution, a careful monitoring of the steps of production, a better trained and motivated workforce, and a more highly capitalized and better managed enterprise (Fruin, 1992, p. 117). Despite these elements of a more modern firm, textile companies integrated and coordinated activities in a decentralized way. For instance, in 1914, one of the largest industrial enterprises in Japan, Toyobo textiles, had a total personnel of 36 694, out of which only 479 (or 1.3 per cent) were located in the head office to carry out administrative, accounting and personnel functions. 'Toyobo's example of not centralising operations was widely imitated, following a wave of mergers in the textile industry during the recession years of 1907–14' (Fruin, 1992, p. 118).

Although *zaibatsu* firms played a central role in Japanese economic development, the independent enterprises were more numerous, widespread and diversified in terms of their activities. 'They were also more important given their estimated two-thirds to three-quarters contribution to the nation's domestic manufactured product' (Fruin, 1992, p. 119).

2.3.2 Small Focal Firms and Interfirm Networks

Both the Japanese economy and Japanese firms grew stronger and stronger, especially after the auspicious conditions of World War I (increased demand of Japanese products by combating countries). The products of the Second Industrial Revolution started penetrating the Japanese markets in an aggressive way. The demand for new products, coupled with new producers and new technologies, accelerated substantially at the time of the Great War, primarily because the country was not part of the European conflict and, as a result, Japanese firms filled prodigious orders for virtually all combatants (for many types of goods). Thus, without the natural resources of many Western nations, through the impetus provided by World War I, Japan was going to further reinforce its existing strategy of creating this unique organizational system of its own. One of the main messages from Fruin's detailed study (1992, especially ch. 5) is that Japanese firms did not grow very large compared with Western companies, they specialized their production in mainly one or two products at the most, and they concentrated their activities within the factories, thus avoiding centralization and accentuating interfirm networks.

Further evidence of the product-focused interfirm networks is the degree of outsourcing as a percentage of manufacturing value added. A government publication of 1936 based on 1932 data (Fruin, 1992, p. 134) showed that this proportion varied between 5 per cent and 65 per cent. In the automobile, textile weaving and electrical equipment industries, at least 20 per cent of the manufacturing value of all products came from suppliers. These figures strongly indicate that vertical integration was low and that a large network of product-based firms was already in place in Japan during the inter-war period. This network was more obvious in the case of *zaibatsus*. For example, Mitsui & Co., the main trading firm for the Mitsui group, held a direct controlling interest in several manufacturing ventures. By 1940, Mitsui & Co. held equity shares in 253 companies covering 14 industries for a total of 275 million yen (ibid., p. 140). Horizontal integration, through mergers and acquisitions, especially in textiles, paper, agricultural chemicals and machinery led to a concentration of resources at production sites and decentralization.

Decentralization of management decisions amounted to the development of what Fruin calls the focal factories. Fruin's work on focal firms in Japan is referred to by other researchers, for instance Boyce and Ville (2002, p. 159):

> Within focal factories, a panoply of corporate functions could be found: quality-assurance offices, marketing and sales staff, research facilities, and even personnel departments. Factory managers, like company presidents, were enveloped by a hive of clerical and technical specialists … Focal factories were charged not only with labour management but also with technology transfer, product and process innovation, engineering, manufacturing, cost accounting, new personnel policies, regional distribution, and sales coordination … (Fruin, 1992, p. 136)

Thus, a focal firm is a factory, relatively small in size (in comparison with American firms), focused in certain activities (thus not diversified), organizationally interdependent within the network it belongs to, and with all distinct functions, such as technology research, localized at the factory level. In Chapter 4 Table 4.2 contains the main characteristics of focal firms according to Fruin's work.[3]

Why has this focal factories system evolved so strongly in Japan? Several reasons can be detected from Fruin's analysis. First, the end of the 19th century witnessed a process of trial-and-error adaptation of Western knowledge and technology, which found a fecund area for development in small, decentralized local and autonomous firms. The time and economic uncertainty constraints made this process even more necessary for such firms.

Second, the economic downturn following World War I and defence cuts caused widespread lay-offs in military arsenals and civilian shipbuilders. This large pool of skilled and semi-skilled underemployed or unemployed workers became the pool of many small subcontractors who became suppliers to large firms between the mid-1920s and the beginning of the 1930s. A survey of Tokyo factories published in 1932 reported that two-thirds of the 322 small metalworking shops in that city were founded between 1922 and 1930. Third, and as a corollary of the second reason, larger firms used the smaller subcontractors because of their lower wages.

Fourth, factories were not producing for the national market but for regional or international markets. Fifth, cost accounting methods were not well developed. Sixth, the well-established regional traders and middlemen prevented companies from centralizing and coordinating relevant functions. This was especially true in the textile, cement, food and beverages industries, which, considered together, accounted for 53 per cent of the 200 largest industrial firms in 1930. Seventh, there was no anti-trust legislation to prohibit interlocking business alliances and, hence, small local firms could exist independently and yet be at the same time part of a bigger group such as the *zaibatsu*.

Eighth, alliances in manufacturing and distribution of small focal firms minimized risks while taking advantage of the production and distribution resources that other firms commanded. 'A strategy of maximising inter-firm economies of scope through group-driven cooperative transactions made better sense than the pursuit of internal production and allocative efficiencies through vertical integration and product diversification' (Fruin, 1992, p. 157). Ninth, as the performance of focal factories was strongly correlated with the existence and intensity of interfirm networks, there were substantial savings in transaction costs and diseconomies of managerial control.

What are some of the other consequences of this dual system of focal factories cum interfirm networks? First, the narrow specialization in products of Japanese firms made these firms – even the larger ones – small by Western standards (for instance, in terms of employment or output). Second, the difference in relative size of Japanese firms created a dual regime of wages and security, where these were higher for larger firms, especially those that were part of a *zaibatsu*, but lower for smaller firms, and especially for those that were subcontractors. Third, the upper management, and to some extent the middle management, was relatively unimportant. Finally, Japanese firms aimed not to just increase the number of different products manufactured in-house, but also to increase their value-added contribution to the manufacturing process.

2.4 OIs IN THE USA AFTER WORLD WAR II

2.4.1 Diversification, Multidivisionalism and Conglomerates

The American economy was already dominating the world economy before World War II. When the latter ended, the American economy was even stronger and totally in control of the non-communist globe. It was not until the late 1960s, and especially the 1970s, that the European and Japanese nations started threatening this American supremacy. It is again worth referring to Chandler in order to assess the situation:

> The successors of the 1950s managers – the third generation – were those who had to meet the challenges of intense inter-industry and international competition that began in the 1960s, the most serious competitive challenge yet faced by US managers ... But in the 1960s, firms from abroad and from related industries began to crowd their markets. Excess capacity increased. Prices fell, costs rose, ROI dropped ... (Chandler, 1994, p. 16)

How did these managers respond to the fierce competition developing during the 1960s and 1970s? Chandler (1990; 1994), and other researchers have suggested that the American firms decided to expand through the process of diversification either to a minor extent into related industries, or, to a major extent, for the first time, into unrelated ones. One of the reasons, although not the most important, for this new way of expanding business, was the continuous passing of anti-trust laws by the American government. Thus, the Celler Kefauver Act in 1950 further tightened restrictions on firm growth through horizontal and vertical integration. However, there are other reasons.[4] For instance Penrose (1959) said that diversification is a solution either to specific problems such as cyclical fluctuations, or a solution to more growth and expansion.

It seems that the latter solution was the dominant one in the 1960s in the USA. The emergence of many conglomerates was an innovation of the 1960s (Chandler and Tedlow, 1985, p. 739) for the USA. These authors (ibid., p. 737) say that:

> In the 1960s, a new kind of business enterprise played a key role in mergers and acquisitions. This was the conglomerate type. Eleven of the top 25 acquiring companies in the 1960s were classified as conglomerates. These 11 firms acquired over 500 companies between 1961 and 1968. And these acquired companies represented over 92 per cent of the acquired firms' assets by the latter years ...

Classic conglomerates such as ITT, Litton and Textron developed a new administrative structure: in brief, central offices were small and included mainly finance, acquisitions and control. Hence, the purchasing, production, research and development (R&D) and marketing were left to separate divisions, namely the former firms. Consequently, in conglomerates the role and importance of the divisions was enhanced.

During the third most important merger movement in American economic history (in the 1960s) the number of mergers and acquisitions rose from about 2000 in 1965 to more than 6000 in 1969, falling back to 2861 in 1974 (Chandler, 1990, p. 622). Overall, American managers, via this merger movement in the 1960s, chose an easy way of making extra profits, or at least of conserving existing ones, but they did so not by lowering unit costs of various products via economies of scale as they had done, so successfully, during the Second Industrial Revolution. However, as Schleifer and Vishny (1994) conclude, in the 1980s there was a tendency to consolidation and specialization, and away from unrelated diversification. The same authors debated as to whether or not both periods examined (the 1960s and 1980s) were based on the principle of efficient diversification, and their conclusion was that the 1980s move by the managers was a move to correct the inefficiencies created in the 1960s.

Another aspect, which ought to be emphasized in this brief discussion, is that diversification and multidivisionalization (M-form of firms) go hand in hand. The continuous process of vertical and horizontal integration as well as product extension into additional industries during more than 100 years of American economic development has ended up in more and more multidivisionalization. A measure of the extent of diversification can be obtained by examining the number of different Standard Industrial Classification (SIC) categories in which the largest firms operated. Thus, in 1948 relatively few of those firms operated in as many as five three-digit industries (see Appendix 6.2), but by 1973 a sizeable number were operating in 10 or more such industries (Chandler, 1990, p. 619).

From the so-called U-form structure of the American firm of the 1900s (which started in the 1890s and continued into the 1910s) which was centralized, and functionally departmentalized, a new structure, the M-form, started emerging from the 1920s which was multidivisional in nature. As Chandler emphasized in his numerous studies (for example, in his *Strategy and Structure*, 1962), not all industries were keen to adopt the M-form. Basically, the latter was adopted by the most technologically and capital-intensive industries, such as Du Pont and General Electric. Rumelt (cited in Scott, 1973, p. 139) notes that the proportion of multidivisional firms within the Fortune 500 grew from about 15 per cent in 1950 to over 75 per cent in 1969. At the same time, in 1969, numerous companies were operating 40 to 70 divisions, whereas before World War II the corporate office of many large firms rarely managed more than 10 divisions (Chandler 1994).

Some general remarks will close this sub-section. First, as Chandler suggested (1990; 1994), following the continuous diversification into related and unrelated industries, there has been a separation between top managers at the corporate office and the middle managers of the divisions. This separation created a problem regarding the statistical and cost accounting tools used by managers to evaluate projects and division performance. Second, after the Second World War, in parallel with decentralized profit-planning, the system of capital budgeting was developed in order to organize data and make relevant decisions. As Baldwin and Clark (1994, p. 80) indicate, the resulting cash inflow-outflow view of investment 'remains the theoretical core of most capital-budgeting systems in the early 1990s'. However, as these authors testify, such systems, though useful in some aspects of efficiency evaluation, led companies to invest vigorously but in the wrong projects.

2.4.2 Fordism, Neo-Fordism and Flexible Specialization

Since the introduction in various ways of the Tayloristic subdivision of labour at the beginning of the 20th century and the mass production of cars achieved by Ford which was also based on applications of scientific management, a growing part of American manufacturing became gradually 'Fordist'. Fordism can be extended beyond its relevance to the production process to become 'a social system combining a Tayloristic organisation of work with a Keynesian role of the state and with mass consumption of standardised industrial products' (Jaeger and Ernste, 1989, p. 163). Ford had actually initiated this parallel mass consumption by substantially increasing wages and promoting his cars.

The new information technology introduced in the 1950s (with the development of the electronics industry) expanded in the 1960s, and really took off in the 1970s. It allowed the appearance of flexible and programmable machines, which can be used not just for one task, but for a whole series of tasks. These machines can extend Tayloristic organization of labour from mass production to the production of small lots and even single products. Consequently, 'Corporations would be run as networks of establishments often considerably smaller than the classical Fordist factory. Also, many forms of subcontracting, franchising and the like are used to a considerable extent to create formally independent units well integrated in such networks'(Jaeger and Ernste, 1989, p. 173). In this respect, Chandler (1990, p. 607) says:

> In the 1960s and 1970s a wide variety of industries shifted from electromechanical to electronically controlled processes of production that began to transform the work place and alter the materials used in production. They realigned the economies of scale and scope, often reducing minimum efficient scale and at the same time expanding the opportunities for exploiting the economies of scope …

Neo-Fordism is the continuation of Fordism in the USA, thus still including the trend of Taylorization and deskilling, but also including the new information technology. As such, neo-Fordism is opposed to post-Fordism, which is more related to the system of flexible specialization. The latter can be defined in broad terms as a vertical disintegration of some core industries or as 'the establishment of a much more independent network of small plants based on a work organisation which as a complement to flexibility and specialisation explicitly emphasises professionalisation' (Jaeger and Ernste, 1989, p. 176).

Overall, neo-Fordism has not introduced new OIs, as Jaeger and Ernste (1989) or McLoughlin and Clark (1994) testify. Post-Fordism on the contrary has, but not in the USA, at least not up to the 1980s. Post-Fordism is more linked to the OIs introduced in Japan after the 1950s (McLoughlin and Clark, 1994); this will be explored in the next section and chapters, though the concept of post-Fordism *per se* will not be analysed (except on an ad hoc basis, for example, in Chapter 4). However, in the USA, during the last 20 years firms in all sectors have been in the process of imitating (American firms were becoming less internationally competitive contrary to Japanese firms) a strong element of post-Fordism, namely the JIT/QC process, which initially had been developed in Japan. This topic regarding the American imitation of the JIT/QC philosophy will be the object of Chapter 7.

2.5 JAPANESE OIs AFTER WORLD WAR II

2.5.1 The Just-in-Time System

Though the JIT cum quality control system is the object of exploration in several sections and chapters of this study, the present sub-section is a necessary introduction to this system, mainly seen through the analysis of Best (1990) (numerous other references will be used in subsequent sections). This author states:

> It was not until the 1970s that Japanese firms developed the plant flexibility to produce a range of products on the same production lines without driving up indirect costs. The new plant flexibility emerged out of the focused system and the refinement of associated production activities which together established the organisational preconditions for the 'just-in-time' (JIT) production system. JIT is not simply an inventory management system … (ibid., p. 143)

Best (1990) suggested that the 'New Competition' (as it is elaborated below) emerging from Japanese firms is not sufficiently explained by strategy and culture alone, but also by new production innovations such as the JIT system. The latter is closely linked to the concept of process efficiency, which includes the productive time (per worker or per machine) as the basis for operational efficiency, as well as the unproductive time, which eventually generates process efficiency. The unproductive time is the time materials spend in inventory, handling, moving, inspecting, recording, batching, reworking, chasing, counting and repacking. Emphasis on process efficiency, and hence unproductive time, is justified when one observes that the amount of time workers are actually transforming material is a small percentage of total production time (estimated as being as low as 1 per cent to as high as 20 per cent) (Best, 1990, p. 148).

Another way to measure process efficiency is the 'work-in-process' (WIP) turn, which is the ratio of annual sales to WIP (the value of inventories excluding inventories of produced goods waiting to be shipped). Abegglen and Stalk (1985, pp. 113–14) provide the following relevant information. In the late 1970s most Western vehicle companies had WIP turns of around 10, but Toyota's (the leader in that respect) WIP turn was greater than 300 per year and Mazda's WIP turn went up from 12 times in 1976 to over 50 times in 1981. The same authors calculated that every doubling of the WIP turn increases labour productivity by 38 per cent.

Process efficiency is not only related to the inventories side of the production process but also to the production itself. Thus, it is inherently linked with short production runs (flexible production reducing the cycle from start to finish; for example, the time to start a new car model and finish its production is considerably diminished under the regime of short runs). These runs reduce the time that materials are stored and level out production runs in supplier firms (see Atkinson, 1990 for a brief description of the close links between short runs, JIT, work cells, customer responsiveness, and so on). In order to find alternatives to long production runs, the solution was to drive down changeover times so that small batches could be run with minimal interference with flows. This necessitated the invention and innovation of 'single-minute exchange of dyes' that are necessary in the production of cars, or other products (Best, 1990, p. 152).

The organization of production in Japan was revolutionized in at least three other related ways (Best, 1990). First, machines were laid out in U-shapes with a single worker overseeing a group of machines around her/him, in lieu of the usual straight-line layout. This organizational innovation also increased productivity. Second, workers became problem-solvers as opposed to merely minders, since their tasks were transformed into detecting the problem of the self-stopped machines and developing a solution. Third, a degree of automation became necessary, though as Abbeglen and Stalk (1985, p. 116) remarked:

> Many Japanese companies view the adoption of JIT as the prelude to full factory automation to further reduce costs. As one recent executive of a robot manufacturer said, 'If you want to know what is wrong with your process, try putting a robot in it'. Factories must be running efficiently before they can be automated ...

Another aspect closely related to the implementation of the JIT system is the subcontracting process. Parent or generally larger companies have established with suppliers, or generally smaller firms, two types of subcontracting, informal or *kyoryokukai* according to which written contracts are non-existent, and formal. A set of 'shared network norms' is established over time in the former with the objective of achieving a trusting relationship. Within the subcontracting system special emphasis must be given to component design, which, depending on the relative independence of the smaller firms, enables technological innovations to be maximized.

Finally, three more aspects related to the JIT system will be briefly mentioned here. First, Western researchers on the miracle of the modern Japanese economy, from 1950 to 1990, repeatedly observed that there is cooperation at all levels of Japan's society. For instance between workers and managers, between government agencies and firms, and so on ' ... Japan's physical and geographical characteristics give social reinforcement to its social history of groupism, interdependence, and sense of ON and GIRI-debt and obligation' (McMillan, 1984, p. 23). As Caves and Uekusa (1976, p. 59) have written:

> Japan is a group-oriented society, and its economy exhibits a startling variety of groupings of firms that extend beyond well-defined commodity markets. These groups in turn are cemented by a variety of commercial linkages: between buyer and seller of goods, lender and borrower, shareholder and issuer of equity ...

This spirit of cooperation can be especially detected in firm networks. Imai (1994) identified four different types of enterprise groups (*zaibatsu*, independent, fission and network types[5]), which can be combined into many other hybrid types in the real world in Japan. Even in the apparently least connected groupings, the fission type, the sense of cooperation is paramount. As cited in Imai (1994, p. 136), the President of the fission-type group Maekawa said: 'Each company strives to have its own character, but the more clearly defined that character becomes, the clearer it becomes that it is part of the whole, and that it must collaborate with other group members which have other characters ... '.

Second, regarding cost systems, 'The Japanese do not let the accounting procedures determine how they measure and control their organisational activities. There are relatively fewer accounting and finance personnel in Japanese firms compared to similar US companies' (Giffi *et al.*, 1990, p. 155). Also, direct labour cost is not used to allocate overhead expenses. The Japanese treat direct labour cost as a fixed asset (ibid.). Third, although the tendency for diversification and multidivisionalism has followed an upward trend since World War II, 'M-form organisation has developed less extensively in Japan than in America, Britain and West Germany, yet still accounted for over 40 per cent of the country's leading companies in the early 1970s' (Cable and Yasuki, 1985, p. 417).

2.5.2 Quality Products, Quality Control and *Kaizen*

The story about the quality issue in Japan is a fascinating one and lies at the heart of the Japanese economic miracle. Before World War II, Japanese products were of low quality (Best, 1990, p. 137). After that war things changed rapidly, mainly because of some suggestions made by American scholars such as Deming (1986) who, in turn, was influenced by his mentor Shewhart, who wrote, in 1931, a book entitled *The Economic Control of Quality of Manufactured Product*. In 1950, the Japanese Union of Scientists and Engineers (JUSE) asked Deming to give a series of lectures on statistical quality control, because they (JUSE) 'saw potential in such methods for addressing the Japanese problem of low quality products' (Best, 1990, p. 159). At the same time Deming met many Japanese managers. 'The rest is history, as the methods (Deming's) came to be adopted by virtually all of Japanese industry' (ibid.).

Following Deming, Duran and Feigenbaum also carried the same message to Japan where any type of quality control soon became an obsession at all levels of production, from the hourly workers to market researchers, to product designers, to materials acquisitions employees and others. Quality control, especially based on statistical methods, became priority number one in all business activities. Overall, Japanese managers and industrialists systematically followed Deming's main ideas, which are summarized in his 14 famous points (Deming, 1986; or Evans and Lindsay, 2002, p. 92), for example: refuse to accept defects; cease dependence on mass inspection; find problems and continually and forever make improvements; provide vigorous and ongoing education and retraining; clearly demonstrate management commitment to all the 14 points every day.

Deming's ideas can also be summarized in the following way: when a worker is deprived of his/her right to do good work and to be proud of himself or herself, this may be the single most important contribution of management to poor quality (Deming, 1986). This is exactly what the Japanese worker was not deprived of, as after World War II the quality movement in Japan has generated a work environment such that the Japanese worker became dedicated to and proud of his/her work, for example via the creation of numerous quality control circles (for more on these circles see Chapter

4). It is relevant to see what Best (1990) has to say in this respect: 'The point is that Japanese companies have not been successful because of cultural characteristics peculiar to the Japanese but because of the organisational characteristics that have tapped human energies presumed, at least by scientific management, not to exist in working people' (ibid., p. 161). Initially, QC was applied in heavy industries such as steel, because they especially required instrumentation control. At a later stage, other industries started to introduce QC for such products as consumer durables and home appliances (Imai, 1991, p. 12). Overall quality control in Japan has been growing at an exponential rate since 1950, as Bhote (1988, p. 28) has shown. This exponential evolution can be compared with the historical expansion of quality control in Japan as Table 2.1 summarizes.

Table 2.1 Initial steps of quality control in Japan

Date	Event
March 1950	The JUSE started publishing its magazine *Statistical Quality Control*.
July 1950	Deming's first visit to Japan invited by the JUSE.
1950s	Several visits by Deming.
July 1954	J.M. Juran invited to Japan to conduct a JUSE seminar on quality control management, the first from an overall management perspective.
1956	Japan Shortwave Radio included a course on quality control.
November 1960	The first national quality month was inaugurated.
1960	Q-marks and Q-flags were formally adopted.
April 1962	The magazine *Quality Control for the Foreman* was launched.

Source: Based on Imai (1991), pp. 10–12.

All these points clearly show at least one issue: a continuous dynamic search for and implementation of improvement. This is inherently related to *kaizen*, which means continuous improvement in all aspects of life. According to Imai (1991) *kaizen* is one of the most commonly used words in Japan.

Kaizen is more of a philosophy, or a strategy, than an actual technique or innovation, and as such it embraces everything else analysed so far. Imai (1991, p. 4) puts under the 'umbrella of *Kaizen*' the following concepts: total quality control (TQC), robotics, QC circles, suggestion systems, customer orientation, automation, discipline in the workplace, total productive maintenance, *kanban*, quality improvement, JIT, zero defects, small-group activities, cooperative labour–management relations, productivity improvement and new product development. *Kaizen* is especially linked with TQC as it is already obvious from the fact that continuous improvement is inherently related to quality control. As Imai (1991, p. 13) has stressed 'Japan has developed an elaborate system of *Kaizen* strategies as management tools within the TQC movement'.

2.6 JAPANESE AND AMERICAN EVOLUTIONS OF FIRMS AND OIs: A BRIEF COMPARISON

From this very brief historical overview of firms and OI developments in Japan and the USA some conclusions can now be readily made (see also Sanidas, 2005a). In Japan, in parallel with only a handful of large conglomerates, the trend has been smaller and more numerous firms from the beginning of the country's industrialization process. This was due to the early development of the focal type of company given the factors of limited resources and technology. Hence the search for cooperation and, consequently, economies of scope became the norm, to the detriment of economies of scale. In the USA a contrary evolution took place, in which, from the beginning of the country's industrialization, there was a tendency for the creation of big business, diversification of products, and eventually multidivisional organization. This was due to a natural abundance of resources, human and non-human, plus a series of major and minor technical and organizational breakthroughs (see subsequent chapters for an exploration of this statement). This divergent historical evolution of the firm in the two countries demonstrates that there is not a unique way to evaluate competition and the role of firms, but this evaluation depends on the historical context of factors such as availability of resources, technology (TIs) and organization (OIs). In other words, what is good for the USA might not be good for Japan or what is good for Japan might not be good for Canada, and so on. In addition, this historical and spatial dependency is part of the contingency approach in management and related sciences.

Also, in Japan, many organizational innovations such as flexible production, JIT, quality control, and Japanese-style subcontracting enhanced and sustained the development of small and medium enterprises (SMEs). On the contrary only recently in the USA, from the late 1980s, has there been a reversal in the continuous expansion of big business; thus in the last 20–25 years in the USA big firms have been shrinking and SMEs expanding in number (for example, cf. Stewart, 1993, or Brown, 2000). This is to a significant extent explained by the reorganization of American firms in terms of the

JIT/QC system as this will be demonstrated in subsequent chapters and especially in Chapter 7. Also note that no other major organizational innovation took place in this country after World War II except for an intensification of product diversification and multidivisionalism.

In Japan, the conglomerate type of business existed from the start of the country's industrialization process and has continued its development up to the present. However, in the USA it only appeared in the mid-1960s, but its development was reversed in the 1980s. Again, this antithesis between Japanese and American firm evolutions can be – at least partly – explained by differences in OIs and TIs, resource endowments, as well as historical circumstances and culture. In subsequent chapters, and especially in Chapter 4, a more detailed comparison of developments in each of these countries will be carried out.

2.7 CONCLUSIONS

In this chapter a historical overview of the USA and Japan uncovered the major OIs as described by eminent scholars and in particular by economic historians. A summary of these OIs and their chronology is provided in Table 1.1 of Chapter 1 (although the list is not exhaustive). In the USA most of the major OIs appeared during the Second Industrial Revolution, between 1880 and 1920: mass production and mass distribution and their integration, mergers, diversification, scientific management, Fordism and so on. In the same country, no major OI was discovered after World War II, except for the adoption of the JIT/QC system during the last 20 years via the introduction of flexible specialization and post-Fordism. On the contrary, in Japan some major OIs appeared after the 1930s and especially after the 1950s: focal firms, JIT/QC, quality control, *kaizen*, and so on.

These OIs, as related to industrial and economic growth, will be extensively scrutinized in the remainder of this study, first from a theoretical point of view and then from an empirical viewpoint. For example, in Chapter 4, from the various and numerous OIs identified in several historical periods some patterns and common ground will emerge.

NOTES

1. Most of this sub-section draws from Chandler (1977), unless otherwise indicated.
2. Sklar (1988, p. 1) says: 'The period 1890–1916 in United States history, encompassing what is commonly called the Progressive Era, was both an age of reform and the age of the corporate reconstruction of American capitalism'.
3. It is worth noting that a similar system of industrial organization to the Japanese focal

factories, also called *territorial administrative systems*, can be found in Italy, France and Germany, where industrial local districts are highly viable (see for instance, Best, 1990, for the Italian case, Piore and Sabel,1984, for the French and Italian cases).

4. Discussions as to why firms diversify are found in Grant (1998), and Rumelt et al. (1994).
5. Examples: Honda Motor for the independent type, Maekawa for the fission type and pharmaceutical sales firms for the network type.

3. The role of OIs in the theory of economic growth

3.1 INTRODUCTION

In Chapter 2, a historical review of the economies of the USA and Japan revealed the nature and the importance of various OIs. The first question to answer subsequent to this review is how OIs fit into the preoccupations of economists and other scholars and, consequently, how OIs fit into the process of economic growth. In this study, the latter is defined as changes in the rate of either real output, or productivity (total factor or labour productivity[1]). When necessary, reference will be made to the growth in employment or other related variables, although it is not a primary concern in this study how these variables (including employment) are related to economic growth.[2]

In this chapter the foundations for the inclusion of the impact of OIs on the process of industrial growth are laid. Although economic growth is not the same as economic growth in general (the latter includes growth of the services, agriculture and mining sectors), most of the analysis carried out in this chapter is related to economic growth (whereas industrial growth will be more explicitly dealt with in Chapter 4). In this chapter, some 'hidden' mechanisms of economic growth and development are explored. In brief, there are some scholars who, for example, have expressed the need to consider or emphasize the impact of OIs *per se*; this is reviewed in the second section together with some other relevant economic concepts. The third section examines the relationship between OIs and TIs in terms of whether technology is embodied or not, or in terms of the simultaneity of impact of OIs and TIs, and so on. It emphasizes the need to examine OIs separately. The fourth section shows why and how OIs can be incorporated more rigorously in a production function (in a non-mathematical way).

3.2 LITERATURE REVIEW ABOUT OIs

3.2.1 The Focus of Scholarly Concern

It is seldom the preoccupation of economists to analyse the impact of OIs on economic growth. This sub-section attempts to collect the writings of scholars who expressed some concern about it.

Marshall was amongst the first economists to have included OIs in his remarkable analysis more than 100 years ago. The fourth book of his *Principles of Economics* (1890, 8th edition reprinted in 1949) is entitled 'The agents of production, land, labour, capital, and organisation', and contains 13 chapters. Thus, he recognizes that 'it seems best sometimes to reckon Organisation apart as a distinct agent of production' (p. 115). More precisely, Marshall suggested (ibid., p. 221) that:

> We may divide the economies arising from an increase in the scale of production of any kind of goods, into two classes – firstly, those dependent on the general development of the industry; and, secondly, those dependent on the resources of the individual houses of business engaged in it, on their organisation and the efficiency of their management. We may call the former external economies, and the latter internal economies …

In relation to the 'internal economies', he explored the relations between machines and labour, and the division of labour in general. In relation to 'external economies' he included the particular localities of specialized industries, and a comparison between large and small firms. He also explored the role of managers, entrepreneurship, leadership and the various legal types of firms, such as the primitive handicraftsmen, cottage factories, professionals, joint-stock companies and so on. Finally, he related improved organization with increasing returns of profitability (ibid., p. 265). To summarize Marshall's contribution it is worth quoting Coase (1937, p. 387) in this respect, 'If a workman moves from department Y to department X, he does not go because of a change in relative prices, but because he is ordered to do so … This fact has not been ignored by economists. Marshall introduces organisation as a fourth factor of production'.

More recently, Maddison (1982) enumerated several factors that have an impact on economic development. These factors are to some extent based on Schumpeter's work, as his comments show:

> Schumpeter described the nature of economic development as the 'carrying out of new combinations', which he defined rather widely as follows (in fact, only the first two of these represents what is conventionally included in the notion of technical progress):

a. Introduction of new goods;
b. Introduction of new methods of production;
c. Opening a new market;
d. Conquest of a new supply of raw materials;
e. *New organisation of an industry* … (Maddison, 1982, p. 20, emphasis added)

Schumpeter (1934) said that OIs are important in determining economic growth, although he was more preoccupied with the role of innovative entrepreneurs and their leadership in combining the five factors mentioned by Maddison (see above quote) than any one of these factors in particular, for example the 'new organisation of an industry, like the creation of a monopoly position (for example through trustification) or the breaking up of a monopoly position' (Schumpeter, 1934, p. 66).

The definition of 'technical progress' in the above quote by Maddison very often assumes that organizational or managerial elements are not included, and if somehow the latter are at least recognized as being a separate conceptual and practical entity they are very quickly eliminated from further analysis. Two examples illustrate this point. First, Jackson (1998, p. 15) has split process innovation into two parts: 'technological change' encompassing change in materials, plant, equipment and so on, and 'change in technique' encompassing change in work group organization (for example, assembly line) and change in individual work (for example, parts of scientific management). However, on the next page, Jackson says: 'Change in technique is a large but somewhat specialised subject and will not be further discussed'. Second, although Hall (1994) includes 'the organisation of production' in the definition of 'technological change', he rarely refers to this organization in his entire book on innovation, economics and evolution.

Also, Abramovitz (1993, p. 237), in his search for the areas of ignorance about the sources of growth, said: 'There is still far too much that is poorly understood about the influence of relative factor costs, about the evolution of science and technology, and about the political and economic institutions and *modes of organisation* on which the discovery or acquisition of new knowledge depend' (emphasis added). This new knowledge is, for Abramovitz, one of the interdependent sources of economic growth.

Along similar lines to those of Abramovitz, Nelson (1996, p. 119) suggested in a much more concrete way: 'From the perspective of evolutionary theory, firm diversity is an essential aspect of the processes that create economic progress'. This author, though much more preoccupied with technological innovations (meaning technical innovations as per our terminology) as the key to economic growth, regards the organizational differences between firms as the key to the appropriate exploitation of technology. In an earlier article Nelson (1994, p. 237) remarked: 'There has been far more study of the new technology advances than there has been of the way firm organisation has evolved'. Nelson, in a more recent article (1997), besides his main point that Abramovitz in 1952 had already elaborated on many more contemporaneous issues of exogenous and endogenous growth models, repeatedly emphasized that organization of firms and institutions should be more on the agenda for academic research, instead of just technology.

In his work Nelson (1996) made it clear that OIs are paramount in determining productivity changes at the firm level. Thus, he explicitly states:

> This calls attention to complications repressed in the neoclassical theory of production. One is the need for mechanism to coordinate action. Given division of labour, jobs must be compatibly designed and appropriately meshed. A network of information flow is needed so that the whole job gets done smoothly. This problem of organisation would exist even if all individuals shared the goals of top management … (Nelson, 1996, p. 20)

Following this statement, and basing his suggestions partly on the work of other scholars, he distinguishes between 'classical' and contemporary views of organizational theory of the firm; in the former he includes Taylor's scientific management (see below), or a machine with human elements; and in the latter he includes the Hawthorne works experiment or social system norms (Nelson, 1996, pp. 20–24). More precisely, Nelson said: 'Management cannot effectively "choose" what is to be done in any detailed way, and has only broad control over what is done, and how well. Only a small portion of what people actually do on a job can be monitored in detail' (ibid., p. 21). This 'dichotomy' between the machine model and the social system model of firm organization will be one of the main themes of analysis in this study in later chapters when the Fordism cum scientific management axis versus the lean production system will be extensively explored.

Despite Nelson's (and other evolutionary economists') recognition of the role of OIs in positively impacting on the growth of firms, Nelson was predominantly preoccupied with TIs and their impact on economic growth. Thus, he explicitly admitted, 'These and the other recent evolutionary models do not address the complexities of internal firm organisation or of individual or social psychology' (ibid., p. 51). Nevertheless, it remains true that eminent scholars such as Nelson have been actively exploring the intricacies of OIs in economic theories.[3]

Hunt (1997) in linking evolutionary economics, endogenous growth models, and his resource-advantage (R-A) theory said: 'R-A theory is an evolutionary, disequilibrium-provoking, process theory of competition, in which innovation and organisational learning are endogenous and in which entrepreneurship and institutions affect economic performance' (p. 437). The same author quoted another writer, namely McNulty, who in 1968 'pointed out three decades ago, that "the essence of industrialisation and economic growth is a changing production function and the development of new products, techniques, and forms of business organisations"' (p. 429). Also, Tisdell (1996, p. 107) remarked: 'A part of productivity progress may be due to ... improvements in organisation of the business firm as it learns how to best adjust to innovation and change in its production processes'. Another well-known scholar, Porter (1990) stressed the importance of strategies, skills and organizational arrangements in determining the 'competitive advantage of nations'. Sayer and Walker (1992) extensively analysed the importance of OIs and often used the same terms, namely 'OIs'.

Rosenberg (1976) has been one of the main proponents of the theory that technological change is a major force in shaping economies and histories. However, overwhelmingly, he analysed TIs and ignored OIs, although he acknowledged the importance of the latter as well. Thus, he said:

> The economy's output may be raised not only by increasing the supply of inputs (movements along an existing production function) or by technological change (shifts in the production function) but also by numerous kinds of alterations in the qualities of the inputs of a sort which typically escape the scrutiny of the economic theorist. It is apparent that economic development is associated with important qualitative changes in the human agent as a factor of production. These improvements take such forms as changes in knowledge, technical skills, organisational and managerial abilities, levels of economic aspiration, responsiveness to economic incentives, capacity to undertake and to adapt to innovation, etc ... (Rosenberg, 1976, p. 86)

Two more recent relevant survey articles have 'discovered' the need to consider more closely organizational issues. The first, compiled by Fuhrer and Little (1996), was based on an economic conference organized by the Federal Reserve Bank of Boston, where several well-known economists such as Solow, Jorgenson, Grossman, Rosenberg, A. Greenspan and others participated: 'Most participants were sympathetic with the need to decompose technology into its constituent parts – innovation, development and diffusion – and, to include intangibles like organisational structure, management skills and culture in the package labeled technology' (p. 4). The second survey by Ruttan (1998) reached a similar conclusion.

The review of the literature has so far not been exhaustive, and more references on this topic will be referred to later in this study. Some other references could also be made, including those that are related to institutional issues, such as Dosi et al. (1992). A supplementary review of the literature and relevant analysis will take place in Chapter 4 for a more concrete elaboration of the issues at hand. Overall, despite these sporadic concerns about the impact of OIs on economic growth, not many scholars have consistently demonstrated this impact, at least qualitatively or even to any extent quantitatively. Perhaps the most outstanding out of these exceptions are Chandler and Lazonick.

Chandler's work is one of the strongest or closest points of reference one can get regarding the impact of OIs on industrial growth. His historical analysis of the American industrial evolution was systematically cited in the previous chapter (based on his books of 1962, 1977, 1990). Three remarks will be made here. First, Chandler's work is a continuous interplay between OIs and TIs, sometimes the former preceding the latter, and sometimes vice versa. Second, he analysed the organization of business mainly in terms of management functions and strategies. Third, he very often referred to the changes that took place in the organization of the shop floor and design of factory works. The organizational features of big business are at the heart of Chandler's work: 'Such institutional linkages may help to explain why Britain and France missed out on what has been called the Second Industrial Revolution … In Britain and France technically trained salesmen, production engineers, laboratory technicians, and university scientists were not coordinated through the instrument of the large business enterprise' (Chandler, 1981, p. 169). On the contrary, in the USA the latter were the moving force of the American 'managerial capitalism'; they replaced to some extent market mechanisms to establish a 'visible hand' on economic performance. Finally, to summarize Chandler's position, the following quote from his 1994 article is useful: 'The actual economies of scale and scope, as measured by throughput, are organisational. Such economies depend on knowledge, skill, experience, and teamwork – on the organised human capabilities essential to exploit the potential of technological progress' (p. 81).

Lazonick's work is again one of the strongest or closest points of reference someone can get regarding the impact of OIs on industrial growth. Effectively, Chandler and Lazonick complement each other in several ways. In his 1991 book, Lazonick thoroughly examined both the American and the Japanese economies, thus extending Chandler's analysis even further. His main points can be summarized in four propositions. First, he used many economic concepts of various scholars, such as Marshall and Williamson, to emphasize that Chandler's 'managerial capitalism' promoted economic growth (for example cf. Lazonick, 1990, p. 291). Second, 'The Japanese have not rejected managerial capitalism, but have elaborated it into a set of institutional relationships that I call collective capitalism', which enabled private sector enterprises to generate high-quality products at low costs (Lazonick, 1991, pp. 15, 36–45). Third, he presented a dynamic theory of the business organization as an engine of economic growth (ibid., ch. 3), based on both TIs and OIs, and by using the concept of the 'high-fixed-cost strategy'. Finally, he combined the 'innovative business organization' with transaction cost theory in a different manner from Williamson: 'By adopting an "organizational success" perspective rather than a "market failure" perspective on the economic institutions of capitalism, one can argue that, when innovation is the issue, the market cannot mimic the high-powered incentives of collective organizations' (ibid., p. 227).

In his 1990 book, Lazonick extensively analysed the organizational transformations that took place on the shop floor during the Second Industrial Revolution in the USA. Thus, he recognizes the immense impact of internal labour structure as a means to advance economic development.

In closing this brief review, reference is made to an article written in an organization-oriented journal. Lillrark (1995) emphasized that OIs such as JIT and TQM become a productive factor in their own right besides capital and labour, and 'the very factor that differentiates successful economies from lesser ones' (p. 971). Such assertions are quite rare to find in economics-oriented journals.

Finally, it is worth adding some comments about the relationship between OIs and entrepreneurship. There is a wide variety of relevant works in the literature, mainly since Schumpeter's work. Some economists have taken a proactive stand as to the important role that entrepreneurs play in shaping economic development (for instance, Lydall, 1998). This indirectly supports the importance of OIs as well, because the latter needs the ideas and actions of entrepreneurs for the implementation of OIs. Baldwin (1972), for example, argues along the same lines, although instead of OIs he refers to TIs. Kirchhoff (1994) suggests that entrepreneurship is the key to dynamic capitalism through the process of business firm formation and growth. Although entrepreneurship is usually related to innovative individuals (mostly creators and owners of new firms), as Schumpeter (1934), Baumol (1995) and others have proposed, entrepreneurship can also exist within large organized companies as McGee (1995) counter-proposed. This enlarged concept of the process of entrepreneurship can then be better related to the process of TIs and OIs (for a more recent and holistic approach to entrepreneurship and OIs, see Sanidas, 2004c).

3.2.2 Link between OIs and Relevant Economic Concepts

Although scholarly concern has not very often been focused on the relationship between OIs and economic growth as the previous sub-section has demonstrated, some economists have developed concepts that can potentially be used to describe this relationship. As a first approximation of the model proposed in this study, Figure 3.1 shows the main elements discussed in this sub-section based on economic concepts and theories suggested by various scholars. This model is only an attempt to put together these various, apparently unrelated, elements.[4] Each one of these elements will now be briefly analysed.[5]

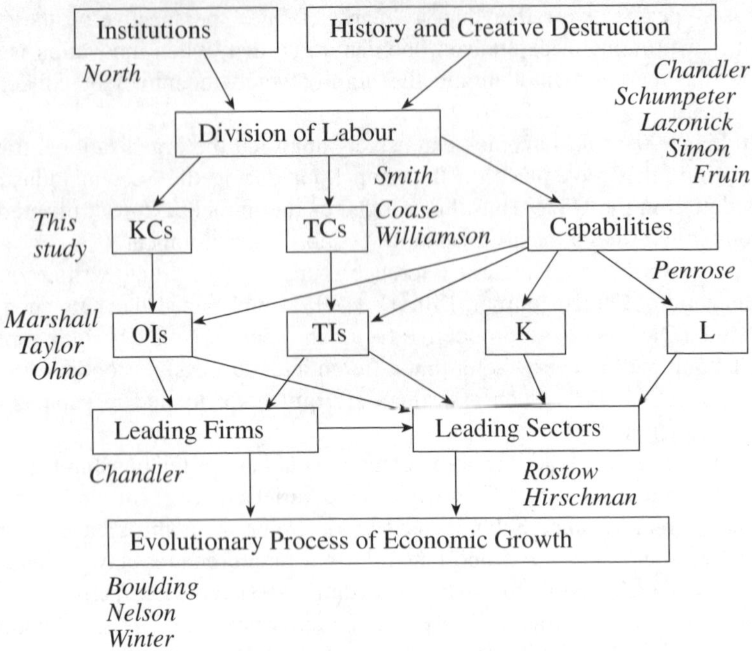

Notes: KCs: kinetic costs (see Chapter 5); TCs: transaction costs; K: capital; L: labour.

Figure 3.1 OIs and economic growth

Coase (1992), in his Nobel Prize speech, outlined some of the main issues related to his contribution to the economic theory of transaction costs (TCs),[6] and the existence of firms and institutions (mainly through his 1937 and 1960 publications). It is appropriate to quote some of his writings in that speech in order to capture the essence of his ideas.

> What happens in between the purchase of the factors of production and the sale of the goods that are produced by these factors is largely ignored … The firm in mainstream economic theory has often been described as a 'black box'. And so it is. This is very extraordinary given that most resources in a modern economic system are employed within firms … I argued in 'The Nature of the Firm' that the existence of transaction costs leads to the emergence of the firm … Not to include transaction costs in the theory leaves many aspects of the working of the economic system unexplained … To have an efficient economic system it is necessary not only to have markets but also areas of planning within organisations of the appropriate size … (Coase, 1992, pp. 713–19)

From Coase's seminal papers many other scholars have substantially analysed the impact of transaction costs on economic theories. Williamson, Demsetz and Cheung are three researchers, among others, as Coase (1992) himself recognized, who have made the notion and idea of 'transactions costs' a new area of fundamental foundation in economic thought. These costs have gradually become a generic name for many other similar or complementary notions, such as governance, opportunism and so on. Many applications of the existence of transactions costs have confirmed its validity. For example, the decision as to whether a firm should outsource some of its inputs or vertically integrate depends on the magnitude of transaction costs involved in these two alternatives.

As North (1992) made clear, when transaction costs are positive, institutions[7] matter. Cheung (1998) suggested that TCs should actually be called institution costs. North (1992) has combined the theory of transaction costs with the evolution of institutions and organizations. He summarized most of his theoretical suggestions as follows:

> In fact, we have incomplete information and limited mental capacity by which to process information. Human beings, in consequence, impose constraints on human interaction in order to structure exchange. There is no implication that the consequent institutions are efficient. In such a world ideas and ideologies play a major role in choices and transaction costs result in imperfect markets … Institutions and the way they evolve shape performance of economies over time … Institutions are the rules of the game of a society[8] … Organisations are the players … The continuous interaction between institutions and organisations in the economic setting of scarcity and hence competition is the key to institutional change … The economies of scope, complementarities, and network externalities of an institutional change are overwhelmingly incremental and path dependent … (ibid., pp. 3–6)

Wallis and North (1986) found that 45 per cent of American national income in 1970 was devoted to transacting. This confirms the existence of transaction costs and hence the existence of the role of institutions and organizations in the economy. However, North's remark in 1992 that: 'We still know all too little about the dynamics of institutional change and particularly the interplay between economic and political markets' (p. 6), adequately describes the state of knowledge in this respect.

In addition, a recent article by Bouttes and Hamamdjian (1997) has shed some extra light on the TCs inside organizations: 'Although coordinating economic activities within firms is very common, it remains a mystery from a theoretical point of view … Our objective is to describe the mechanisms by which operators and activities are coordinated within large corporations' (p. 59). First, these two authors elaborated on Williamson's (1985, ch. 10) typology of industrial relations, itself based on four alternatives of combinations of asset specificity and technological non-separability. Thus, they, for example, formalized the concept of 'residual rights' and their allocation or implementation. Second, they highlighted the need to resolve the coordination issue inside firms due to the problems of communication and bounded rationality, and to satisfy this need they explored three types of contracts: complete, project and residual rights related.

Thus, coordination of operations within firms is inherently linked with numerous aspects of the three types of contracts. Failure of such coordination leads to decreases in firm productivity. Furthermore, the basic elements of the theory of TCs are interrelated with the three types of contracts, and the process of intrafirm coordination. For example, routines are essential in 'complete contracts' that depend on qualifications and professional experience, but they do not exist in 'project contracts'; or, 'discretionary power' and 'information links' are weak in complete contracts but more important in the other two types of contracts. Also, if coordination of the firm's operations fails due to the domination of one team, the residual rights of each team are eventually redefined and hence new contracts are made. Bouttes and Hamamdjian's article (1997) considerably enhances the importance of TCs, since it extends TCs to an important sphere of OIs, namely those that take place inside firms.[9]

Transaction costs in general and the evolution of institutions and organizations are not sufficient to explain economic growth by themselves. We also need the ample and extensive evidence provided by Chandler, Lazonick and Fruin[10] whose historical studies of firm evolutions in several countries support and complement North's and Coase's writings. The complementarity comes from elements such as strategy, sectoral and particular firms' leading role in economic development. It also comes from other elements as Simon (1997) has highlighted: 'The attempts of the new institutional economics to explain organisational behavior solely in terms of agency, asymmetric information, transaction costs, opportunism, and other concepts drawn from neoclassical economics ignore the key organisational mechanisms like authority, identification, and coordination, and hence are seriously incomplete' (p. 237).

We also need Schumpeter's creative destruction and the role of entrepreneurs in shaping booms and recessions, and historical economic development. Entrepreneurs introduce OIs and TIs in a dynamic environment characterized by continuous disequilibria. Nelson (1996, p. 109) nicely summarizes the strong link between Schumpeter's creative destruction and evolutionary economics: 'In our book, *An Evolutionary Theory of Economic Change* (1982), Winter and I spent quite a bit of space presenting a "theory of the firm" which is consistent with, and motivates, a Schumpeterian or evolutionary theoretic view of economic process and economic change'. Thus, within the boundaries of this view, diversity of firms is an expected phenomenon (Nelson, 1996, p. 113).

To further understand the impact of TCs and institutions, and of historical creative destruction on the evolutionary process of economic growth, we also need to emphasize the bounded rationality of economic activities. Simon

(1997) criticized the concept of rational economic man from as early as the 1950s. He emphasized that individuals have limited knowledge; they generate choice alternatives by satisficing and not maximizing their utility. Overall their rationality is bounded as a rule, due to 'the limits of human cognitive capacity for discovering alternatives, computing their consequences under certainty or uncertainty, and making comparisons among them' (ibid., p. 291). This bounded rationality impinges upon many economic activities such as organizational identification (ibid., p. 203). How are these concepts related to economic growth? It seems that the existence of such rationality always creates room for improvements in organizational issues; such improvements in turn increase productivities that eventually speed up economic growth. In other words, the existence of bounded rationality and satisficing behaviour in decisions of production entails that firms, and hence the economy as a whole, are always situated inside their production possibility frontiers[11] (this point will be further analysed later).

The evolution of historical creative destruction and institutions are related to the theories of specialization and division of labour (Smith, 1776; Young, 1928), which constitute strong moving forces for economic growth, especially via the mechanism of comparative advantage. As Cheung (1998, p. 517) remarked, 'Transaction costs as a proportion of gains from specialisation provides a critical measure which, in my view, very largely explains observed differences in the wealth of nations'. Furthermore, division of labour influences capabilities and competences of the firm (and vice versa). Penrose (1959) started a new wave of awareness about the role of firm resources,[12] which in turn became the basis of the impact of capabilities and competences on the growth of firms. We need to know what slows down that growth so that we better understand the overall mechanism of sub-sectoral and macro levels of growth. Penrose remarked that limitations to the growth of firms depend on managerial ability (related to conditions within the firm), product or factor markets, and uncertainty and risk. Consequently, internal resources, and managerial ability, should be included in our models of economic growth. Penrose generated a new stream of thought regarding firm competences. Langlois (1992), for example summarized the importance of these competences:

> Although one can find versions of the idea in Smith, Marshall, and elsewhere, the modern discussion of the capabilities of organisations probably begins with Edith Penrose (1959), who suggested viewing the firm as a "pool of resources" ... ' (p. 286). ' ... In a metaphoric sense, at least, the capabilities of the organisation are more than the sum (whatever the means) of the skills of the individuals in the organisation. In addition to the "skill" of the firm's physical capital, there is also the matter of organisation. How the firm is organised – how the routines of the humans and machines are linked together – is also part of a firm's capabilities ... ' (p. 287).

In addition, as Langlois (1992) again suggests, capabilities, governance and transactions costs are all intrinsically interrelated. This theme of the link between firm competences and transaction costs will be further analysed in the next two chapters, although it is impossible in this study to include most of the substantial research conducted on the firms' capabilities mainly by management scholars. Nonetheless, it is worth mentioning here the substantial contribution of Langlois and Robertson (1995) in this area of research, because it is a significant move to bridge many issues related to both economics and management (see Chapter 5).

Economic growth is affected by OIs as the following short account also shows. The pioneering innovative entrepreneurs Ford, Toyoda and Ohno (discussed in more detail in later chapters) introduced and implemented new modes of production, thus having destroyed the old ones. These three revolutionaries and their organizational systems of production – Fordism and JIT – are a strong point of evidence for Schumpeter's thesis. Taylor's introduction to scientific management provides another piece of fundamental evidence that there can be a positive interplay between institutions and organizations through a reduction in relevant transaction costs. Taylor's concepts became the basis for Fordism and later for JIT, as scientific management was the starting point of an era of rationalization of all intrafirm and interfirm activities. Connected with the JIT system is quality control. Deming (1986) and Juran (1995) theoretically developed the connection between quality control and productivity increases. The continuous improvement in the quality of produced goods is directly linked to a reduction in both production costs and transaction costs (for example, via customers' satisfaction). In turn, QC is inherently linked with JIT, and hence it becomes a continuation of scientific management.

As is shown in Figure 3.1, leading sectors are also important in explaining economic growth. Effectively, Rostow (1960, 1990) has elaborated on the role of leading sectors in the economy and its stages of growth by 1960 (though his ideas on these leading sectors were expressed in earlier publications) and expanded on it in 1990. Again let the author explain some of his main points:

> It is possible to isolate empirically certain leading sectors, at early stages of their evolution, whose rapid rate of expansion plays an essential direct and indirect role in maintaining the overall momentum of the economy … The overall rate of growth of an economy must be regarded in the first instance as the consequence of differing growth rates in particular sectors of the economy … The behavior of sectors during the take-off is merely a special version of the growth process in general; or, put another way, growth proceeds by repeating endlessly, in different patterns, with different leading sectors, the experience of take-off … (Rostow, 1960 [1971], pp. 14, 52 and 53)

> Initially, of course, the growth-rate curve of the leading sector greatly exceeded that of the industrial index. After a time the rate of growth of the leading sector settled down (or fell below) the rate of growth of the overall index ... (Rostow, 1990, p. 459)

Hirschman's (1958) theory of leading sectors complements Rostow's contributions, in the sense that the micro and macro levels of the economy are linked together through backward and forward linkages. In addition, this author has emphasized the importance of the administration and management of firms as being a crucial element to these linkages (Chapter 8 of his 1958 work). The leading sectors in many countries (for example, Great Britain, the USA, Japan), at the beginning of their development, or take-off, were the

Table 3.1 Concepts related to the impact of OIs on economic growth

Scholars	Concepts
Chandler	History, big business, strategies and managerial organizations
Coase	The black box and transactions costs
Deming	Quality control
Ford	Mass production system
Fruin	Focal firms, interfirm economies of scope
Hirschman	Unbalanced growth and leading sectors
Juran	Quality control
Lazonick	History, big business, shop-floor changes
Marshall	Organization as a distinct input of production
Nelson, Winter, Boulding	Evolutionary economics, emphasis on disequilibrium economics
North	History in economics, institutions and organizations
Ohno	JIT process
Penrose	Firm's resources and capabilities
Rostow	Leading sectors, stages of economic development
Schumpeter	Creative destruction and innovative entrepreneurship (disequilibrium economics)
Simon	Bounded rationality and satisficing
Smith	Economies from specialization, division of labour
Taylor	Scientific management
Toyoda	Lean production
Williamson	Transaction costs
Young	Economies from specialization, division of labour

Note: On dates of publications, see text and references.

textiles and clothing and railway industries. The duration of their positive effect on the overall economy was about 50 years (Rostow, 1990). More recently, the leading sectors were the vehicles and electronics industries for Japan, and the chemicals and electronics industries for the USA. However, leading sectors must be linked to the concept of leading firms. As will be further analysed in the next two chapters, leading firms always play a leading role in shaping leading sectors, for example IBM in computers, Ford in cars, and so on.

In summary, the pioneers and the economic foundations of the relationship between OIs and economic growth as they have briefly been described in this section can be summarized in Table 3.1. These foundations will be further explored and used in the remainder of this study in a more specific way whenever necessary.

3.3 OIs AND TIs

In this section a necessary preliminary discussion takes place to clarify the important issue of technology, which has almost always been regarded as substantially contributing to the process of economic growth. Technology is not only TIs, or embodied technology (as mentioned in Chapter 1), but also OIs, or disembodied technology. In this chapter, and the following two, the role of OIs in this process is thoroughly explored, as the focus of this study is disembodied technology, or OIs, in order to emphasize the organizational character of this type of technology. However, the distinction embodied/disembodied has not always been very clear, as the following paragraphs will briefly demonstrate, amongst other issues. Thus, when writers refer to 'technology', they usually implicitly mean the embodied type, or they sometimes mean both TIs and OIs, unless they specify it. Consequently, care is required to differentiate the two types of technology as much as possible, since the object of this research is primarily related to OIs.

3.3.1 Distinction between Embodied/Disembodied Technological Innovations

Several scholars have made the distinction between embodied and disembodied technology. For example, Nelson (1996, p. 29) provides a similar distinction though using different terms: 'To the extent that firm structures and decision-making style are important variables influencing productivity, this fact in itself cautions against thinking of the determinants of labour productivity simply in terms of the quantity of complementary inputs, and technology … '.

One of the original sources relating to embodied/disembodied technology is Solow (1957; 1960), who made this distinction in the late 1950s. In his 1957 article, he says: 'It will be seen that I am using the phrase "technical change" as a shorthand expression for *any kind of shift* in the production function. Thus slowdowns, speedups, improvements in the education of the labour force, and all sorts of things will appear as "technical change"' (p. 312, original emphasis). Further, in the same article, Solow says: 'Obviously, much, perhaps nearly all, innovation must be embodied in new plant and equipment to be realised at all' (p. 316). The same author in his 1960 article is more specific:

> It is *as if* all technical progress were something like time-and-motion study, a way of improving the organisation and operation of inputs without reference to the nature of the inputs themselves. The striking assumption is that old and new capital equipment participate equally in technical change. This conflicts with the casual observation that many if not most innovations need to be embodied in new kinds of durable equipment before they can be made effective … (Solow, 1960, pp. 90–91, emphasis added).

As expected, Solow did not talk about the 'time-and-motion' (cited in the above passage), or 'a way to improve organisation' any more in the remainder of his article, as his preoccupation was to distinguish capital equipment of different dates of construction or vintages. Brown (1968) analysed more extensively the embodied/disembodied distinction, mainly based on Solow's work. He recognizes its usefulness in theory and policy, and amongst other points he reminds us: 'It has been shown that in the long run the embodied and disembodied models yield the same rate of growth when the elasticity of substitution between labour and capital is unity' (Brown, 1968, p. 78).

The distinction between embodied/disembodied technology and its relationship with OIs can be further explored by referring to Peters (1980, Ch. 1). According to this author's discussion, disembodied technology can also be hardware capital that becomes 'embodied' at the first year of operation but becomes 'disembodied' in subsequent years when it has an additional impact on productivity (hence, this definition is more akin to Solow's definition and his calculations of vintages, as was mentioned in the previous paragraph). However, disembodied technology is also due to non-hardware factors such as organizational techniques, thus this definition being similar to the one given by UNCTC (1985, as quoted in Chapter 1). In this respect, Peters refers to three distinguished authors to support his arguments (and hence this study):

1. 'Technological change is the advance of technology, such advance often taking the form of … and new techniques of organisation, marketing, and management' (Mansfield, 1968, p. 11).

2. 'The advance of productivity is sometimes discussed as if it were very largely the result of new technological knowledge. But surely this is wrong. Dramatic advances in management and organisational techniques and in architectural layout are visible to the naked eye' (Denison, 1962, p. 232).
3. 'Denison's emphasis on the importance of "software" is an important corrective to a widespread tendency to think of technological change in purely "hardware" terms' (Rosenberg, 1972, p. 7n).

Then Peters (1980, p. 17) says: '"Scientific management" is a clear example of disembodied change'. This is exactly the argument of this study; OIs such as scientific management are disembodied technology that enhances productivity. Therefore, OIs (or 'software' innovations) must be explored separately from TIs (or 'hardware' innovations). From the cognitive point of view, a priori, there is no reason to assert that both OIs and TIs grow together or in parallel or even with lags, or that they are complementary. On the contrary, there is no deterministic pattern between the two types of innovations. One obvious reason for this view is that it does not have to be the same people who introduce both types, or if it is the same people then there could be delays or lags between the implementation of the two types of innovations.

3.3.2 Non-contemporaneous Impact of OIs and TIs

The non-parallel way of development in OIs and TIs is one of the main reasons why both impacts – that of the OIs and that of the TIs – on economic growth should be analysed. None of these two types of innovations is sufficient for sustained long-term growth. We have a good historical example, that of the former USSR, which shows that TIs alone were not sufficient for their continuous economic development. The former USSR's rigidity of institutional and organizational structures over a period of 70 years finally had a negative impact on this country despite its brilliant record of technological achievements (TIs related) (for example, cf. Nafziger, 1995; Thurow, 1996).

Other examples of the non-parallel development in OIs and TIs are those that occurred in the USA during the period 1900–30 and in Japan during 1950–70. In the first case a substantial improvement in OIs occurred without such a parallel progress in TIs. These OIs included the integration of mass production and mass distribution, vertical integration, scientific management and Fordism. Evidence of this case is also provided by Freeman (1984, p. 66) in a graph depicting significant inventions (patents), in which it is shown that, overall, there was a trough in inventions during the period 1900–20 in the USA.

This finding is also confirmed by the historical exposition of OIs in Chapter 2. In addition, even if we consider the impact that electricity (as an important TI) had on the American economy at the beginning of the 20th century, these conclusions remain the same. According to figures and other aspects provided by various researchers and discussed by Mowery and Rosenberg (1998), residential use of electricity increased more than threefold from 1920 to 1930 (perhaps as a response to a 19 per cent decrease in price) and this happened mainly in urban areas. It is also in the 1920s that the electrical appliances of radio, refrigerator and electric water heater were introduced. The use of electricity in business rose from about 5 per cent in 1899 (electric motors as a percentage of total horsepower) to 39 per cent in 1914 and 55 per cent by 1919 but jumped to about 82 per cent by 1929. However, as Mowery and Rosenberg (1998) remark:

> The effects of industrial applications of electric power on measured productivity growth are difficult to detect until the 1920s … The restructuring of factories, including the reorganisation of the flow of work on the factory floor, new work arrangements, and the development of the necessary new patterns of specialisation on the part of both workers and management took decades of experimentation and learning … (ibid., p. 116)

This comment by Mowery and Rosenberg (1998) provides extra evidence that the introduction of an important technology (TI) is not sufficient to guarantee economic growth but must be accompanied by OIs. The latter can vary from firm to firm and from sector to sector given a new technology (TI). It is these variations, which create differences in growth for firms, for sectors and for economies, given the same technology (TI).

In addition, Table 3.2 summarizes the main technical innovations according to some important periods from the 19th century up to the 1970s. Although the exact dates of innovations (in contrast with the dates of invention) are often approximate, Table 3.2 is a rough but pertinent indication as to when most innovations took place. They did appear mainly during the period 1921 to 1950, and especially in the 1930s and 1940s, and to a lesser extent during the period 1951 to 1975. Admittedly, it is also difficult to judge the impact of TIs by examining their mere number; it is possible that fewer TIs during a given period might be more important than more TIs during another period. Nonetheless, it is also possible that each period has some major TIs to offer,

Table 3.2 Technical innovations

Up to 1865	1866 to 1895	1896 to 1920	1921 to 1950	1951 to 1975
9	27	25	62	41

Source: Freeman (1984); the figures were compiled from his Tables 3.4 and 5.1.

such as, for example, telephone and telegraph as well as steel processes during 1866–95, and the impact of electricity and cars during the 1920s.

In the case of Japan during 1950–70 some leading sectors, such as vehicles, electronics and machines, grew in a substantial manner, not because of any significant TIs (these were mostly imitated from the West), but because of some major institutional and organizational innovations such as cooperative business networks, quality control, JIT and governmental planning assistance (these OIs will be further explored in subsequent chapters).

Sustained productivity growth without technical change has been known since the early 1960s as the 'Horndal effect', introduced in 1961 by the Swedish economist Lundberg (Lazonick, 1992). Lazonick (1992) discusses this effect in order to provide evidence that it is due to other factors such as the impact of management–worker relations on work intensity, and not due to Arrow's 'learning by doing' effect (as Arrow himself has suggested in order to explain the case analysed by Lazonick).

Overall, the suggested theme of this section of a not necessarily parallel development of OIs and TIs is not new in the relevant literature. For example, Wood (1989, p. 457) said:

> This article reinforces the point made by Nelson, Lazonick, and Coombs that we need to clearly separate innovations in management from technological advances. We must also avoid the danger of assuming that particular changes in technology require changes in management methods, or that changes in work organisation must follow technological innovations …

Consequently, once the need to separately examine the OIs is recognized, then in a second stage, more evidence is also needed to examine the evolutionary paths of OIs and TIs.

3.3.3 Links between OIs and TIs: Further Discussion

Nelson (1996, p. 118)[13] was very precise as to how he saw the relation between OIs and TIs:

> From one point of view it is technological advance that has been the key force that has driven economic growth over the past two centuries, with organisational change a handmaiden. But from another perspective, we would not have got that technological advance without development of new ways of organisation that can guide and support R&D and enable firms to profit from these investments …

Whether OIs are just a 'handmaiden' or something more cannot be verified in this study; however, there is some evidence that TIs by themselves are not a sufficient condition for improving productivity. For example, Boer (1994, p. 101) argues that TIs related to the flexible manufacturing system (FMS) can lead to market success providing that there is 'company-wide involvement in the design and management of organizational innovation'. Substantial evidence that OIs are determined first, and then TIs, is provided by Womack and Jones (1996) who explored many firms in revealing detail. Mintzberg and Quinn (1991) in their examination of the Taurus project for Ford cars also provided evidence that OIs were determined first and TIs followed; in addition, 'Petersen attributed 85 per cent of Taurus' gains to "managing smarter" and only 15 per cent to the new technology' (ibid., p. 496). Also, Fujimoto (1999, p. 117) mentions that, 'Kim Clark and I, for example, pointed out that the main source of interfirm difference in the frequency of new product introductions was not the difference in R&D resource inputs but how efficiently given R&D inputs were used', thus inferring that OIs are necessary for the efficient use of R&D. Overall, the message is that we cannot equate OIs with TIs and examine only TIs without examining OIs (or vice versa). This study pays more attention to the OIs; nevertheless, more work is needed to explore more closely the links between OIs and TIs.

Two significant examples will be given here in order to indicate that often TIs and OIs are both present in organizing production in important new ways, but it is difficult to separate their distinct influence without any additional research. The first example is related to the development of mass production as introduced by Ford through the adoption of the moving assembly line. Womack et al. (1990, ch. 2) emphasized that designing new tools that could cut hardened metal and stamp sheet with absolute precision was the key to interchangeable parts, which in turn assisted in developing the system successfully. The same authors (ibid., ch. 3) provided another example; they remarked that the introduction of new dyes that were easy and fast to change was a cornerstone in developing the lean production system in Toyota. Blackburn et al. (1985, ch. 5) seem to adopt the stance that there is a one-to-one relationship between stages of mechanization (namely, control, transfer and transformation) and stages of production process (namely, craftwork, Taylorism, Fordism, neo-Fordism). However, as it will be seen in subsequent chapters, this one-to-one relationship seems to break down when the passage is made from neo-Fordism to post-Fordism (the latter includes elements of the JIT/QC system). The links between TIs and OIs will be further explored in other sections of this study as the need arises.

Chandler and Lazonick filled many pages of research on the co-evolution of these two types of innovations. For Lazonick, it seems that internal management of the firm is necessary to coordinate a specialized division of labour, which includes all employed people in the firm in order to produce high-volume and high-quality goods; technology will assist in producing these goods (1991, pp. 100–103). Furthermore, 'The innovative organization enhances its value-creating capabilities organizationally by unbounding its cognitive competence and by transforming the behavior of its participants, and technologically by committing itself to the development and utilization of organization specific assets' (ibid., p. 229).

As for Chandler, let Lazonick summarize his ideas on the matter:

> Chandler's key contribution to economic analysis is a conceptual framework that can comprehend how organizational capability transforms an innovative investment strategy into low unit costs and competitive advantage … By imposing a transaction cost interpretation on Chandler's historical material, Williamson has failed to comprehend both the causes and consequences of the dynamic interaction between organization and technology that is central to Chandler's approach … (Lazonick, 1991, p. 230)

Whether Williamson (for example, 1985) failed to comprehend the dynamic interaction between organization and technology will not be debated here. However, it is certain that Williamson thoroughly attempted to link organization and technology through the examination of the concepts of asset specificity, governance, transaction costs and so on. Perhaps for this scholar, technological factors taken as being exogenous influenced the organizational features of firms. Nonetheless, the important point for this analysis is that it is not wise to ignore OIs and just examine TIs. Both types of innovations are needed for a comprehensive analysis of the process of economic growth, on a firm, sub-sectoral and macro levels.

Langlois and Robertson (1995) extensively analysed primarily the TIs and to a lesser extent the OIs in relation to each other through using the concepts of competences, routines, transaction costs and so on. For instance, they attempted to find out which firm networks are more or less conducive to technologies (both systemic and autonomous), or why some firms and industries are immersed into inertia in terms of TIs; factors such as institutions, routines and capabilities, learning, strategies of market dominance and others were scrutinized. In their conclusions section of their sixth chapter, Langlois and Robertson (1995, pp. 118–19) said:

> The model presented here is intended as only one explanation among many for the co-existence of a variety of competing technologies and organisational forms even in nations at similar stages of development … It is nevertheless clear that institutional factors, especially those embodied in capabilities and routines, can both improve the ability of a firm to exploit an existing technology and make it more difficult to innovate by generating an inertia that is hard to overcome …

In an attempt to be more precise about the links between OIs and TIs, reference will now be made to Breschi and Malerba (1997). One of the main preoccupations of these two authors was to first link technological regimes (TRs) (predominantly TIs oriented) with sectoral innovation systems (SIS),[14] and second, to quantitatively (by using the proxy of patent applications) link five types of SIS with Schumpeter Mark I and Schumpeter Mark II[15] models in six countries. A TR can be broadly defined by the particular combination of four fundamental factors: opportunity conditions (likelihood of innovations), appropriability conditions (possibilities of protecting innovations from imitation), cumulativeness of technological knowledge (serial correlation among innovations) and nature of the relevant knowledge base (properties of knowledge upon which firms' innovative activities are based). These four factors are defined and analysed in relation to TIs. However, an attempt will be made by the present author to relate the various elements of these factors to OIs. Table 3.3 shows the results of this attempt.

The aim of Table 3.3 is to see whether TR as defined by many authors (for example, Breschi and Malerba, 1997; Malerba and Orsenigo, 1996; Nelson and Winter, 1982) can also be used to describe OIs or to relate OIs and TIs. The conclusion is quite clear (the 'yes' answer to the links with OIs column shows that): TRs can be extended to include OIs.[16] However, the situation is not so simple as is expressed by the answer 'yes'. Organizational innovations, and especially the major ones such as Fordism cum scientific management, have their own characteristics: for example, for JIT/QC, appropriability, cumulativeness and knowledge base are very high, but the JIT/QC system does not need to be patented in order to generate high profits and to penetrate markets. This system generates higher productivity and economic growth due to its connectivity with all layers and aspects of the firm that adopts it (although adopting it takes a lot of time as will be seen in subsequent chapters). Thus, OIs are related, besides the traits appropriability, cumulativeness and knowledge base, to specific elements of complex evolutionary systems[17] such as information and entropy. As Clark (1988, p. 524) remarks: 'Thus the greater the amount of information in any given system the smaller will be its entropy, or the greater will be its "negentropy", and, equally, the greater the flow of information into any system the greater the degree of organisation such a system will exhibit'.

Consequently OIs are characterized by their degree of negentropy. Furthermore, another characteristic of OIs is their ability to support TIs and other types of innovations, hence their degree of supportiveness. Overall, it is useful to adopt the conclusion reached in Edquist et al. (2001, p. 17): 'organisational innovations take place under different circumstances and are governed by different determinants as compared to technological innovation, even though they are also closely related to technological innovation in their development'.

Returning to Breschi and Malerba' s (1997) paper, another way of applying their quantitative analysis to the present discussion is to observe that the sectors belonging to the Schumpeter Mark II model (ibid., Table 6.4, p. 153),

Table 3.3 Technological regimes (TIs oriented): links with OIs

Opportunity conditions	Links with OIs	Appropriability conditions	Links with OIs	Cumulativeness	Links with OIs	Knowledge base	Links with OIs
Level	No	Levels	Yes	Technological level	Yes	Nature of knowledge	Yes for specific knowledge
Variety	No	Means	Yes	Firm level	Yes		Yes for degree of tacitness
Pervasiveness	Yes			Sectoral level	Yes		Yes for degree of complexity
Sources	No			Local level	Yes		Yes for degree of independence
						Means of knowledge transmission	Yes

Source: Based on Breschi and Malerba (1997) and extended by the author.

are predominantly those that have invented and propagated the JIT/QC system (an important OI) not only in Japan but also in the USA (Chapters 6 and 7 provide evidence for this assertion). Thus, the Schumpeter Mark II model (of TIs) is linked with OIs in order to generate economic growth. Malerba and Orsenigo (1996, p. 42) in their attempt to answer the question, 'are firm behavior and organization related to the specific technological environment in which firms are operating?', have examined technological (TIs-related) strategies and specialization/diversification in relation to the main characteristics of TRs. Thus, these authors explicitly state: 'this chapter does not address the detailed features and articulation of firm strategies, the specific types of firm organizations, the relationship between capability and firm behavior, the changes in behavior and strategies during the evolution of an industry and the dynamic interaction between firm behavior and technological regimes' (ibid., p. 43). In particular, capabilities or specific competences are not considered by these two authors except specialization/diversification. On the contrary, in this book, competences and related concepts as defined by these two authors, are the main objects of analysis:[18] 'The concept of competence is intrinsically an organizational one, in that it necessarily implies a set of organizational rules (routines) for the communication and implementation of individual knowledge' (ibid., p. 48).

The overall conclusion of this brief discussion so far is that TIs are only part of the whole impact of innovations on economic growth. More research is needed to more exactly pinpoint the links between OIs and TIs (and other types of innovations as suggested in Chapter 5 of this study) *in relation to economic growth.*[19]

Some scholars recently started more systematically to 'involve' OIs in their exploration of TIs and generally innovations. In Van de Ven (1995), the role of entrepreneurship and institutional arrangements are two of the links between OIs and TIs that are more thoroughly examined. Ehrnberg and Jakobson (1997) include 'firm specific organizational routines and competence' as one of the determinants of the outcome of a technological discontinuity and related inertia. Andersen and Lundvall (1997) explicitly take into account the division of labour in analysing national innovation systems.[20] Davenport (1993) in his study of process innovation (TI oriented) has recognized the need to introduce complementary organizational changes. Rycroft and Kash (1999) in their attempt to show the complexity of technological innovations (mainly TI oriented) examine the relevant organizational complexity and organizational learning. Teece (1998) examines the following factors affecting the development and direction of TIs: monopoly power, managerial hierarchy, scope of diversification, vertical integration, organizational culture and values, governance modes and external linkages. Fujimoto (1999) in his detailed analysis of the Toyota manufacturing system often examines OIs and

TIs together. For example, this author says: 'JIT principles, which had prevailed at Toyota by the 1960s, were applied to such manufacturing activities (dye making, prototyping and startups) in product development' (ibid., p. 195).[21] Antonelli (1999, p. 231) concluded in his book 'the laws governing the creation of new knowledge and the introduction of technological and organisational innovations are central to understanding the dynamics of economic systems'. Dawson (1997, p. 111) remarked: 'The design of new technology and the development of new forms of work organization is an important and generally under-investigated area which may provide useful insight to our current understanding of the process by which these new innovations unfold in practice'.[22]

This brief review of recent literature brings up the following issues: how is the process of innovations (TIs or OIs) triggered and how is it developed or evolved? Especially, how is the process of innovations linked with the continuous flow of organizational change taking place inside firms? What is the dynamic interplay between OIs and TIs (and other types of innovations, as described in Chapter 5)? This study provides only partial answer to all these questions and instead attempts to answer one question more extensively: how do major OIs impact on economic growth? In this context, it is important to emphasize the following 'axiom' of one of the relationships between OIs and TIs as expressed by Kogut (1996, p. 273): 'This essay has traced the coevolution of technological innovations that spawned a wave of new firms, and the concomitant organisational innovations that developed to manage their growth'.

This 'axiom' implies that though TIs might be a necessary condition for the development of new products, the subsequent growth of firms that develop these products depend on OIs. Nonetheless, there are cases where OIs can be the major factor in the growth of a firm (a good example is that of Dell computers as analysed in Chapter 6), thus without the concomitant significant contribution of TIs. Again only precise studies can demonstrate the exact relationship between TIs and OIs. For example, Malerba (1985) in his detailed analysis of the semiconductor sector in the USA, Japan and Europe shows how vertical integration (an OI) depends on TIs and vice versa (within a more general context of demand, supply and public policy issues). Also, Bresnahan and Malerba (1999, p. 124) in their detailed study of the world computer industry conclude 'in each coevolutionary process, a different appropriate model of the firm emerged, with its own competences, organisation and strategy'. Thus, for mainframes, a Chandlerian integrated firm became quite successful due to major and continuous R&D efforts and investments in management, production and marketing (ibid.). However, these two authors did not fully explain why Dell computers became increasingly the major firm in personal computers (PCs) in the last 15 years, except that Dell had marketing or distribution advantages (ibid., p. 111). This particular point will be taken up in Chapter 6, where it will be indicated that internal organization, based on the JIT/QC system, played a substantial role in Dell's success. In order to be more precise in the relationship between TIs and OIs, Chapter 5 of this study explores this relationship from the internal fundamental operations of the firm point of view.

3.4 INCLUSION OF OIs IN THE PRODUCTION FUNCTION

Simon (1997) in examining differences in productivity between car manufacturers said:

> The causes for these enormous differences in efficiency *have almost nothing to do with the classical physical production function.* They also appear to have little to do with cultural differences at the blue-collar level. They seem to have nothing to do, either, with material reward structures, which are not significantly different in the various plants. They must be attributed in large part to differences in management practices (for example, quality control practices, and inventory policies), perhaps bolstered by differences in management attitudes and motivations … (ibid., p. 232, emphasis added)

Mefford (1986) has attempted to test empirically the impact of these 'management practices' in about 30 plants around the world. These practices consisted of three criteria: output goal attainment, cost (factory budget) over or under fulfilment, and output quality level. The general production function estimated was $Q= f(K^*, L^*, MGMT)$, where K^* stands for quality adjusted capital, L^* is quality-adjusted labour and management (MGMT) stands for the management variables mentioned above. 'Management is shown to be an important input to the production process in these plants regardless of the particular functional form estimated' (ibid., p. 96). As Mefford recognized, several other studies were carried out before his own, the general conclusion out of them being similar to his own. The same author also refers to the X-inefficiency theory of Leibenstein (1966) for which he says:

> His theory of X-inefficiency makes the case that allocation of resources is usually less important than effectiveness of use of these resources in determining the amount of output produced. The effects that make-up X-inefficiency are the organisation of the work process, motivation and supervision of employees, and the monitoring and controlling of the operation … (Mefford, 1986, p. 96)

However, management practices as briefly analysed in the above paragraph are not the only OI that can be found to be quantitatively significant in explaining economic growth. Kogut (1996, p. 269) provides both qualitative and quantitative evidence that during the early decades of the 20th century: 'The critical factor was not the organizational capabilities of managerial hierarchies, but the American national strength in the rationalization of work and the standardization of production'. Also, Antonelli (1999) has quantitatively linked sectoral growth in several countries and the concomitant impact of TIs and OIs. In the present book, more quantitative relevant evidence will be provided in Chapter 6 and especially in Chapter 7.

Edquist et al. (2001) in their detailed analysis of the relationship between innovation and employment explicitly take into account organizational process innovations in explaining employment and indirectly growth in the context of a general approach of systems of innovation that looks at an economic system as a whole, but without referring to general equilibrium theories, 'since innovation is the enemy of equilibrium' (ibid., p. 8). These authors distinguish four types of innovations: product innovations in goods, product innovations in services, technological process innovations and organizational process innovations (such as JIT production, TQM and lean production). The relationship between product and process innovations (thus between OIs and TIs) is also discussed (for example, ibid., s. 5.3). Thus, these authors say that consumer products (TIs-related) and intermediate products (TIs-related) never become process innovations (OIs are included in this category). Overall, 'the study of organisational innovations is neglected compared to the study of technological innovations' (ibid., p. 12).

Edquist (1997, p. 23) mentions as an argument favouring organizational change[23] the Japanese auto assembly plants in the USA. These plants though using process technologies similar to their US counterparts, are more productive than domestic plants because of their organizational innovations such as the JIT/QC system (this argument is fully explored in Chapter 7). Edquist (1997, p. 24) proposes three important reasons for including OIs in the concept of innovation, and as an extension in the discussion about the relationship between innovations and employment (and hence growth):

> Organizational changes are important sources of productivity growth and competitiveness and they might also strongly influence employment.
>
> Organizational and technological changes are closely related and intertwined in the real world, and organizational change is often a requirement for technological process innovation to be successful.
>
> All technologies are created by human beings; they are in this sense 'socially shaped', and this is achieved within the framework of specific organizational forms …

Edquist et al. (2001) in their detailed discussion of the impact of their four types of innovation, including OIs, show that effectively, OIs are a part of the production process of firms, sectors and economies. Thus, in their hypothesis five, they argue that one category of OIs is similar to TIs, that is, they are labour saving; another class is capital saving with direct effects on employment levels (ibid., p. 35). These propositions are further discussed in the context of the empirical analysis conducted in Chapter 7 of this book. Gregersen and Johnson (2000) in their review of various approaches in economics as to how innovations affect economic growth, first examine theories such as the neoclassical, growth accounting, new growth, and catch-up growth theories which all use a 'simplistic' innovation concept. These authors then recommend that innovation should be at the centre of analysis within an evolutionary process. Consequently, they propose the 'techno-economic paradigm' according to which it is not only technology, but also a constellation of technological, organizational and institutional factors that form the growth model of a period. In particular, the organizational modes of firms become increasingly important in order to enhance learning capabilities that feed into the processes of innovation. Finally, the 'techno-economic paradigm' can be extended into the system of innovations (SI) approach (Chapter 5 further examines this issue as it is part of the process of growth suggested in this study).

Given this sporadic inclusion of OIs in the process of economic growth by some scholars as a background, the purpose of the remainder of this subsection is to append onto the 'classical physical production function' the variable OIs. In summary, the following equations show the broad relation between the major OIs that have been described so far and elements of economic analysis linked to economic growth.

Output growth (macro or micro) is a function of:

TIs +
Quantity of capital (abbreviated as K) + Quality of capital (QK) +
Quantity of labour (L) + Quality of labour (QL) +
Quantitative relation between K and L (KtgrL) + (3.1)
Organizational relation between K and L (KogrL)+
Organizational relation between L and L (LogrL) +
Organizational relation between K and K (KogrK)

The inputs used in equation 3.1 are defined as follows:

1. K: simple quantity of capital (the usual stock of capital)
2. QK: quality in steel, machines, tools and so on (cf. Jorgenson, 1990)
3. L: simple quantity of labour (as hours or number of employees)
4. QL: the so-called human capital (education and so on) (for example, as number of years in schooling)

5. KtgrL: the classical technological or technical relation or quantitative substitutability between K and L, for example KtgrL could be represented by a Cobb-Douglas function (tgr stands for technical relation)
6. KogrL: the organizational relation between K and L, which is different from KtgrL, that is, altering KogrL does not necessarily change KtgrL (ogr stands for organizational relation)
7. LogrL: the organizational relation between L and L, which is different from KtgrL, that is, altering LogrL does not necessarily change KtgrL (ogr stands for organizational relation)
8. KogrK: the organizational relation between K and K, which is different from KtgrL, that is, altering KogrK does not necessarily change KtgrK (ogr stands for organizational relation).

The inclusion of the last three inputs (KogrL, LogrL and KogrK) explicitly in the production process (or function) is the outcome of this section. These three inputs represent OIs, as these have been described in Chapter 2 and will be further explored in the next two chapters. Their inclusion in the production function is very rare. For instance, Menard (1996) remarked, 'Thus, the production function should include three related inputs: capital, labour, and organizational capabilities: Q = Q (K, L, O)' (p. 291). In addition, Lydall (1998) says, 'The productive enterprise needs to make use of many technologies besides the physical. These may be classified under the headings of commercial, financial, and organisational technologies' (p. 33). An extensive review of the relevant literature was provided in previous sub-sections of this chapter.

A general way of incorporating the above organizational inputs in the production function is now proposed.

$$Q = f(M, K, L, O, T) \tag{3.2}$$

In this production function, all inputs (M: materials, K: capital, L: labour, O: OIs, T: TIs) participate in the output Q in an autonomous way. Thus, changes in O entail changes in Q, even if the other inputs do not vary. In addition, if we adopt the usual calculus results regarding the marginal products of output with respect to each one of the inputs, we can obtain from Lagrangian maximization of Q the cost or shadow price of each input. For example, the marginal product of Q with respect to L (MPL) will provide us with the optimum cost of labour or wages: P. MPL = w, where P is the price (as taken by the firm) in the output market and 'w' is wages/salaries (cf. Nicholson, 2002). In the same way, the marginal product of Q with respect to O (MPO) will provide us the optimum cost of organization: P.MPO = o, where 'o' is organizational costs. These latter costs include transaction costs between the various agents and principals of the firm. But they also include other types of costs, such as informational and pure organizational costs. This point will be re-examined at a later stage, in the next two chapters, and especially in Chapter 5 where the 'kinetic costs' (KCs) will be introduced.

A particular form of equation 3.2 is:

$$Q = f(O_1M, O_2K, O_3L, O_4T, \ldots) \quad (3.3)$$

In this production function, the O_i variable plays the role of a special factor having an impact on each input separately and directly. Figures 3.2 and 3.3 show this type of representation in the case of three factors (K, L and O).

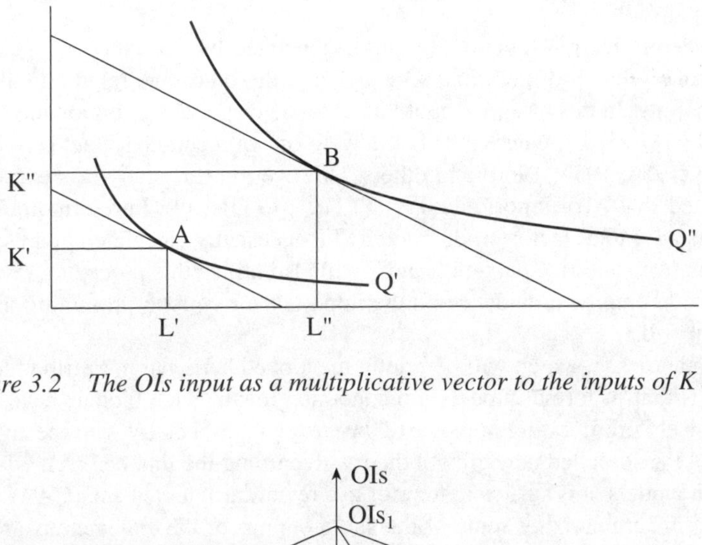

Figure 3.2 The OIs input as a multiplicative vector to the inputs of K and L

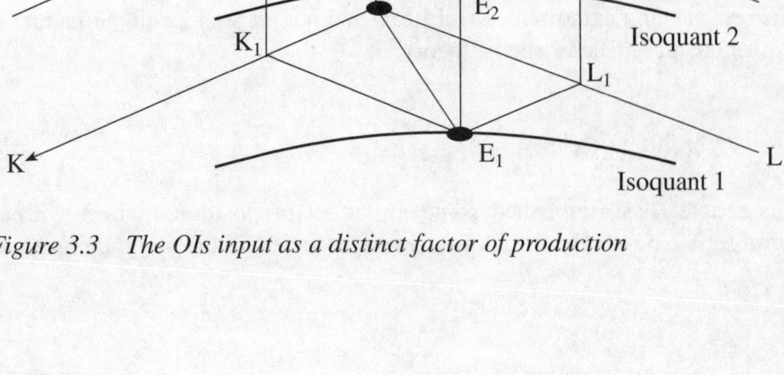

Figure 3.3 The OIs input as a distinct factor of production

In Figure 3.2, the factor O_2 of equation 3.3 when multiplied with K′ increases K′ to K″ and the factor O_3 when multiplied with L′ increases L′ to L″, so that the initial point A on the isoquant Q′ becomes point B on the isoquant Q″, hence generating economic growth. Thus, O increases the effective K and L, although the actual stock of K and L remain K′ and L′. On a three-dimensional Figure 3.3, the isoquant 2 is above isoquant 1 and hence the productivity effects of OIs (the variable O) can be easily discerned. A good example of how equation 3.3 can be applied in the real world is the effect of layout reorganization within the JIT/QC system from a linear layout to a U-shaped layout.[24]

Finally another way of expressing (3.2) is:

$$Q = f_1(O, T)\, f_2\,(M, K, L) \tag{3.4}$$

In this production function, output is generated by two independent functions; one related to the inputs O and T and the other one related to all the other inputs. In this case, we could include the classical Cobb-Douglas function $Q = A(t).K^aL^b$, where A(t) is the well-known technology factor used by Solow (1956; 1957; 1960) and others. However, in this study it is explicitly indicated that A(t) is not only TIs (T) but also OIs (O). Furthermore, from equation 3.4 total factor productivity (TFP) is usually calculated and used in quantitative studies. Chapters 6 and 7 will also utilize this procedure in order to provide empirical evidence on the importance of OIs in promoting industrial growth.

Another issue, which will be briefly mentioned here, but not further developed, is that of formal models of economic growth. Such models have been extensively proliferating in the last 20 years or so, especially with the appearance of the so-called new growth theory. Regarding the link between OIs and growth models, it is sufficient to refer to a review article by Pack (1994). This author, in summarizing some of the shortcomings of the endogenous growth type models (for example, see Romer, 1994), said that factors such as management practices, labour relations, innovativeness in devising new systems such as JIT, quality circles and flexible new models of consumer products are not taken into account in the endogenous growth theory; in addition, too much reliance on externalities and R&D and not on organizational factors is another characteristic of such a theory.

3.5 CONCLUSIONS

This chapter has established some important propositions, which will be useful for the remainder of this study. First, a review of the literature revealed that OIs, though sporadically suggested to be explicitly taken into account in the growth process of production by some eminent scholars, ought to have their special place in the agenda of research, especially because the separate role of OIs in this process has not as yet been consistently explored. Several concepts and theories that are pertinent to the inclusion of OIs in the growth process were explored in this chapter, for example the transaction costs theory and its extensions, the role of leading sectors, the role of competences and capabilities, and so on.

Second, a comparison between OIs and TIs by various scholars has established that the link between these two types of innovations is not straightforward and requires further investigation. A priori, there is no reason for assuming that OIs and TIs have a parallel existence, or that they are complementary, or that OIs are the 'handmaiden' of the TIs or vice versa. Hence there is ample room for a thorough exploration of their mutual relationship. In addition, some concrete examples of OIs shed more light into the rather autonomous existence of OIs.

Third, the inclusion of OIs in a general framework of economic growth, and indeed in a production function, was suggested. The inputs of such a production function explicitly include the OIs in a more detailed and concrete way but not mathematically. In particular, the distinction is made between the quantitative relation between capital and labour and the organizational relation between K and L, or between L and L, or between K and K.

However, this chapter is only an introduction of the various issues involved. In Chapter 4, the close and important relationship between the various concrete OIs and industrial growth will be more extensively scrutinized. In subsequent chapters, the principles established in this chapter will be further justified.

NOTES

1. In Chapter 7 all the types of productivity are linked and explained.
2. Edquist et al. (2001) discuss in detail this relationship. For instance, they talk about the phenomenon of 'jobless growth' in relation to TIs and OIs. See also, how in this book this discussion is briefly extended in subsequent chapters and especially in Chapter 7.
3. Nelson's views on the relation between OIs and TIs is discussed further in section 3.3 of this chapter.
4. Note that not all interactions between the different factors and hence not all relevant arrows are shown in this figure. The inclusion of all the relevant arrows would have shown the links of all aspects in reciprocal ways in order to indicate the endogenous character of these factors of economic growth. Here, the mathematical models developed by Solow (1956) and Romer (1994) are pertinent for showing the axis of exogeneity versus endogeneity of economic growth.
5. In Chapters 4 and 5 a more comprehensive study of all these elements plus necessary extensions will be undertaken. Especially in Chapter 5, a model will be proposed that

takes into account new issues about OIs. However, some aspects are not discussed in this study. For example, the role of demand and consumers can influence economic growth through events such as wars, governmental activities and business cycles, or tastes, preferences, introduction of modular products (cf., for instance, Langlois and Robertson, 1995; Sanchez, 2000) and saturation of consumption (cf. Engel curves).

6. Cheung (1998, p. 515) defined TCs 'to be all the costs which do not exist in a Robinson Crusoe economy'; they cover both costs associated with 'exchanges' and those associated with 'commands from higher authority'. Transaction costs will be more extensively analysed in the next two chapters.
7. Cheung (1998, p. 515) defined institution as 'any arrangement used to conduct economic activities which involve two or more individuals'.
8. Some other relevant studies are worth mentioning here: North (1990a) examined the importance of constitutional rules in countries such as England and Holland through history. North and Thomas (1973) and North et al. (1971) did similarly for Britain and North America, whereas Best (1990) did for Japan. Regarding the special role of culture in Japan see Von Tunzelmann (1995, p. 328) and Watanabe (1991, p. 75); and in the USA see Roy (1997).
9. However, TCs are not sufficient to explain the complex process of economic growth. For instance, the topic of 'kinetic costs' (KCs) (and their relationship with the TCs) will be extensively explored in Chapter 5. The insufficiency of TCs is pinpointed by many scholars such as Perrow (1986) and Stiles and Mick (1997). The last two authors say for example, 'As Perrow (1986) has pointed out, proponents of transaction cost economics often forget that intra-organisational decision making, control, and enforcement are themselves complex and costly enterprises' (Stiles and Mick, 1997, p. 211). In addition, in the above analysis by Bouttes and Hamamdjian (1997), it becomes apparent that contracts depend on how the firm is organized, and hence this dependence must be further explored in Chapter 5.
10. These three authors' work has already been referred to in previous sections, and it will be further explored in subsequent sections and chapters.
11. Other holistic analyses such as that by Arrow (1974) are required to complement our overall knowledge.
12. Sanidas (2005a) shows through the tool of linear programming (Baumol, 1977) that American and Japanese firms evolve historically in opposite directions because of the relative availability of national resources. Thus, the Japanese evolution of the firm has been influenced by the limited resources, (technical, human and organizational), from the beginning of its industrial development a century ago, and consequently, it still depends on the myriad of smaller firms for strong economic growth. On the contrary, the American firm has evolved from the beginning of its industrial take-off from big business and relies much less on the peripheral smaller firms, because resources have always been relatively abundant in the USA (in Chapter 2 this point was more extensively analysed, especially by referring to Chandler's and Fruin's work).
13. A similar comment by Nelson and Wright (1992, p. 1940) shows perhaps how strong is the belief that TIs are the primary moving force in this discussion: 'Here again, we should not think of organisational strength as an *alternative* but as a complement to advanced technology … '.
14. A sectoral analysis of economic growth was used, for example, by Kuznets (1966; 1971), especially by examining the technological (TIs) impact on various sectors (the impact of OIs, as in his 1966 book, was much less explored). However such a sectoral analysis is not the same as that achieved by examining SIS (see also Chapter 5).
15. An SIS is composed of those firms that are active in the innovative activities of a sector. Schumpeter Mark II is characterized by a small and rather stable population of innovators, thus creative accumulation is emphasized, whereas Schumpeter Mark I is characterized by a large and highly turbulent population of innovators, thus creative destruction is emphasized (Breschi and Malerba, 1997; Malerba, 2002).
16. A detailed examination of this extension will not take place in this study as it is beyond its scope.

17. These systems will be further explored in Chapter 5.
18. The exact relationship between OIs and competences will be indicated in Chapter 5.
19. The important point to stress is how the relationship between OIs and TIs affect economic growth, and not just how OIs affect TIs or vice versa.
20. A remark is relevant here: most of these studies refer to either small or overall incremental changes in firm organization in relation to TIs or to specific firm organization issues related to the promotion of R&D. The OIs treated in this study are major OIs as determined in Chapter 2.
21. More details on this are found in chapter 6 (for example p. 186) in Fujimoto's (1999) analysis.
22. McLoughlin and Harris (1997) is a collection of papers, such as that by Dawson, that explore the theme of the relations between TIs and OIs.
23. Organizational change is the topic of numerous papers and books in the management discipline literature (for example, see Dawson, 1994). However, the present study examines OIs in relation to economic growth expressed in terms of productivity, real output and so on.
24. See Chapters 4 and 5 for more details on the layout factor.

4. Contribution of specific OIs to economic growth and the three axes of OIs

4.1 INTRODUCTION

In the previous chapter it was established that there is a need to include OIs in the process of economic growth and, by extension, into the production function. In that chapter OIs were examined as a group. In this chapter, the analysis becomes more specific. Thus, each major OI is explored in the USA or Japan in order to show that each OI contributes to industrial growth in the light of some major economic theories. Emphasis is placed on the factors that increase productivity, for example, elimination of economic wastes (such as unnecessary inventories). For instance, the benefits of the division of labour are evoked to explain the advent of Taylor's scientific management cum Fordism system (TFS) in the USA and their impact on productivity increases. The concepts of focal firm and the continuation of the TFS in Japan are used in order to clarify the JIT/QC system and its usefulness. Overall, the unit of analysis in this chapter is the firm, and the exploration conducted seeks to identify the deeper reasons for firm expansion,[1] which, if transmitted to sectors and then to the economy, leads to industrial growth. The link between firms, sectors and macro economies will be further analysed in the next chapter. Finally, mainly the industrial or manufacturing sector will be considered in this chapter, thus overall excluding the primary and tertiary sectors of the economy.

Although some specific OIs are included under the umbrella of the USA, and others under Japan, all OIs exist to some extent in both countries. Thus, for instance, the inside subcontracting system was also part of the Japanese economy, although this is mainly analysed in the American context. Lazonick (1990, p. 293) can provide us with a concrete example.

> In the late nineteenth century, Japan, like the United States, had skilled workers, often employed as internal subcontractors, who exercised considerable control over hiring, firing, and the organization of work on the shop floor. But there was one critical difference between US and Japanese shop-floor relations in this era. Japan's skilled workers did not form collective organizations that sought to enforce craft control on the shop floor ...

The remainder of this chapter proceeds as follows. The second section deals with the OIs that took place in the USA and their role in promoting productivity and growth. These OIs are the replacement of the putting-out system by the factory system, the advent of big business and vertical integration, the advent of scientific management and Fordism, and, finally, mergers, product diversification and conglomerates. The third section concentrates on the development of the JIT/QC system in Japan. Finally the fourth section synthesizes all OIs into three fundamental groups, or axes so that the whole study becomes easier to undertake.[2]

4.2 AMERICAN FIRM ORGANIZATION AND ECONOMIC GROWTH

4.2.1 The Putting-out System Replaced by the Factory System of Production

The domestic production system was prevalent in the USA for a long period up to the 1850s, and beyond to a lesser extent. The factory system started its inroad slowly and accelerated after the 1850s. At the same time several other OIs took place in the USA, as analysed extensively in Chapter 2. The domestic system itself had replaced the handicraft system from the 16th century by relying on a greater division of labour under the coordination of a central, usually merchant, coordinator. Dietrich (1994) suggested that the main advantage of the putting-out system, compared to the handicraft system was the possible development of a division of labour, whereas the disadvantages included embezzlement, cheating, quality control and large inventories. These disadvantages were also the cause for the gradual domination of the factory system over the putting-out system as will be briefly shown in this first subsection.

Jones (1999) explained the emergence of the putting-out system during the 17th and 18th centuries in Britain as being due to the following factors: divisible processes of production in certain industries, simple technology, cheap rural labour surpluses and cheap transport of semi-finished material. Thus fixed costs were low, division of labour could be practised and, at times of depression, the merchants who hired domestic workers could easily reduce production. In other industries, such as iron, porter, glass, paper and sea-salt, which were characterized by technological non-separabilities, plants were much bigger and closer in nature to the factory system. The latter, despite 'the advantages of a reduction in embezzlement, an improvement in scheduling, better quality control and the evident profitability of the Crowley empire'[3] did not take off until later, from the 1780s, with the emergence of new technologies in the silk and cotton industries (Jones, 1999, p. 23).

New technology seems to be the main factor influencing the appearance of factories. This involved new machines, sometimes larger and more automatic than before, and a central power source such as a steam powered machine. Furthermore, as Langlois (1999, p. 55) says, 'obviously, if increasing extent of the market led to what I have called the volume effect – more highly-skilled machines embodying more durable 'dyes' – then there is likely some connection between the extent of the market and the factory system'. With more of this technology inside factories, fixed costs rose and, at the same time, the marginal productivity of labour rose as effort levels increased. This usually meant higher wages for factory workers than for the outside and inside contractors (cf. Jones, 1999; Langlois, 1999). However, despite this difference in wages, the employers still made a profit because, in the factory system, supervision could be done more effectively, discipline was generally easier and hence the workers intensified their efforts to produce the required output (cf. Marglin, 1974). Whether a better or more effective supervision was the critical factor in deciding in favour of the factory system cannot be easily demonstrated. Other factors could also have been as essential; for instance, at the same time, as demand grew substantially, supply bottlenecks induced more inventions and innovations which in turn drove the cost of capital down further, thus increasing the capital stock and consequently necessitating the presence of all workers under the same roof to operate more machines and capital equipment.

The factory system replaced the previous systems very slowly; perhaps the whole 19th century was needed for the complete transition to the factory system in both Britain and the USA. This is not surprising considering the complexity of the situation as described in the previous two paragraphs. Certainly, both technical and organizational innovations played a simultaneous role in this historical transformation. The division of labour concentrated in a single location meant some sort of deskilling of work, more specialization and, at the same time, entailed new methods of supervision and management. It also entailed more entrepreneurial possibilities and innovations (cf. Pollard, 1968, for more details on this topic).

Langlois (1999) distinguishes three organizational alternatives, which adequately describe the various production modes relevant to this sub-section: the factory system, the inside contracting and the putting-out system. Each of these alternatives has a different set of transaction costs. In this respect Langlois (1999, p. 46) said that writers like Williamson and North explained the arrival of the factory system in terms of greater efficiency based on lower TCs, such as those due to embezzlement, coordination and supervision, and monitoring of product quality. Thus, for example, Williamson (1980) argues that the factory economized on buffer stocks (work-in-process inventories) relative to an inside or outside contracting system. However, the TCs approach has been heavily criticized by other scholars (for example, Dietrich, 1993a, remarked that both costs and benefits are important in determining the final outcome), especially the economic historians (cf. Jones, 1999). Yet, as Jones (1999, p. 41) remarks, together with the paramount impact of technology, there is evidence supporting the TCs arguments.

The factory system emerged during the so-called 'industrial revolution' in Britain, and coincided with an acceleration of economic growth in that country. The same process took place in the USA about fifty years later. This process was closely related to the emergence of new dynamic sectors such as the mechanical ones, which were more inclined to quickly adopt the factory system. These leading sectors supported the existing and prospering textiles industries already fully-fledged in the accelerating economy. More will be said about leading sectors in subsequent chapters.

4.2.2 Big Business and Vertical Integration

The factory system was expanding during the 19th century in all directions. In the USA, endowed with a huge array of resources including usable plant space, factories kept growing in size. In Chapter 2, where this process was briefly described, it was shown that from the 1880s the systems of mass distribution and mass production began. Also, the integration of mass distribution and production with a parallel development of vertical and horizontal integrations took place that finally led to the big business we know today. At the same time there has been a tendency for the creation of monopolies or oligopolies in many industries within the American economy. What were the underlying economic reasons for such changes, which ended up with the appearance of the so-called big business?

Technology (as TIs) seems to have played a preponderant role. As Von Tunzelman (1995) described, using Rosenberg's work (1969; 1972), the American manufacturing system was, from the beginning, based on the routine assembly process, which in turn was based on the development of interchangeable parts and the corresponding industry of machine tools. Von Tunzelmann (1995, p. 203) mentions five advantages of using interchangeable parts: first, the ability to replace any defective part with another equally usable one; second, dispensing with expensive and time-consuming handicraft labour; third, dispensing with one-off designs through standardization; fourth, huge possibilities of further developing machines and tools; and, fifth, an obvious impact on the increase of speed of throughput. In due course, interchangeable parts were linked to two main organizational directions: first, to the assembly line and continuous processing introduced in some industries in the last 30 years or so of the 19th century (for example, production of cigarettes, or beverages) and epitomized with the production of cars by Ford; and, second, to using the equipment much more flexibly to produce various products, which eventually led to the Japanese so-called 'lean' system (to be described in the next section).

Despite obvious historical trends for firms to become bigger and bigger in the USA from the second half of the 19th century until the 1980s, economists, in general, considered the issue of firms' organization as non-existent, or as simply a black box. Influenced by the strong analytical tools of the neoclassical theories, economists ignored the content[4] of this black box until Coase (1937) and Williamson (1975; 1985) among others disturbed the calm waters of these theories. Demsetz (1997, p. 426) summarized the latter as follows:

> Neoclassical theory's objective is to understand price-guided, not management-guided, resource allocation. The firm does not play a central role in the theory. It is that well-known 'black-box' into which resources go and out of which goods come, with little attention paid to how this transformation is accomplished. In the theory's core model, perfect competition, the transformation accords with the dictates of known technology and prices. Management has no real influence …

Without entering into rather fruitless debates as to whether we should or should not discard neoclassical theories, mention is now made of some of the developments in the theory of the firm since Coase's seminal contribution in 1937. According to this author, and many others who followed him, a firm exists because market transaction costs are higher than the organizational costs[5] involved in producing a profitable quantity within the boundaries of the firm. Also a comparison between these two costs in terms of marginal dimensions will determine to a great extent the degree of vertical integration. Bounded rationality (cf. Simon, 1997) and uncertainty (cf. Knight, 1921) are necessary conditions for the role of TCs to be effective. In addition, the establishment of property rights is a prerequisite for TCs to exist.

The internalization of transactions into a firm enables the exploitation of economies of scale and scope, which in turn determines the size of that firm. Related to this internalization are the elements of opportunism and asset specificity. The former expresses the tendency that employees have to take advantage of the imperfections of employment contracts and hence to 'cheat' in the production process. The latter refers either to physical or human elements in the transaction and involves assets, which are not freely available for other uses. In general, the greater the degree of opportunism and asset specificity, the more likely it will be that internal governance will replace market governance. Thus, the most integrated firms also have the most specific assets in their possession. All this is a brief summary of Williamson's work (1975; 1985).

However, Hodgson (1998) suggested that the TC-based view of the firm can and should be complemented by the competence-based view. For instance he says (ibid., p. 189): 'An important but not exclusive factor explaining the existence, boundaries, nature and development of the firm is the capacity of such an organization to protect and develop the competencies of the groups and individuals contained within it, in a changing environment … '. These competencies create a culture, an institution, trust and loyalty, none of which is translated as a contract but, rather, becomes tacit knowledge difficult or impossible to buy and sell. This cultural transmission facilitates group and individual learning, and eventually increases firm productivity, thus perpetuating the existence and development of firms. Finally, it should be emphasized that all these elements are path-dependent and hence historical analysis is relevant.

This last point also implies that the competence-based approaches of the firm are part of, or complementary to, the more general evolutionary theory. As Hodgson states, these approaches 'fit more closely to the philosophy and methodology of the evolutionary paradigm pioneered by Nelson, Winter and others' (ibid., p. 192). These authors, Nelson and Winter (1982), extensively explored the close link between capabilities and routines. Nelson (1996, p. 110) includes core capabilities as one of the three strongly related features of an adequately described firm, the other two features being strategy and structure. He then summarized his work with Winter:

> Winter and I have proposed that well-working firms can be understood in terms of a hierarchy of practiced organizational routines, which define lower-order decision procedures for choosing what is to be done at lower levels. The notion of a hierarchy of organizational routines is the key building block under our concept of core organizational capabilities … (Nelson, 1996, p. 111)

In addition, Chandler (1992, pp. 92–3) argued that American and German firms (as opposed to British ones) grew fast in world markets because they were, 'driven much less by the desire to reduce transaction, agency and other information costs and much more by a wish to utilize the competitive advantages created by the coordinated learned routines in production, distribution, marketing, and improving existing products and processes'. In other words, organizational capabilities developed in a dynamic firm are based on routines and procedures, which can be defined as what an organization actually does every day. Thus a firm's routines are a subset of its capabilities. One could say that the latter are future or potential routines. Finally, Coriat and Dosi (1998) argued that Taylor had the pioneering understanding that the organization of production is equivalent to questions of know-how and competence. Hence, Tayloristic practices represent a 'coevolution between forms of incentive governance, routines, competences, under circumstances of acute interest conflict' (ibid., p. 114).

All the above theories[6] explain why,[7] in parallel with the development of big business from the end of the 19th century, we also have the phenomenon of vertical integration either of production or distribution or of both.[8] This trend continued unhampered until the 1960s and to some extent the 1970s.

> This amounted to the replacement of the market economy by the corporate economy. In 1969, J. K. Galbraith predicted that the inherent advantages of firms over markets in allocating resources and permitting long-term planning would result in increasing dominance of capitalist economies by a small number of giant corporations. During the 1980s and the 1990s, these predictions were refuted by a sharp reversal of the trend toward increased aggregate concentration … (Grant, 1998, p. 319)

Thus, the trend is now to disintegrate,[9] to specialize again, to be more flexible, to collaborate with other firms and to reinforce core competencies (ibid.). However, this applies to the USA and some other countries only, since Japan had followed a different path from the beginning (for example, through focal, flexible firms as this is explained in the next section).

Vertical integration led to the appearance of the M-form of internal managerial organization, which gradually replaced the functional structure of large firms. Did this continuous process of multidivisionalization contribute to economic growth in the USA? There exists some evidence, which suggests that the M-form firms are better performers than other type of firms. For instance Rumelt (cited in Scott, 1973, p. 143) has shown that from 1950 to 1970, the product divisional firms scored higher on all the conventional financial measures (for example, sales growth, earnings growth and so on) than the other type of companies. Also, as Chandler (1969, p. 278) mentions, and based on a governmental study in 1963, 'In 1947, the 200 largest companies, many of which had not yet become fully diversified and decentralised, accounted for 30 per cent of the value added by manufacturing. By 1963, when most of them had adopted the new strategy and structure, they accounted for 41 per cent of value added … '.

From a more theoretical point of view, Williamson (1994, especially pp. 372–9) mentions the following arguments for a better economic performance of multidivisionals than other types of firms:

- They enjoy more disciplined internal organization in terms of stronger incentives, better accountability, and superior internal resource allocation.
- Value creation (entrepreneurship) and loss reduction (administrative function by top managers) are both realized.
- Internal organization allows the squeezing out of managerial discretion.

- Economization on bounded rationality and attenuation of opportunism.

However, Hill (1994) makes a valuable remark: although he agrees with the fact that the M-form of firms is a superior type of governance (according to Williamson's thesis), 'for a firm to fully realize value from hierarchical governance, the correct internal organizational structure and control systems must be in place' (p. 307). This precondition is essential as the theoretical extensions in Chapter 5 will show. There it will be suggested that the 'process of movements' (POM) (akin to the internal organizational structure of Hill's remark) is at the heart of firm operations and thus the 'process of contracts' (akin to the hierarchical governance of Hill's remark) depends on the POM.

The M-form of governance is closely related to diversification. Scholars have attempted to explain a diversified firm with the aid of transaction costs theory. If the TCs of using the market mechanism exceed the bureaucratic costs of a management hierarchy, then the firm will internalize these costs and thus diversify its production into new products, and hence there will be horizontal integration. This argument resembles that explaining the process of vertical integration. However, there is a difference, in that a diversified firm can realize economies of scope whereas a vertically integrated firm can realize economies of scale. Furthermore, Hill (1994) says that economies of scope through diversification emanate from unutilized excess capacity in human capital as Penrose had already claimed in 1959. However, a TCs approach to understanding firm expansion through diversification is not sufficient. As Milgrom and Roberts (1992) remark 'firms seeking profitable expansion are most likely to be successful by identifying the areas of their own special competencies, investing to build those competencies, and introducing products where the competencies give them a cost advantage or a marketing advantage' (p. 570). In particular, for mergers, Milgrom and Roberts (1992) mention the following transaction costs to be the reasons for potential failure: 'conflicts of corporate culture, political battles leading to influence costs, or misbehavior by the central office of the acquiring firm, which reneges on promises made to managers and employees of the formerly independent subsidiary' (p. 574).

4.2.3 The Role of Scientific Management and Fordism

The American system of manufacturing as it existed up to the early 1980s started at least 100–130 years earlier (cf. Locke, 1996). The mass production system based on scientific management and its extensions, and on Fordist methods, provided the American economy with an edge over all other economies during that century. This sub-section links elements of these methods to some important theoretical issues in economics, and analyses the implications for productivity and economic growth.

Before analysing this new order in business management that had slowly started by the 1880s, a brief reminder of the old order which was based in the more sophisticated industries such as the mechanical ones, on the inside contracting system, is presented.

> contractors were prosperous entrepreneurs who engaged skilled people to do vital shop floor production work. But the system also encouraged 'scams' that resulted, as Frederick Winslow Taylor described it, in 'a jungle of restrictive practices, ca'canny, hanging-out, or soldiering', none of which induced efficient production in standardized American process manufacturing … (Locke, 1996, p. 19)

On the other hand, scientific management was based on different premises. George (1972, p. 93) summarized Taylor's ideas as follows:

> First: Develop a science for each element of a man's work, which replaces the old rule-of-thumb method.
>
> Second: Scientifically select and then train, teach, and develop the workman, whereas in the past he chose his own work and trained himself as best he could.
>
> Third: Heartily cooperate with the men so as to insure all of the work is being done in accordance with the principles of the science which has been developed.
>
> Fourth: There is an almost equal division of the work and the responsibilities between the management and the workmen. The management take over all work for which they are better fitted than the workmen, while in the past almost all of the work and the greater part of the responsibility were thrown upon the men …

In this quote, it is worth noting the contrast between the old paradigm of organization and the new one proposed by Taylor. However, there is more to it (George, 1972, pp. 96–9):

- 'Scientific management involves a complete mental revolution'[10] (Taylor, 1947, p. 27)
- Systematic research to determine all aspects of work
- Establishment and use of standards in every phase of management
- Systematic planning and control
- Conservation, not waste in effort, materials, time and so on.

On the last point, the Taylor Society in its early days enunciated the following aim: 'To earn through a waste-saving management and processing technique, a larger income from a given expenditure of human and material energies, which shall be shared through increased wages and profits by workers and management' (quoted in George, 1972, p. 97). This aim is obviously related to increasing productivity. Hoxie, an early critic of Taylor's ideas and suggestions, also emphasized the purpose of scientific management as being the elimination of 'avoidable wastes' in the process of production (quoted in Nyland, 1996, p. 986).

What were the consequences of the implementation of scientific management? George (1972, p. 96),[11] in his extensive historical account, suggests the following positive outcomes (regarding other scientific management proponents, see also further below):

- not only improvements on the shop floor, but also in other parts of the firm such as sales, general administration and so on
- more effective utilization of equipment, labour and materials
- more accurate controls, routing and planning, mainly through measurement of time requirements of carrying out various tasks
- for labour, better placement, more opportunities for advancement, higher wages, better working conditions and so on
- a more reliable product
- a better workforce
- a better understood customer.

Scientific management was not only expressed by Taylor, but also by several other scholars after and before him.[12] According to George (1972, p. 96), 'Babbage, among others, had traveled much the same road', before Taylor. Gilbreth was a contemporaneous 'competitor' with Taylor, and he introduced the complementary motion study to Taylor's time study. The scholars who extended the Taylor cum Gilbreth ideas and actions were Gantt (the Gantt chart, service as an objective), Munsterberg (industrial psychology), Scott (applied psychology to motivation and advertising), Emerson (efficiency principles, objectives of a firm), Person (emphasis on total scope of management) and Fayol (first comprehensive theory of management) (George, 1972, ch. 7). It is worth noting about Emerson, that his concept of efficiency was based on the elimination of waste. All these pioneers of scientific management, plus some others such as Church, Diemer, Duncan, Brandeis, Hartness, Bloomfield, Hoxie, Drury, Tead, Cooke, Parsons, Leffingwell, and Herford, Hildage and Jenkins (George, 1972), were all active in the 1910s and/or 1920s (or even beyond). The movement of the new management science, of course, continued later with many other scholars such as Sheldon and Mayo.

Overall, during the period of the 1910s to 1920s, social and economic conditions favoured the diffusion of rationalization, standardization and simplification; thus, the Americans achieved a significant lead in the organization of work and accelerated productivity. In this respect, Kogut (1996, p. 269) said: 'The critical factor was not the organizational capabilities of managerial hierarchies, but the American national strength in the rationalization of work and the standardization of production … '. The same author remarks that Taylorist ideas had spread widely by the 1920s; thus, by 1935, 34 per cent of all manufacturing firms with more than 250 employees reported the use of time-and-motion studies.

However, scientific management was not the only revolution in organizational matters that took place in the 1910s. There was also the semi-automatic assembly-line process engineered and introduced by Ford. Although American factories had already introduced continuous-process machinery on the floor to increase production throughput from the 1870s (Chandler, 1977; Locke, 1996), Ford's assembly line, in operation by 1913, epitomized this mass production process. Fordism,[13] when introduced in the 1910s and 1920s, needed Taylor's principles about standardization, work codification and time studies. Above all, it most likely needed Gilbreth's motion studies as Kijne (1996) correctly remarked.

Was Taylorism complementary to Fordism, or were these two systems rivals? Although these dilemmas will not be fully resolved in this study, the impact of both philosophies, one way or another, has been profound and positive on American organization of businesses, be it small or big. The principles of Taylorism cum Fordism and their various applications are of primary importance for this study. In this respect, the stand suggested by Aglietta (1976) and by Blackburn et al. (1985) is adopted: 'Fordism took up the principles of Taylorism and put them more effectively into practice, to obtain an ever greater intensification of labour. Fordism further developed the mechanization of labour, increased the intensity of work, radicalized the separation between manual and mental labour' (Aglietta, 1976, p. 117); and 'Taylorism, then, differs from Fordism in that the former describes a set of techniques for the management of labour and the latter a particular configuration of a production system seen as a combination of process and product technologies with human labour in a specific form of work organization' (Blackburn et al. 1985, p. 44).

It is now time to ask the question: how is the Taylorist-Fordism system related to some fundamental economic theories?[14] The following interdependent propositions are suggested and briefly discussed below:[15]

- The more meticulous is the division of labour organized by the TFS, the more waste and transaction costs are reduced.
- The introduction of standards by the TFS creates a better institutional framework.
- The whole enterprise is affected by the TFS.

Antonelli (1998) remarked that economic analysis has recently appreciated the role of standards as institutions (rules) that shape both markets and organizations. They are non-pure private goods that contain technical, commercial and procedural information, are part of technological and organizational changes, have an impact on the competition process, and 'affect radically the division of labour and the organizational setup of firms' (ibid., p. 79).[16]

Since standards are institutions, they provide the structure for exchange that determines the costs of transacting and the costs of transformation (North, 1990; 1992). Standards reduce the variety of asset specifications, increase the frequency of transactions, reduce market and intrafirm uncertainties, increase the number of parties involved in the marketplace or within the firm, reduce opportunistic behaviour, and reduce the size of governance structures (Antonelli, 1998).[17]

Another main feature of the TFS, as described above, is the relentless pursuit of continually reducing waste in the production process. This reduction in waste is, for instance, translated into reduction in inventories, reduction in unproductive work (due to a more meticulous and adequate division of labour) and so on.[18] This pursuit means that the production possibility frontier (PPF) is not attainable in the short run; in the long run, reduction in waste would drive the firm (or the economy in general) closer to its PPF. However, in the long run, at the same time, the PPF is displaced upwards due to technical and organizational innovations, thus the firm is continuously situated below and inside the PPF (points A and B in Figure 4.1). In addition, there is a continuous interplay between OIs (and/or TIs) and the elimination of waste, that is, waste is reduced because of OIs (and/or TIs) and, on the other hand, OIs (and/or TIs) are introduced because of the existence of waste. Consequently, there are many possibilities as to where the firm is situated in the first place and where it dynamically relocates itself.

The TFS strives, through its organizational changes of work on the shop floor and everywhere in the firm, to shift the current point A (in Figure 4.1) of production towards the ever moving outwards PPF line because it reduces waste and uses resources and inputs more effectively. Here reference can also

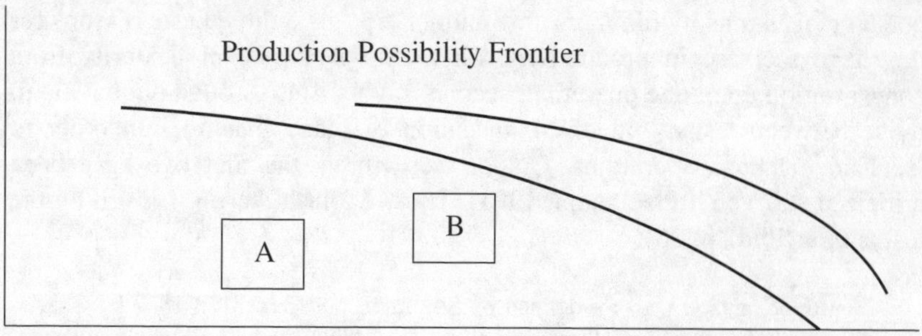

Figure 4.1 Taylorist-Fordism system and the production possibility frontier

be made to the X-inefficiency theory of Leibenstein (1966), according to which one of the main reasons for not having non-allocative efficiency is an incomplete specification of labour contracts (cf. also Tisdell, 1996, p. 238). A study by Kilby (as cited in Frantz, 1997) showed that simple alterations in the physical organization of a plant such as work flows throughout the plant, changes in plant layout, and waste control increased labour productivity substantially. Kilby's pioneering study is used by Frantz (1997) as quantitative evidence of Liebenstein's X-efficiency theory, but it can also be used to support the arguments in the present book.[19]

The issue of reducing waste and its impact on increasing productivity is a major issue in this study and will be taken up again on several occasions, for example, in this chapter where the JIT/QC system is introduced, in Chapter 5 where waste will be more specifically defined and in Chapter 7 where waste reduction through its decrease in inventories will be quantitatively tested against increases in sectoral productivity. It is also worth noting that the TFS is very often related to mass production and big business of a monopolistic or oligopolistic nature. In such a case, Schumpeter's (1942) creative destruction through the existence of imperfect markets generates more economic growth, not through Paretian efficiency (Tisdell, 1996, p. 237) but through X-inefficiency (my addition).

The more meticulous division of labour according to the TFS will now be briefly discussed. The first known reference to the paramount role of the division of labour and hence specialization was made by Plato (as quoted in George, 1972, p. 16), when he said: 'One makes men's shoes, another women's, one lives entirely by the stitching of the shoe, another by cutting the leather … A man whose work is confined to such a limited task must necessarily excel at it' (Plato's *Republic* contains a substantial account of this topic of specialization and division of labour). In modern times Smith (1776) became famous with his detailed analysis of the pin factory where he showed that 18 different operations within that factory raised labour productivity dramatically, much above the level of productivity of the case when only one person makes pins by him/herself. Smith (1776) gave three basic reasons for this huge increase in productivity. First, improved worker dexterity from concentration upon one operation; second, saving time through not having to move between tasks; and, third, invention of many machines in order to mechanize labour operations. Out of these three, the first two have been criticized (cf. Von Tunzelmann, 1995). Thus, a fourth reason is given by the last author (ibid., p. 36):

> A fourth advantage was not defined by Smith but put forward in 1832 by Charles Babbage … Babbage argued that the division of labour would allow slotting individual workers into vocations for which they were best fitted, particularly those

most closely matching their skills, thus providing what he referred to as a mental division of labour. This fourth factor became famous (or notorious) as the foundation for 'scientific management' along Taylorist lines in the 20th Century …

Furthermore, the more meticulous division of labour and specialization brought about by the TFS can be linked to another important relationship affecting economic growth, as established by Smith (1776) again. This states that the market expands more and more as there is gradually more specialization in the economy and vice versa, that the enlargement of the market permits greater division of labour and that the link of this two-way relationship is the growth in productivity. This is a very important conclusion to be drawn from the analysis because it fully supports the present study, which emphasizes the role that organizational changes play in accelerating economic growth.

Two articles by Yang and Borland (1991) and Borland and Yang (1992) have given more evidence about this close link between division of labour and economic growth, especially through a dynamic general equilibrium model of endogenous growth based on the issue of specialization. These two authors, by introducing the somehow vague notion of 'transaction efficiency' and allowing for increasing returns to scale, concluded that 'As long as the division of labour has evolved to a sufficient degree and the potential for further division of labour remains, the rate of growth of per capita income will increase over time' (Yang and Borland, 1991, p. 462). Also, as the two authors confirm, their conclusions have captured Young's (1928) insight that increasing returns to scale in the economy depends on the progressive division of labour.

The final point examined in this sub-section is the relationship between the TFS and the theory of transaction costs applied in intrafirm cases and not in interfirm cases. Within the organization of firms, the points made in the first pages of this sub-section suggest that the actual use of the TFS principles involves the following TCs-related traits. First, a more scientific way of determining each employee's job tasks reduces TCs since the division of labour becomes more detailed and precise. At the same time the adjacent opportunism is also lessened, thus providing more effective governance. Second, the introduction of planning and control reduces uncertainty about the future, thus again reducing TCs and opportunism as well. Third, the shift of supervision from shop-floor foremen (through the elimination of inside contractors) to middle managers reduces the TCs within the echelons of overall management of the firm.[20] Furthermore, the TFS is not only related to TCs but also to other aspects of production such as learning aspects, or time-and-motion studies and so on. In Chapter 5, OIs, and hence the TFS, will be more precisely linked with the full set of firm operations.

4.2.4 Summary of the Role of OIs in Promoting Industrial Growth

In this brief sub-section a summary of the findings of the previous sub-sections will be presented. Although the objective was to follow a historical sequence in explaining OIs since 1850 in the USA, it has perhaps become apparent that there is a considerable overlap between theories, OIs and periods of time. For example, the division of labour issues developed in sub-section 4.2.3 is easily applicable to the other two sub-sections, or the TFS is also related to the development of big business and integration outlined in sub-section 4.2.2. This is an important remark in so far as the economic growth phenomenon is concerned. However, from each sub-section the elements which are conducive in promoting economic growth in each historical sub period can be extracted. Table 4.1 summarizes these findings.

The underlying reason for choosing the appropriate elements in each column of Table 4.1 is the potential increase in any kind of productivity (for example, labour productivity or total factor productivity). In Table 4.1, the putting-out system was not included, as it would make the comparison even more difficult. In addition Table 4.1 mainly includes those elements that are more closely linked with the study, that is, OIs that are a significant driving force of economic and, in particular, industrial growth. The quantification of the links between elements of each production system and productivity is rather non-existent on a sectoral or macro level; on a firm basis there is, generally, more empirical work. For a quantitative example of how the original scientific management applications increased productivity in a particular firm see Peters (1980). The empirical analysis in Chapter 7 regarding the impact of the JIT/QC system on American industrial growth will indicate the limited empirical evidence in this respect (however, on a firm basis more quantitative evidence regarding the JIT/QC is provided in the next section of this chapter).

4.3 JAPANESE FIRM ORGANIZATION AND INDUSTRIAL GROWTH

Throughout their long-term development Japanese firms have followed their own distinct path, as discussed in Chapter 2. Aoki (1990), amongst other scholars, has extensively analysed many aspects of the Japanese firm, especially from the governance point of view. In this section particular attention is paid to the OIs of JIT and QC (through TQM or TQC), which have been at the centre of Japanese management innovations since the end of World War II. The implementation of TQC/JIT has been the necessary condition for the high and sustained economic growth of Japan for most of the years since

Table 4.1 Characteristics of the main production systems

	Factory system	Mass production/big business system	Scientific management and Fordism
1	Transaction costs reduced	Transaction costs reduced	Transaction costs reduced
2	Better quality control for a larger production	Development of core competences	Quality of products increased
3	More specialization	Division between entrepreneurs and managers	Greater division of labour on the shop floor
4	Inventories decreased	Transport costs reduced	Reduction in wasteful efforts, materials, time
5	Improvements in scheduling	Moral hazard reduced	Cooperation between employees increased
6	Increased coordination	Increased coordination	Increased coordination
7	Intensification of efforts	Organizational and technical non-separabilities	More control of labour by management via planning
8	Improved supervision	Supervision more difficult	Improved supervision
9		Creation of externalities	Establishment of standards
10		Greater size of markets	Develop a science for each element of a person's work
11	New TIs	New TIs	Variety of asset specifications decreased

World War II. This section links JIT/QC with theoretical economic concepts and with other OIs such as networking. First, however, emphasis is placed upon showing how Japanese firms became ready to accept and develop the holistic JIT/QC system.

4.3.1 Preconditions: Taylorism, Fordism and Focal Firms

Chapter 2 analysed the historical evolution of the Japanese firm during the last 150 years or so. A summary of this evolution can be found in Fruin (1992). Thus, the modern industrial enterprise was developed from 1890 to 1920. This type of firm became large during 1920 to 1955 and built up a factory with organizational capability for scale and some scope economies. Finally, from 1955 to date, interdependent, large modern industrial enterprises were formed, which focused on task forces and products. These firms were reciprocity oriented, and became focal factories with organizational capabilities for scale, scope, learning, transaction-cost economies and decentralized coordination.

The catching-up efforts of the Japanese society to reach Western standards have been paramount in explaining many of the country's economic features. For example, 'traditional labour-management practices, such as the internal contracting system, disappeared within several decades of the introduction of Western models of industrial relations' (Fruin, 1992, p. 59). The focal factories, though widely spread out after World War II, 'first appeared during the inter-war period when the level of economic development did not permit nation-wide coordination of output but it did demand local and regional centralization of business functions. Balkanization of markets yielded smallish corporate head offices, incomplete divisionalization, and robust manufacturing works' (ibid., p. 210). In this sub-section it is worth summarizing the characteristics of Fruin's 'focal factories' because they are the place where the lean production system took place with all its traits of JIT/QC, networks and industrial cooperation, which will be the object of the next few sub-sections. Table 4.2 summarizes the salient characteristics of these factories.

The features of the Japanese focal factory (especially those of organizational and managerial interdependence, localized resources and focused efforts in well-defined activities) immensely helped introduce quality control and just-in-time philosophies after World War II (see sub-sections below). However, the latter were also linked to the impact of scientific management on Japanese business thinking from as early as the 1910s. An article by Nakagawa (1996) explains how this happened. This author tells us that in 1911, Yasunari published an article introducing scientific management (SM), hence before Taylor's own book. In 1913, Ikeda published a book entitled *The Secret of*

Table 4.2 Characteristics of focal factories

Importance	Feature
High	Management-intensive
High	Learning is emphasized
Low	Similar to strategic business units in functions and significance
Low	Similar to subsidiaries and affiliates of big American firms
High	Link between production and distribution
Very high	Linked with corporations and interfirm networks
Very high	All distinct functions are localized: applied technology research, product and process research, market research, product design and planning, design engineering, trial manufacture, sales coordination and so on
Low	Unbalanced product specialization (a number of products are produced in varying amounts)
Medium	Interactive, feedback process
Very high	Focus efforts in well-defined activities
High	Complementary activities between focal factories
High	Organizational interdependence, hence permeable boundaries joining factory, firm and network
High	Usually low-volume manufacturing (due to imitation of foreign technology, a recently formed mass consumption society)
High	Often a cluster of focal factories orbiting the core firm, but having their own dynamic centres of technological competence
Medium	If the focal factory is part of a core firm (for example, Toshiba) then strategic planning remains in the hands of divisional and corporate-level managers
High	The number of product families is limited
Very high	Intrafirm and interfirm economies of scope

Source: Information gathered from Fruin (1992).

Saving Lost Motion, which was heavily based on Taylor's work and which sold more than 1.5 million copies. In addition, several Japanese scholars spent time with Taylor before he died in 1915. In June 1925, the first overseas branch of the Taylor Society was established at the Institute of Industrial Efficiency in Tokyo. By 1917, there were six Japanese factories practising SM, as against 169 in the USA.

The pioneering of SM applications in Japan took place in the repair works of the Japanese National Railways (JNR). This company introduced the entire system of SM step by step, whereas many other firms did it only tentatively and fragmentarily. For example, when the concept of work-study was introduced to the employees of JNR, great care was taken to obtain their consent and cooperation. Also, the chief person who led the SM effort at JNR, Yamashita, had said: '*Kaizen* should not be produced only by the engineers and managers. If all the men in the shop do not engage in operations to achieve *kaizen*, essence of shop management would not be achieved' (quoted in Nakagawa, 1996, p. 172). Overall, JNR became the progenitor of all SM installations in Japan. Finally, during the period 1931 to 1945, in other words during the industrial rationalization movement and wartime production, several industries became large enough and had sufficient technological sophistication to be ready for an SM approach.

Post-World War II witnessed the development of the Toyota or lean production system, which will be the underlying topic of many sub-sections below. As Ohno, the man who conceptualized the LPS, suggested, all the methods and concepts embodied in the LPS are also part of the original SM approach as invented by Taylor:

> For example, JIT (just-in-time) and the Kanban system were derived from the concepts of standardization, routing and line balancing, key concepts of scientific management. Supplier relations are seen as an integral part of total production process … Having flexible job structures that promote job multiskilling is not in conflict with the principle that every worker should use a standard operating procedure as determined by time and motion study and as described by his job card … (Nakagawa, 1996, p. 164)

However, the common points between the LPS and SM did not preclude some specific developments in Japanese society. For instance, as Littler (1982, p. 157) said: 'Taylorism in the West involved the rigid separation of planning and doing, resulting in the downgrading of the foremen's role. Whereas in Japan the foremen/oyakata retained considerable production planning responsibilities … ' The JIT/QC philosophy developed in Japan is a unique system, a Japanese invention, which nonetheless has its roots in the principles of SM.

Other scholars suggested that the two systems are not unrelated to each other. For example, McMillan (1996, p. 285) remarked:

> As it turns out, the Toyota system has theoretical origins in scientific management. More than fifty years ago, Taylor's student and assistant, Henry Gantt, developed planning tools which today look at the total production sequence in an attempt to develop assembly balancing techniques. The kanban system requires higher levels of fixed costs for additional tooling, materials handling, and factory layout, but operating costs are substantially lower than the traditional mass assembly system …

Fujimoto (1998) tells us a similar story: 'The Toyota system has adopted various elements of the Ford system and hybridised them with their own indigenous system and original ideas. There is an obvious continuity between the two systems' (p. 22). Thus, whereas Ford introduced an extensive standardization of jobs with a parallel vertical separation between single-skilled workers and elite industrial engineers, Toyota's standardization emphasized continuous improvements on the shop floor through a flexible and multiskilled workforce. Another aspect that is worth mentioning here is that in some respects there is a similar evolution from a pre-Tayloristic system to a Tayloristic one in both the USA and Japan. Thus, as Taylor reacted to the inside-contracting system and proposed the takeover of the shop-floor management by engineer-planners, in the same way Ohno in Toyota replaced the foremen-craftsmen by the flexible system, now known as JIT/QC, hence emphasizing the multiskilling modality of the workforce (Fujimoto, 1998).

Is the JIT/QC system as developed in Japan also related to the focal factory system (FFS) described earlier? One of the main themes of the latter is the focus on well-defined activities being complementary between factories under a decentralized organizational interdependence. This theme has greatly facilitated the JIT process and outsourcing, as the efficiency of such a process greatly depends on the close links and cooperation between focal factories. Another central theme of the FFS is the emphasis on learning to continuously improve imported technologies (TIs). This theme has greatly facilitated the QC process, which is interrelated with the JIT system. It is worth quoting Krajewski and Ritzman (1999, p. 744) who used the concept of 'focused factories' (similar to the FFS) and directly related it to the implementation of the JIT pull system in a company case. The sub-sections below will further analyse the close link between the FFS and the LPS.

4.3.2 Just-in-Time Manufacturing: A Basic Description

Although there are some suggestions that the JIT system was introduced in the 1920s in the USA through the 'hand-to-mouth' purchasing approach (Langlois and Robertson, 1989), this system became known to the world as such through Japanese initiatives (Harrison, A., 1992; 1994; Imai, 1991; 1997; Schonberger, 1982; 1986). Historically, the JIT system became known to the Western world through Toyota's rigorous implementation of JIT principles, as developed by the two Japanese pioneers Taichi Ohno and Shigeo Shingo from the mid-1950s to the late 1970s. However, as Schonberger (1982, p. 17) remarked, according to his own sources, the shipbuilding industry initiated the JIT idea with inventories '20 years ago' and subsequently this idea spread to other Japanese companies; and all this took place before Ohno and others started writing about JIT in the mid-1970s. Nonetheless, the story told by Fujimoto (1998) indicates that it was Toyota's founder who introduced the term 'just-in-time':

> The idea and slogan of 'Just-in-Time' was created and advocated by Toyota's founder-entrepreneur Kiichiro Toyoda during the 1930s … When Kiichiro started the automobile business, he first posted the words 'Just in Time' on the walls, and told his subordinates to receive just twenty engine-blocks in the morning and no more if only twenty were needed that day. Kiichiro was frequently walking about the factory and threw away anything above what was needed … (p. 23)

Furthermore, the proper JIT process started in the 1950s with the introduction of the formal mechanism *kanban* by Ohno, who also took many ideas from the textile industry where he had previously worked. Thus, Toyota was the pioneer of the JIT system, and it was mainly through Toyota's experience (and the other car producers) in the 1970s, that the JIT system spread rapidly to other industries. For instance,

> In the electrical industry, Matsushita (a much larger but less well-known company than Sony) developed its own version. Shingo thought that the Matsushita production system was better than Toyota's. Instead of using *kanban* to signal the need for more parts between separated operations, Matsushita concentrated on placing operations next to each other so that there was no need for signaling … (Harrison, A., 1994, p. 180)

Furthermore, it must be emphasized that as QC in general started in Japan immediately after the end of World War II, many Japanese companies were ready to adopt the complementary system of JIT later in their evolution. Already in 1951 Toyota implemented the 'Idea Suggestion System' to continuously reduce production costs (Fujimoto, 1998); this subsequently became the basis for *kaizen* and TQC.

A first succinct but holistic definition of JIT is given by A. Harrison (1994, p. 175). This author distinguishes three fronts in the quest for excellence regarding the JIT philosophy or, as it is also called, 'Lean Production', or 'World Class Manufacturing':

- Techniques are systematically put in place to attack all sources and causes of waste.
- Everybody is included and participates in the JIT process and management.
- Continuous improvement searches for the ideal case of zero scrap, defects and inventories.

Strictly speaking, the JIT system is only a major part of the 'lean' system; however, the former can only be successful if it becomes holistic in its nature and thus it becomes equivalent to the latter. Womack et al. (1990, p. 13) shed some light on the definition of the lean production system:

> Lean production (a term coined by IMVP researcher John Krafcik) is 'lean' because it uses less of everything compared with mass production – half the human effort in the factory, half the manufacturing space, half the investment in tools, half the engineering hours to develop a new product in half the time. Also, it requires keeping far less than half the needed inventory on site, results in many fewer defects, and produces a greater and ever growing variety of products …

Note that in the just mentioned IMVP (International Motor Vehicle Program) researcher's article (Krafcik, 1989), it becomes evident that lean production, flexible production and minimum inventories through the JIT process are all intrinsically interrelated.

Briefly the aim of JIT is to meet demand instantaneously with perfect quality and no waste. Hence, TQM or TQC and JIT are complementary strategies in order to excel in manufacturing (see for example, Flynn et al., 1995). Since total quality primarily means that the customer is the next process (Ishikawa, 1985, p. 104), the JIT/QC course of action involves many areas of the production chain, especially the design part, sales/distribution and the supply component. According to A. Harrison (1994), there are three elements in JIT/QC management: elimination of wastes, total quality control and importance of people. The first two elements are further explored in various parts of this book, as they are fundamental in the analysis and synthesis of relevant concepts.[21] The element of importance of people is less systematically analysed, as it is indirectly included in other concepts. The elimination of waste can be expressed in terms of the following factors: avoiding overproduction, eliminating waiting times, minimizing transport, aiming at zero inventories, reducing motion, cleaning the process and eliminating any defective goods (see ibid., pp. 182–3).

A few comments are worth mentioning regarding several techniques that are major features of JIT companies (ibid.). The 'focus' technique is related to the product focus. This supports the ample evidence provided by Fruin's work that the Japanese firm is basically a 'focal' firm. In relation to the 'layout and flow' technique, it is well known that the 'U' pattern of arranging machines substantially improves the JIT/QC process (see, for example, Abegglen and Stalk, 1985; Donchess, 1990). In addition, there is another element of layout relevant to this process, that is, the product-focused machine layout that often has the U-shape pattern. Thus, Ohno, in his implementation of the JIT system in Toyota, converted the engine-machining, transmission and suspension factories into product-focused machine layouts (Fujimoto, 1998).[22] Regarding the 'flow scheduling' technique, the aim is to synchronize all operations throughout the factory so that accurate inventory timing takes place. This is achieved with the use of *kanban* cards (for example, each container must have a *kanban* card). Thus, as the parts of production are always pulled by the succeeding process (the user or customer), no parts are started without a *kanban* card, containers of parts must never be removed from a storage area without a *kanban* first being posted, and so on (cf. Harrison, A., 1994; Krajewski and Ritzman, 1999).

A cornerstone of the JIT/QC process is the simultaneous continuous improvement of all operations and the continuous reduction in inventories. The situation is explained with a picture by Krajewski and Ritzman (1999, p. 741):[23]

> The rocks represent problems encountered in manufacturing or service delivery. When the water surface is high enough, the boat passes over the rocks because the high level of inventory covers up problems. As inventory shrinks, rocks are exposed. Ultimately, the boat will hit a rock if the water surface falls far enough. Through JIT systems, workers, supervisors, engineers, and analysts apply methods for continuous improvement to demolish the exposed rock …

This reduction of the 'sea of inventory' is the very visible result that any efficient JIT/QC company achieves. Thus, it makes sense to use this gauge (the target to reduce inventories) in order to measure the efficiency of firms, sectors, industries and the whole economy (the importance of inventories reduction is shown in various other passages of this study, as it is a crucial element of the analysis).

4.3.3 The Links of the JIT/QC Process with Other Organizational Issues

One of the experts on JIT/QC and operations management Schonberger (1986, p. 4) has described the following three major events during the history of American manufacturing management:

1. The beginning of scientific management around 1900 with the suggestions and experiments of Taylor, Gilbert and others.
2. The Hawthorne Studies of motivation at Western Electric, around 1930.
3. The beginning of the JIT/QC implementation in some pioneering companies in the USA in the 1980s.

When Schonberger wrote this in 1986, it was still difficult to assess the importance of the JIT/QC system. However, a few years later, as will be shown in Chapter 7, that system played a leading role in reviving and strengthening most industrial sectors in the USA, as well as accelerating economic growth in that country during the 1990s.

The impact of JIT/QC on manufacturing management can also be seen by examining alternative organization pyramids as shown in A. Harrison (1994, p. 196). Thus for a traditional organizational pyramid, the basis is one of its sides, as usual, with top-down command and a structure of many layers of hierarchy. On the contrary, for a total quality company, the basis of the pyramid is one of its apexes. In this case, there are fewer layers of management whose role is to support and lead at the 'coal face', or the top of the pyramid (its side) where the value-added work is done. This new organizational pyramid in a JIT/QC company can only work if there is cooperation at all levels of economic activities: between employees and employers, between employees, between managers, between the company and its suppliers, between the firm and other firms in the same industry, or in the same industrial district, or in the same *keiretsu*. With such universal cooperation, the motivation process of everybody increases with the aim of growing as much as possible by producing quality products; at the same time, transaction costs are substantially reduced. Another way of looking at the integration of the JIT/QC system within the overall Japanese management apparatus is suggested by Keys et al. (1994, p. 385). The most relevant aspect of their model is the 'effectiveness' part. As the authors remarked (p. 386): 'If Japanese management practices are superior to Western practices, the advantages should be observable in terms of productivity, product quality, and the utilization of human resources … '. This is exactly what happened after World War II in the USA and Japan (but in different sub-periods and opposite directions). Productivity, product quality and the utilization of human resources were much higher in the latter country than in the former, as will be extensively shown in Chapters 6 and 7.

How does the holistic JIT/QC system compare with systems that do not contain JIT/QC? Table 4.3 summarizes the differences and similarities between the JIT/QC system and the non- JIT/QC system.

Another way of examining the relations between JIT/QC and other OIs is by answering the question: why did JIT/QC take place in Japan and not in the USA? The answer also lies in the very Japanese characteristics of the focal factory analysed by Fruin (1992) and summarized in a previous sub-section, as well as in the tightly integrated system of *keiretsus*, and the whole society. In this respect, Abe (1997, p. 305) said: 'The establishment of hierarchical *keiretsus* (which could be described as multilayered production and distribution networks involving vertical interfirm integration) paved the way for the just-in-time system … '. The JIT/QC was devised and implemented in Japan because of several concurrent factors, such as cooperation between employees and employers, cooperation between SMEs and *keiretsus*, cooperation between the government and the private sector, cooperation between all levels of the managerial hierarchy, relative independence of factory management

Table 4.3 Comparison between the JIT/QC and non-JIT/QC systems

Characteristics	JIT/QC system	Non-JIT/QC system
Labour division	Flexible work teams	Rigid work segmentation
Setting standards	Standardization methods	Standardization methods
Inventories	Low inventories (high stocks are a waste)	High inventories for flexibility
Discipline	Self-discipline of workers	Through strict hierarchy
Production runs	Small batch sizes	Long runs
Planning flow	Last stage first	First stage first
Set-up times	Frequent	Infrequent
Operating control	Decentralized	Centralized
Interdependence	Increased	Lowered

Source: Based on information collected from A. Harrison (1994), McMillan (1996) and Schonberger (1982; 1986; 1996).

within the focal firms, limited natural resources and an adequately educated workforce to imitate foreign technology (TIs).[24]

On the contrary the JIT/QC philosophy overall has not been a feature of the American capitalist system (except in the last 20 years, as will be shown in Chapter 7), because that system, as it was developed in the period between 1880 and 1930, contained elements of OIs that did not agree with the JIT/QC philosophy (for example, tendency for big business, emphasis on TIs and not OIs, and so on). Despite this overall tendency, there were some instances, as in the auto industries in the late 1920s, where practices such as 'hand-to-mouth' purchasing were more closely related to the JIT/QC process (Langlois and Robertson, 1995, p. 62). Overall, mass production and scientific management explicitly separate each worker's functions and each management level's functions in a meticulous way, which excludes cooperation between them in the manner described in the previous paragraph and elsewhere in this report. Also, the dominance of big business over the whole economy and SMEs, and the abundance of natural resources in the USA allowed the American system to fully take advantage of economies of scale and scope (as shown by Chandler and other scholars), and, hence, the JIT/QC philosophy was not necessary until very recently in the development of the American system (see also Sanidas, 2005a).

4.3.4 Benefits from Implementing the JIT/QC Process, and Economic Growth

In Chapter 2 some of the benefits associated with JIT/QC were briefly indicated. In this sub-section a more detailed exploration of these benefits will take place. Imai (1997, p. 45) summarizes the benefits from implementation of JIT/QC as follows (explanation of Japanese words in the quotation are provided below).

> Opportunities for cost reduction on-site may be expressed in terms of muda. The best way to reduce costs in gemba is to eliminate excess use of resources. To reduce costs, the following seven activities should be carried out simultaneously, with quality improvement being the most important. The other six major cost-reduction activities may be regarded as part of the process quality in a broader sense:
> 1. Improve quality.
> 2. Improve productivity.
> 3. Reduce inventory.
> 4. Shorten the production line.
> 5. Reduce machine downtime.
> 6. Reduce space.
> 7. Reduce lead-time.
>
> These efforts to eliminate muda will reduce the overall cost of operations …

Note that *muda* means waste and *gemba* means shop floor or work place in general. Quality refers to both process quality and *gemba* quality. The former includes the quality of work in developing, making and selling products or services. The latter refers to managing resources and it includes the five Ms, namely man, machine, material, method and measurement. Although Imai does not advocate firing employees, the JIT/QC will tend to 'reduce the number of people on the line; the fewer line employees, the better' (ibid., p. 46). As will be shown in Chapter 7, this is exactly what happened to the leading sectors of the American economy in the 1990s, namely, a substantial increase in productivity with a concurrent small increase, steady, or even a decrease in employment (depending on the manufacturing sector).

Furthermore, as Imai (ibid.) explains, the two aims of improving quality and reducing costs are not mutually incompatible, and a JIT system encompasses both issues of cost and delivery. Once *kaizen* is started in *gemba*, shortcomings in upstream management will be identified in the process and eventually a good quality system will be in place, thus embracing everybody and everything in the company and also its suppliers. 'By eliminating all kinds of non-value-adding activities, JIT helps reduce costs' (ibid., p. 49). Further evidence of the substantial benefits the JIT/QC system offers to firms and eventually to the whole economy, is provided by numerous other researchers. The following paragraphs are just a small sample of this evidence. Womack et al. (1990), in their comprehensive study of the LPS related to the world vehicle industry, included several tables that show the higher productivity of the LPS *vis-à-vis* the non-LPS. Thus, Japanese firms in Japan and in North America perform much better in terms of hours of work per vehicle, assembly defects per 100 vehicles, space per vehicle per year, size of repair area, inventories, percentage of workforce in teams, job rotation, suggestions per employee, number of job classes, training of new production workers, absenteeism and automation (ibid., p. 92).

The same authors explain that the LPS's better performance is due to three reasons in terms of OIs:

> The truly lean plant has two key organizational features: It transfers the maximum number of tasks and responsibilities to those workers actually adding value to the car on the line, and it has in place a system for detecting defects that quickly traces every problem, once discovered, to its ultimate cause … So, in the end, it is the dynamic work team that emerges as the heart of the lean factory … (ibid., p. 99)

Kim and Takeda (1996), tested in their study (using 81 questionnaires in Japan) the hypothesis that the implementation of a JIT system will improve the effectiveness of the production control management system and competitiveness, and found that JIT techniques are effective in promoting productivity. Kobu and Greenwood (1991) remarked that increased living standards in industrialized nations, and in some of the developing countries, and increased competition have made customers more selective. One component of this new competition is that many markets are being quality driven rather than price driven. A manufacturing firm implementing a JIT system through a network of suppliers guarantees the highest quality delivery of goods. The same authors then analyse the causes of the decline in US competitiveness in the 1970s and 1980s. They also put forward several critical questions, one of them being: 'Why do we have to live with twice the layers of management, five times the cost of keeping the books, and ten times the inventories of our international competitors?' (ibid., p. 59). Schonberger (1982; 1986; 1996) showed in an extensive and convincing way the benefits of JIT/QC in Japan. The essence of his deductions is that there is higher productivity by using less material, labour and indirect inputs for the same or higher-quality products.

Although JIT and QC are intrinsically interrelated, it is also necessary to examine the benefits of QC more specifically. Watanabe (1991), after having briefly tested some concrete firms and industries, suggested four types of benefits of QC circles (p. 69):

- the development and production of low-cost devices and equipment

- more efficient use of existing equipment, for example, through a modified layout and working procedure
- the development of workers' capabilities
- intangible attitudinal and organizational improvements.

Quality control circles, their link with the suggestion system (millions of suggestions are made every year by millions of workers at their workplaces) and their paramount impact on productivity, are related to the high motivation of employees. The latter is based, according to Watanabe (ibid.), on three factors: the lifetime employment system (mainly in the larger firms), the egalitarian remuneration system and the promotion system that typically prevails among Japanese industries.

It is also worth noting the more abstract analysis in Watanabe's (1991) article regarding the impact of innovations and QC circles. Thus, there are two types of innovations: first, discontinuous dramatic ones (for example, introduction of the latest production techniques or management concepts) and, second, continuous incremental progress achieved through the QC circles. Continuous efforts for improvement are required because any system tends to start deteriorating as soon as it has been established. This spontaneous deterioration is the entropic effect (my term, which will be fully explained in Chapter 5). Thus, for a given firm with only discontinuous dramatic innovations, the expansion path is subsequently negatively affected by the entropic effect. However, with continuous improvement through a QC circle programme, the development path is overall much better than in any other situation. As Watanabe (1991, p. 68) remarks, an American study has observed the improved path (due to the QC circles) followed by Japanese firms: 'As the MIT Commission points out, this (improvement) can be large enough to outweigh the effect of technical breakthroughs, which seems to be exactly what happens in Japanese industries … '.

Abbeglen and Stalk (1985, ch. 5), in a classical study exploring the Japanese firm, provide substantial evidence of the benefits of the JIT process. Their main conclusions are:

- JIT eliminates the substantial amount of productive labour lost in material handling. For example, in the factory of a US hand tool manufacturer, between 40 per cent and 50 per cent of workers' time was spent moving material into and out of their departments. This is achieved by altering the layout of factories and machinery, by creating product areas, by developing multi-machine manning, by using simpler machines focusing on fewer tasks, and so on.
- The factory must first work as efficiently as possible in a manual way, and then automation is introduced.

- Overhead labour is greatly reduced with JIT.
- The adoption of JIT reduces the time required to manufacture a product.
- With JIT the entire factory is more flexible, and it works efficiently even when demand is uncertain.
- The labour productivity improvements of the "new" (JIT) factory are within the factory's four walls. Component suppliers can help achieve additional savings. If suppliers match their deliveries with the JIT needs of the factory, the management of the factory can be simplified even more. More important, if the supplier adopts JIT to provide this flexibility, the supplier will also enjoy savings from increased productivity, and some of these savings will be passed onto the factory' (ibid., p. 108).

The last benefit is a positive externality generated by the JIT system, with cumulative effects throughout the whole economy when JIT/QC expands gradually to all industries.

Krajewski and Ritzman (1999, p. 749) mention the following operational benefits:

- reduction in space
- reduction in all types of inventories
- reduction in manufacturing lead times
- increase in the productivity of direct labour employees, indirect support employees and clerical staff
- increase in equipment utilization
- reduction in paperwork
- setting valid priorities for production scheduling
- encouraging participation by the workforce
- increase in product quality.

It is worth noting that the list of authors who have written about JIT/QC and its benefits is too long to be mentioned in this study. For instance, Sakakibara et al. (1997) pointed out that from 1985 to 1990, over 700 JIT-related articles were published recognizing the importance of the JIT/QC system.

At this stage it is worthwhile summarizing some of the main elements of the analysis so far. According to the evidence presented in this sub-section, once the JIT/QC process is properly implemented in firms, it raises production value-added and productivity, which in turn raises the rate of economic growth. If more and more companies use the JIT/QC process we also have the effect of positive externalities (pecuniary or otherwise) on the whole economy, since JIT/QC involves many people and firms at all levels. In addition, as it has often been remarked, JIT/QC is not an isolated phenomenon within a firm, but it touches upon all aspects of its functioning and working.

In the remainder of this sub-section, the relationship between JIT/QC and economic growth from a more micro point of view is examined by referring to transaction costs theory (TCT), competences and so on. Reference will be mainly made to Dietrich (1994), as his propositions are closer to those advanced in this study than perhaps any other author in this respect (this author emphasizes not only the transaction costs and capabilities of the firm, but also strategic, learning and organizational issues such as those related to the JIT system). Dietrich (1994, in particular in chapter 4) in his attempt to push the TCT beyond its present state, summarizes the inefficiencies of the JIC (just-in-case) system of organization[25] which is the usual non-JIT/QC system. Based on the work of several other scholars, Dietrich groups these inefficiencies under the following headings:

- JIC does not respond to market changes flexibly enough.
- JIC requires expensive information and monitoring systems to allocate intrafirm resources.
- JIC creates large expensive buffer stocks.
- JIC conceals through buffer stocks production and quality problems.
- JIC makes quality testing more expensive than building quality into production. A separate quality control department increases overheads without increasing value added.
- JIC requires a rigid vertical hierarchy for coordination and control.
- JIC inhibits the development of dynamic economic and learning effects.

Dietrich concludes that the only way to reduce the above inefficiencies is to opt for the JIT/QC type of organization, which is superficially consistent with Williamson's 'relational team' (a relational team is efficient when human assets are specific and task non-separabilities are evident). A detailed analysis indicates that JIT/QC's superior efficiency is not explicable in solely transaction cost terms, mainly because the reduction in the ratio of indirect to direct labour (which is a consequence of the JIT/QC process) would reduce the importance of non-separabilities, which in turn would shift organizational characteristics away from the relational team towards an obligational market (according to Williamson's categorization). Unfortunately, for the TCT, this shift is not occurring; indeed, the opposite appears to be the case. Consequently, Dietrich suggests that in order to reconcile Williamson's TCT (and, for instance, Ouchi's 1981 'clan' form of organization, which is also based on the concept of relational team), we should distinguish between technological and organizational non-separabilities.

Once we make this distinction we can adequately use the concepts of 'core competences' (Prahalad and Hamel, 1990) or 'idiosyncratic organizational assets' (Pettigrew and Whipp, 1991), and others, in order to gauge the importance of systems such as JIT/QC. The latter, despite its static efficiency implications, derives its real superiority from dynamic costs and benefits. Thus, to conclude Dietrich's analysis (1994, and his related article in 1993b), it is worthwhile to quote him as follows: 'Orthodox transaction cost reasoning, with its comparative static emphasis, has little to say about these organizational innovations … Clearly, the emphasis in these non transaction cost contributions lies in, what is called here, the benefits of idiosyncratic, and tacit, organizational knowledge' (1994, p. 64); and 'The existence of modern organizations, involving just-in-time and total quality control systems, is dominated by benefit factors: the importance of idiosyncratic knowledge, dynamic advantage, and non-contractual motivation' (1994, p. 66).

The JIT/QC system is, however, not without drawbacks. It, for instance, generates negative externalities such as urban congestion, environmental and recycling costs (Cusumano, 1994). This author also sees the following limitations of the LPS: long geographical distances, overseas locations, stress on suppliers, too much product variety, shortage of blue-collar workers, the high cost of frequent model replacement, the high cost of frequent model line expansion and too much product variety. B. Harrison (1994), in analysing the negative side of flexible production, contends that more and more Americans are facing involuntary part-time or part-year work, low wages and few benefits. Dassbach (1999) strongly criticized the JIT/QC system because it is not a 'win-win' situation, but rather management wins and workers lose. Nonetheless, it remains that, as this section has demonstrated, the JIT/QC philosophy has been providing firms with a new ground for substantial productivity improvements. Thus, for Japan, such a production philosophy has been the basis for substantial economic growth as Dicken (1998, p. 168) reminds us: 'Although the Japanese production system has undoubtedly built heavily upon the new IT-based technologies, including the intensive use of robotic equipment, it can be argued that it is the 'organizational' technologies developed within the Japanese system which have been far more important to its postwar economic success … '.

4.3.5 The Role of Networking Relations and Cooperation

Although networking between firms is not only a characteristic of Japan (for instance, see the case of the Silicon Valley in the USA), the Japanese networks of *zaibatsus* and then *keiretsus* are important OIs to further explore. This sub-section analyses the impact OIs have mainly on the external organization of firms or, in other words, on how firms are interrelated through cooperation with each other. In Chapter 2 the role of firm networks as OIs was briefly mentioned. Here this role will be related to economic theories and concepts. Perry (1999) has made a substantial contribution towards understanding the role of business networks in several countries. His main theoretical conclusions about such networks are as follows.

According to the transaction cost perspective, business networks take place when neither the market nor the internal management hierarchy offers a satisfactory way of minimizing transaction costs. In this case, networks 'combine elements of the two basic ways of managing transactions by replacing the certainty of internal management or legally enforceable contracts by relationships based on trust, reputation and mutual dependence' (Perry, 1999, p. 4). Are networks a third permanent alternative to organizational structure in the economy? According to the modified transaction cost perspective, Perry (1999, p. 5) refers to Powell's (1990) table of attributes to emphasize that networks have an advantage over the other two forms of economic organization in three main areas: improved resource use and risk spreading, flexibility and adaptability to changing market opportunities through specialization, and effective accessing of information and skills.

According to a third perspective analysed by Perry, and based on Granovetter's work (1985), 'All economic transactions are embedded in social relations and entail cooperation as well as competition, implicit relations as well as formal agreements' (Perry, 1999, p. 6). Thus, economic relations in networks are also history dependent and inherited patterns of behaviour may originate even from pre-industrial societies. An extension of this perspective could perhaps be found in the notion of a cohesive power elite created by closely tied networks of corporate leaders, key policy-making bodies and elite social groups. The so formed interorganizational networks 'are a vehicle for enhancing the integration of the power elite and maintaining control of key social institutions' (Perry, 1999, p. 11).

As already mentioned, a key trait of networks is the trust implicated in various transactions (Perry, 1999; Sako, 1992; Zucker, 1986). Is trust a scarce commodity? Probably yes, as it heavily depends on historical relations, the cultural disposition towards cooperation, and the 'extent of commitment to higher goals in society' (Perry, 1999, p. 16). In the opinion of this author, there are certainly differences in all these factors from country to country, Japan being the obvious one where all these factors exist in abundance and hence trust is at a high level, at least until the recession that has occured in the Japanese economy over the last 10 to 15 years. This high density of confidence in Japanese businesses has been one of the main reasons why strong networks originated in Japan. This reasoning agrees with Dore (1987) who linked relational contracting with a cultural predisposition towards mutual indebtedness in networks (cited in Perry, 1999, p. 21).

The origins of networks in Japan have also been linked to the pressures of the economic environment. Perry (1999, p. 21) summarizes this link.

> The small size of the Japanese domestic market ruled out the possibility of pursuing economies of scale in the fashion of Fordist mass production. The entire year's output of the Japanese automobile industry, for example, was then equivalent to one and a half days of production in the United States. A shortage of resources and space for storage further enforced the experimentation with the low-inventory manufacturing methods that developed into the just-in-time system. Among the practices that had to be learned to make this system viable was 'governance by trust' in sub-contracting relations …

All the issues developed in this sub-section are strongly pertinent to the Japanese type of business, as already indicated. Japanese networks take several forms. One of them, the *zaibatsus* and later *keiretsus*, was more extensively described in Chapter 2. There are also industrial districts (Perry, 1999; Whittaker, 1997), small-business cooperatives (Perry, 1999), industry associations (Perry, 1999) and so on. An examination of these networks leads to the following main characteristics (Perry, 1999, especially ch. 5):

- cooperation between individuals and firms
- specialization of production
- a high degree of trust and preference
- minimization of transaction costs
- long-term associations that reduce the risks of changing demand
- increased flexibility for product development and design.

All these characteristics contribute to accelerating economic growth, since they make production more efficient.

Based on other sources, Perry (1999) summarizes some of the above points in the following sentence: 'Specialization results in a high use of market transactions relative to internal management coordination. Transaction uncertainties are contained by managing external relations within networks of interfirm mutual obligation with a long-term focus' (p. 44).

Although subcontracting is not a strong type of networking since it is usually based on independent bilateral relations (whereas networks involve several bilateral and mutual relations, and are thus interdependent), mention is made here of only some basic features of the subcontracting system of an obligational nature (see below) in Japan (again based on Perry's, 1999, pp. 147–54, synthetic work, which itself is based on the contribution of many other scholars' work). First, a series of contracts, related documents and customary practices govern the business relationship; thus, the legal framework is supplemented by the prospective sanction of a bad reputation. Second, both parties are expected to be guided by a commitment to a long-term association through sincerity and patience. Third, buyers expect their suppliers to make progressive reductions in cost each time contracted prices are reviewed (usually twice yearly).

Obligational contracting took place in Japan during the high-growth period of the 1960s. It has the following five features. First, there is high asset specificity in the investments made to facilitate subcontracting. Second, there is utilization of relatively customized inputs with design expertise at least partly conducted by the supplier. Third, there is opportunity to obtain advantage through product differentiation and diversification and reduced product cycles. Fourth, there is the strategy of market stability, in respect of both final sales and the buyer's preparedness to smooth out market fluctuations. Finally, trust becomes relatively inexpensive due to continuous market growth.

Besides Perry's work, other scholars increasingly have been attempting to identify the benefits of networking, for example, Ebers (1997b); also, Edwards and Samimi (1997) who attempted to explain the successful Japanese interfirm networks again with the theory of transaction costs. Ebers (1997a, p. 3) observes that 'in recent years, we have witnessed remarkable growth in various forms of co-operation among organizations', not only in the USA, but also in Europe (Korea and Japan have always been network-inclined in their industries). According to Ebers (1997a, p. 6), the motives of networking are summarized in two aims: increased revenue and decreased costs.

It is also necessary to emphasize again the close relationship between the JIT/QC system and networks. This system, as has already been explained, links the production of a firm to the production of other firms through the buyer–supplier relationships. Consequently, as networks offer a more co-operative version of the market, JIT/QC becomes even more efficient because we have reinforcement of positive externalities in both directions. Networks assist the smooth functioning of JIT/QC, and the latter assists the smooth functioning of networks.

Finally, a short note on clusters is now due. Clusters can be regarded as being part of a certain type of network, and their importance as an important factor for competition, mainly from the geographical point of view, can be further stressed. In this respect, Porter's (for example, 1990) analysis becomes very relevant. Effectively, Porter has advocated the significant contribution clusters make to competition and faster economic growth. In a more recent article, Porter (1998) includes a map of the USA identifying the most eminent clusters within the American economy, such as Silicon Valley, Seattle aircraft and boat building, and others. He also suggests 'clusters increase productivity vis-à-vis outsourcing or vertical integration' (ibid., p. 10) through the following factors:

- improving access to specialized inputs and information
- facilitating complementarities among cluster participants
- improving incentives and performance measurement
- improving the rate and success of innovations
- lowering barriers to new business formation that improve the environment for productivity
- increasing public investments in infrastructure.

Thus, despite the tendency for globalization and easiness of access to local resources, industrial clusters seem to proliferate not only in the USA but also in other countries (for Japan, cf. Perry, 1999, ch. 4; or Whittaker, 1997). Hence, as Porter reminds us, location economics is still very important.

In Perry's (1999, p. 82) synthetic work, we can find five reasons for the strengthening of competition of industrial districts. First, there is a blend of well-established enterprises alongside high rates of new firm formation. Second, there is an extended social division of labour. Third, within the cluster there is intense competition among firms that specialize in the same phase of production but, at the same time, there is a high level of interfirm cooperation, as specialized firms must collaborate with firms in a complementary phase of production. Fourth, rapid innovation diffusion (spillover effect) is realized through the geographical concentration, cooperation, and flow of personnel between firms. Finally, social and economic cohesion helps perpetuate the life of the district.

However, clusters, and generally all types of networks, seem to adjust to their own environment of such factors as capabilities, transaction costs, the scope of TIs, the presence or absence of economies of scale, the stages of product lives and so on. Hence, a Chandlerian firm is not always the 'best' firm, or a *keiretsu* is not always the 'best' network. Alternatively, the TI of computers has been developed in a particular type of 'innovative network', the Silicon Valley type for instance, and not in the context of Japanese or Marshallian, or Italian networks (Langlois and Robertson, 1995, ch. 7). All this shows that competition, efficiency and growth are contingent on many particular aspects of the space and time in which they take place, something that has been advocated throughout this book.

4.4 THE THREE AXES OF OIs

4.4.1 The Main Axis: Scientific Management cum Fordism versus JIT/QC

This sub-section expands on a previous sub-section regarding the importance of scientific management cum Fordism for the evolution of Japanese firms (Sanidas, 2004a is the main source in this section 4.4). In a recent article, Wilson (1996) argued that H. Ford was a JIT pioneer. His comparison between Fordism (or the original model T production system) and JIT/QC is illuminating, as shown in Table 4.4 which contains many of his ideas. Fujimoto (1999, p. 59) also arrives at a similar conclusion: 'Toyota's JIT method and the Ford system of the early days (Henry Ford's era of Highland Park experiments in the 1910s) had much in common … '.

Overall, as Wilson (1996, p. 30) said: 'The differences between them (Ford and Toyota) arise from Ford's restricted product line, and from the very different labour and industrial environments each faced … '. Furthermore, modern Fordism has some differences with the original one as implemented by Ford. For example, in the modern version, staff continue to produce regardless of quality until a batch finishes; also, although original Fordism intended to reduce inventories, modern Fordism accumulates inventories. This is contrary to the JIT/QC system according to which workers stop their routine work until any defects in machines or tools and so on are completely repaired, and hence inventories are minimized since the quality of products is high. Nonetheless, even modern Fordism and JIT/QC still have many of the similarities shown in Table 4.4.

However, the differences between Fordism and the LPS are more substantial if we follow some of the arguments elaborated by other scholars. For example, Fujimoto (1999, p. 60) says: 'Toyota's knowledge transfer from the Ford system was rather selective', for instance, Toyota's production architects emphasized semi-synchronization (for example, *kanban*) with small-lot production and supplier outsourcing. Wood (1989) remarked that if we interpret the Fordism system as implying a complete separation of conception from execution (for example, no cooperation between workers and management), then the LPS would be distinguishable from it. This author concludes that, overall, the LPS with its emphasis on harnessing tacit skills and latent talents of workers substantially differs from Fordism, which is based on disposable labour, on training for immediate needs, on supervision by specialist engineers and on jobs narrowly defined. The Japanese approach of labour management explicitly involves workers in *kaizen* and quality control through which every worker is encouraged to think like an industrial engineer. This entails cooperation and the development of skills, especially those

Table 4.4 Comparison of original Fordism and JIT/QC

Original Fordism	JIT/QC	Common feature
Mass production of model T for almost 20 years made large on-hand stocks unnecessary	Adaptability to rapid responses in markets and decreasing product life cycles	Decreased inventories
Dependence on high quality through machine and travelling inspectors	Dependence on high quality through teamwork and *kaizen*	High quality of products
High levels of staff turnover	Low levels of staff turnover	—
Labour force non-homogeneous, non-well-educated, little loyalty to employers	Labour force well trained, loyal to employers and homogeneous	—
The greatest distance any material has to be trucked is 20 feet	Deliveries directly to the assembly lines	Continuous flow
The central stock room and high work-in-progress eliminated	Minimize inventories, zero if possible	Minimum inventories
Often machines are arranged in a circle	U-shaped layouts	Layout is important
Long-term contracts with suppliers	Long-term contracts with suppliers	Long-term contracts with suppliers
Ford's suppliers were often far away, and rail transport was the only method available	JIT suppliers are often very close to the factory, and there are many ways of transportion	—
Ford's variable costs were often drastically reduced	Variable costs are reduced because less direct labour is required	Variable costs reduced

Source: Information compiled from Wilson (1996).

tacit skills that will allow the avoidance of mistakes and getting it 'right first time'.

According to Wood (1989) there are six elements that greatly facilitated the development of tacit skills, and hence the promotion of the LPS. First, the segmentation of the labour market involves the use of temporary and contract workers, all of whom can act as a buffer for the employment security of the permanent workers. Second, wages are not only determined by the employees' skills, age and seniority, but also by their performance. Third, trade unionism is structured on an enterprise basis. Fourth, the supervisors and team leaders are given a strong training role. Fifth, regular team meetings are held to discuss standards, performance levels and operational problems. Finally, there is job rotation.

The division between Fordism and the LPS has been the object of many studies such as that of Piore and Sabel (1984) who mainly analysed the various systems in a rather distinct way, that is the craft versus the mass versus the flexible specialization types of production. However, these two authors also suggested that the spread of flexible specialization 'amounts to a revival of craft forms of production' (p. 17). As will be emphasized below, we can create paradigms of production, which might not be, in reality, separate entities but somehow mixed. McLoughlin and Clark (1994) have summarized some of Piore and Sabel's ideas. For example, from the product market point of view, flexible specialization operates in a more uncertain environment, and hence it has a variety of differentiated products, which are not standardized as in Fordism. Also, Fordism promotes deskilled and adversarial labour, instead of up-skilled and cooperative labour for flexible firms.

As has been hinted earlier, the various types of production can overlap each other, or evolve in a parallel way, or influence each other. As McLoughlin and Clark (1994, p. 53) said: 'Continuity with the Fordist phase of industrial capitalism, rather than a "radical break" from it, appears to provide a more compelling starting point of analysis … '. Thus, the post-Fordist region, which has as a variant the flexible specialization sub-system, combines a high degree of all three factors in determining the region, that is, a high level of labour responsibility, process variability and product innovation. Also, both the initial Fordist paradigm and the neo-Fordist paradigm contain a low degree of labour responsibility (for more details see McLoughlin and Clark, 1994, p. 55). In addition, it must be emphasized that the LPS, once established in a large part of the economy, made a significant impact on growth in the same way as the initial Fordist evolution made a significant impact on growth a few decades earlier. These impacts will be quantitatively explored in Chapters 6 and, especially, 7.

The distinction between the LPS and the SM system can take other directions. Coriat and Dosi (1998) tell us that, 'Whereas the Tayloristic approach has been aimed to separate the functions of production, maintenance, quality control, planning, etc. and to fragment the tasks required by each function, the Japanese way on the contrary has been to create work stations where the different tasks are to different degrees reaggregated' (p. 119). A similar conclusion is reached by Lazonick (1990, p. 286). However, as the same authors (Coriat and Dosi, 1998) remark, the Taylorist legacy is not abandoned by many Japanese firms but rather it is 'regained'. This is achieved in two ways: first, through flexible work standards and transferable work components; and, second, through the implementation of the so-called 'management by eyes' as elaborated and designed by Ohno, according to which everything can be very easily visible and hence any worker can, for instance, stop the line at any time in order to make sure that quality and quantity of workmanship is absolutely guaranteed. In addition Ohno (1988) emphasized that this procedure should not only apply to the machines but also to the methods of production, the circulation of *kanbans*, the levels of inventories and so on. Furthermore, the famous 'five-dollar day' wage introduced by Ford to boost workers' motivation is matched by a system of individual and collective performance bonuses in the LPS. Coriat and Dosi (1998) summarized their findings as follows: 'Japanese practices may indeed be considered as a profound organizational innovation originally grounded in the local adaptation of Taylorism and Fordism, which eventually led to a distinct archetype of organizational routines for problem solving and governance of industrial relations' (p. 122).

This brief comparison between the two organizational schemes shows that there are two forces that shape them. First, they both (the original American and the Japanese ones) have a common background, namely, the attempt to rationalize the production process,[26] which started mainly with Taylor's principles and Ford's application to mass production. Second, they both diverged to accommodate other national influences such as size of the market, industrial relations, cultural traits and so on. These two points are well summarized in the following conclusion reached by Lillrark (1995, p. 981): 'It should be noted that while the Japanese were, and still are, Frederick Taylor's best students as far as work simplification and rationalization goes, they never adapted the American personnel management principles of Scientific Management, such as piece-rate pay and strict division of labour … '. A similar view is expressed by Sayer and Walker (1992) in their comprehensive study of the relevant concepts of Fordism, flexibility, JIT and so on: 'So, unlike neo-Fordism, JIT arose neither from the problems facing Fordism in the recession of the late 1970s and early 1980s, nor from the possibilities offered by microelectronics, but as an attempt to adapt western manufacturing practices to the context of post-war Japan' (p. 180).

The axis of Taylorism-Fordism versus JIT/QC or LPS is also related to other issues examined in this study. For instance, it is illuminating to quote Coriat and Dosi (1998, p. 116) again.

> The 'Tayloristic Revolution' describes at the level of production routines a process which co-evolves with the reshaping of the organizational structure of the firm, entrusting the general knowledge on coordination and strategies upon professional managers – as described by Chandler. Later on, we shall also argue that the modes of international adaptation of such 'American' (Chandlerian and Tayloristic) corporations have deeply affected for a long period the growth patterns of each country …

To close this sub-section, Tsutsui's (1998) revealing analysis is used to support further the relevant arguments. This author has historically explored the proposition that Japan's LPS or its equivalent JIT/QC are fundamentally based on 'Taylorite' premises and principles: 'Drawing on a common heritage of Taylorism, neither 'Japanese-style management' nor contemporary American practice has transcended the intellectual confines of Scientific Management' (ibid., p. 244). After the introduction of SM in Japan in 1911, the 'efficiency movement' in the 1920s, the 'industrial rationalization' of the depression-era, the mobilization campaigns of World War II, the post-war drive for productivity, the QC initiatives and the drive for minimum production waste of JIT during the after World War II era, all these stages embodied the ideological framework and the concrete shop-floor methodology of SM.

More precisely, 'the ideological backbone of "Toyotism" was patently Taylorite: at the core of the 'lean' model lay a compulsive urge to eradicate all waste within the manufacturing process, reduce costs, and boost the efficiency of labour' (Tsutsui, 1998, p. 180). The eradication of waste (as this was also mentioned in a previous sub-section) was accompanied by a parallel tendency for standardization (ibid., p. 181). Toyota's pioneering production system (that led to the LPS and JIT/QC system) 'was based on the rigorous application of the two Taylorite methodologies most highly developed in Japan, time-and-motion study and layout design'[27] (ibid., p. 183). Regarding the QC process, Tsutsui (ibid., p. 216) concludes that the QC pioneers in Japan regarded the Taylorite rationalisation of the work process as an essential prerequisite to QC. In particular, the motto was that QC starts and ends with standards. In addition, over the last 20 years or so, QC circles (see also previous sub-sections in this chapter) have been regarded as a Japanese OI because they are based on cooperative small-group activities that involve *kaizen* and feedback to the management. However, even QC circles are based on the Taylorite imperative to appropriate craft knowledge from shop-floor workers for managerial staff.

Overall, the above scholars' analyses point to the proposition made in this book that an important axis of OIs is the axis that starts with SM cum

Fordism and extends to the JIT/QC system. There are different variations of the content of this axis according to the historical period and country considered, but this axis can safely represent the internal OIs of firms.[28] Consequently, this axis of Taylorism-Fordism versus JIT/QC or LPS, enriched with Chandlerian elements, can be considered as one of the main driving forces of economic growth (besides the well-known force of TIs and others). Moreover, this axis is the main one of this analysis because it is related to the internal structure of the firm and hence it is more closely linked with OIs, that is, the organization of work, tasks, routines and so on. This is more extensively dealt with in the following chapter.

4.4.2 The Other Axes of OIs

The purpose of this section is to group all OIs under some very broad categories in order to facilitate the analysis and synthesis of this study. In the previous sub-section one such category was identified, namely the 'main axis'. In order to identify more significant categories, all the activities of a firm within the economy are shown, as a system, in Figure 4.2 (see Sanidas, 2004a).[29] The main elements of this system are the value chains of the two products, which are linked with the two factories and their internal structure and organization of firm F_0, and the firms and customers that this firm is linked with. The internal organization of this firm is the main axis; the firms in the network together with F_0 constitute the network axis, whereas parts of the value chain not yet in the production of F_0 constitute a potential vertical integration process axis. Thus, Figure 4.2 shows how these three axes are integrated into a larger system that encompasses the three axes and the value chain of production.

All three axes are dependent on each other. Thus, Fordism is intrinsically related to mass production and vertical integration, whereas JIT/QC is intrinsically linked with flexible or lean production; both underlying axes promote industrial growth in a contingency framework of conditions, that is, under the path dependency of country and history conditions. Also, mass production is linked with a rather thin type of networking (less cooperation) between firms whereas the LPS is linked with a rather dense network of firms (more cooperation).

The relationships between the three axes of OIs[30] can also be shown in Figure 4.3. The y-axis represents the 'main' axis as defined and analysed in the previous sub-section. Hence, between the two pure extremes of either Fordism or JIT/QC there exist systems which have in various degrees elements of the two extremes. Note that these links express tendencies and not absolute laws. Thus, generally, the x-axis as represented by the vertical integration line (1) in Figure 4.3 is positively related to Fordism (the higher

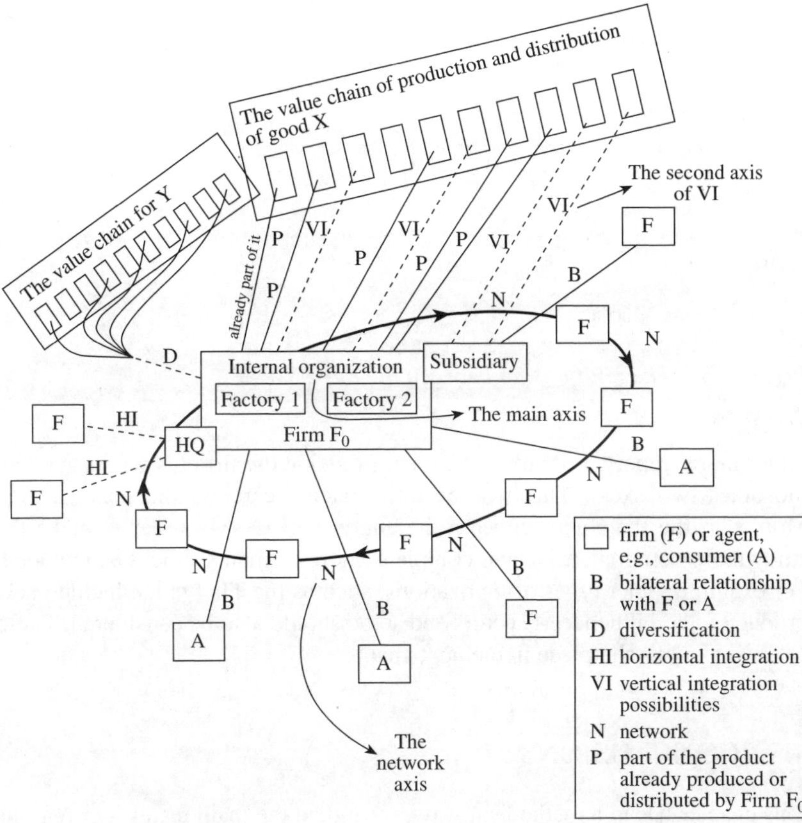

Source: Sanidas (2004a).

Figure 4.2 The three integrated axes

the Fordist state of the firm, the higher the tendency for vertical integration *ceteris paribus*); on the contrary, the x-axis represented by the networking line (2) in Figure 4.3 is negatively related to Fordism (the higher the Fordist state of the firm, the lower the tendency for networking *ceteris paribus*); as the JIT/QC is the other extreme of the Fordism/JIT/QC axis, the above relations are reversed for JIT/QC.

Note that all the OIs identified in Chapter 2 and summarized in Table 1.1 can be interpreted in terms of the three axes in Figure 4.3, or a combination of them. For example, the putting-out system is part of the vertical integration axis, the *keiretsu* is part of the networks axis, and so on. Furthermore, it is proposed that what happens inside the firm (represented by the main axis) is

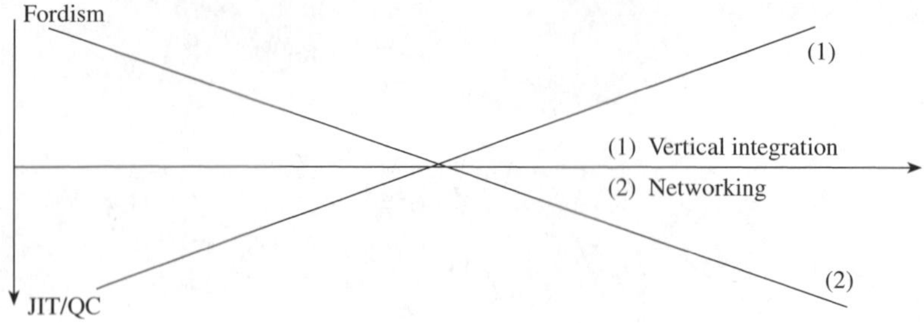

Source: Sanidas (2004a).

Figure 4.3 The relationship between the three axes

rather more important than what happens outside the firm (as represented by the other two axes). This will become much clearer in the next chapter. Moreover, the three axes concern the analysis of OIs. In order to make the study more comprehensive and complete, more elements should be included. For example, other types of innovations, such as the TIs (which include new products), the influence of sectors and so on should also be considered. These extensions will take place in the next chapter.

4.5 CONCLUSIONS

This chapter has in a methodical way determined the main features or reasons of the major OIs that have been the driving forces of productivity and hence industrial growth. Each sub-section identified these features within a more theoretical framework of various salient aspects. The relevant sub-sections summarized the main reasons as to how industrial growth is generated by OIs. These reasons cannot be properly understood unless they are analysed in the context of the particular OI considered, as effectively carried out in this chapter.

In addition, section 4.4 by synthesizing the common traits and background of all OIs, made the underlying reasons for considering OIs as significant forces of growth more evident. Thus, the main axis of Fordism cum JIT/QC was synthesized and briefly compared with the other two axes of OIs, namely vertical integration and networking. The latter provides more credibility to the substantial benefits the JIT/QC system brings to firms and industries. These three axes can also be used to differentiate between the firm evolution in the USA and Japan (thus, in that respect, complementing the brief comparison undertaken at the end of Chapter 2). According to the findings of this chapter, for the USA, the main axis and a high degree of the axis of vertical integration were the two axes on which the American firm evolution was based. For Japan, the main axis and a high degree of the axis of networking were the two axes on which the Japanese firm evolution was based.

Thus, overall, this chapter has scrutinized specific OIs in terms of several concepts and theories that make OIs a substantial moving force for the process of industrial growth (and economic growth as a consequence). The three axes synthesize all the OIs and provide a useful tool for analysing economic growth. However, these axes and, in particular, the main axis and its paramount importance will not be fully understood until after completing Chapter 5. Chapters 6 and 7 will relate the main axis – and to a lesser extent the other two axes – to industrial growth in a more quantitative way.

NOTES

1. Chapter 5 will substantially expand on these deeper reasons.
2. Categorization of concepts is always the first step before further analysis can be undertaken.
3. A. Crowley established in 1691 one of the earliest factories that relied on free labour to manufacture nails. Expansion soon followed to produce numerous items of hardware and ships' chandlery to finally create an enterprise that 'was undoubtedly a giant in the age of pygmies' (Jones, 1999, p. 22).
4. Another part of this content is the role played by the existence, the evolution and the actions of the entrepreneurs cum managers. One of their most remarkable traits is to be able to identify new and hitherto unrecognized market opportunities under conditions of uncertainty. Second, entrepreneurs and managers take risks, lead their companies and instil trust in the functioning of their firms. Third, and not less important, it is to innovate (in this respect Schumpeter's work, 1934 or 1942a/b, is most relevant). These innovations can be TIs, OIs, or both, or other types of innovations (to be introduced in Chapter 5). In the present study, only the business *consequences* of the decisions taken by entrepreneurs cum managers will be explored.
5. These costs include the following elements (Hill, 1994, p. 314): costs of hierarchy (for example, salaries of upper level management, and related incentives); inefficiencies due to information-processing problems within the structure of hierarchies; inefficiencies due to internal politics or influence costs (the latter were introduced by Milgrom and Roberts, 1992, p. 573: 'they arise from attempts to relocate and protect rents and quasi-rents within the organization').
6. In Chapter 5, these theories will be complemented with some important extensions such as the concept of 'kinetic costs' and they will all be put into the framework of system analysis.
7. The existence and growth of firms can also be seen through the lens of some other similar theories. The principal–agent theory was analysed more extensively by Williamson (1985), or Milgrom and Roberts (1992). Similarly the managerial theory of the firm was developed by Marris (1966), Odagiri (1981), Jacobson and Andreosso-O'Callaghan (1996) and Baumol (1967). Jacobson and Andreosso-O'Callaghan (1996) in their chapter about the theory of the firm also include the 'co-operative game theory', whose creator was Aoki (1984). It is also worth mentioning here Porter (1980; 1990), who introduced the five forces of competition, the three generic strategies, the value chain and the national diamonds. Last but not least, it is important to mention Alchian and Demsetz's (1972) contribution with their introduction of technological non-separability.
8. Big businesses are also due to vertical integration, and not just to the mere expansion of firms.

9. More details about the process of integration/disintegration can be found in Grant (1998), Carlton and Perloff (1990) and Williamson (1975; 1985). Also, the vertical integration analysis can include joint ventures, agency agreements and so on, according to combinations of formalization and commitment (see Grant, 1998, p. 325).
10. Nelson's (1992) book contains valuable contributions by several authors on the theme of 'mental revolution'.
11. Similar conclusions were reached by other scholars, for instance Drury (1922), Nelson (1992) and Wren (1979).
12. A recent collection of the contributions of some of these authors is found in Urwick and Brech (1994).
13. The term 'Fordism' is a generic term that encompasses more than what Ford innovated in the first place; for instance, it also includes the further split of the professional division of labour to comprise financial managers or marketing specialists introduced by Sloan of General Motors (Womack et al., 1990).
14. TFS will be further explained through the theoretical extensions of Chapter 5.
15. These propositions are conclusions reached by the scholars in the brief discussion that follows.
16. There are various types of standards, such as product ones and document ones. In this subsection reference is made to the compatibility standards that are related to processes (see Antonelli, 1998, for relevant definitions).
17. However, this author does not include the elements related to intrafirm impacts.
18. For a recent analysis of the meaning of waste and its importance to governance and TCs, see Williamson (1994).
19. Frantz (1997) has included many other quantitative studies to support the X-inefficiency theory.
20. Note that a more detailed account of the type of TCs within a firm was provided in the previous chapter based on the article of Bouttes and Hamamdjian (1997).
21. For example, the elimination of waste process is inherently linked with the 'process of movements' explored in Chapter 5.
22. From the principles point of view, the application of proper layouts can be traced back to other earlier cases (for instance, cf. Chandler, 1977 for steel plants, or Pollard and Robertson, 1979, for shipbuilding plants and so on).
23. Krajewski and Ritzman (1999) also analysed the following traits: the pull method of materials flow, consistently high quality, small lot sizes, uniform workstation loads, standardized components and work methods, close supplier ties, flexible workforce, line-flow strategy (for example, product layout), automated production, preventive maintenance, continuous improvement.
24. The imitation of foreign TIs will be further explored in subsequent sections and chapters.
25. See also Sayer and Walker (1992) and, much earlier, Abernathy et al. (1983) on the comparison between JIT and JIC.
26. Fujimoto (1999, pp. 63–4) mentions that as the indigenous craft-type system persisted in Toyota's production processes during the 1930s and 1940s, the first step that Ohno took was to standardize jobs. Also, Toyota introduced in 1951 (as imported from America) a formal training in 'scientific management' for supervisors (ibid., p. 70).
27. These two concepts and/or practices are part of the process of movements, which is the centre of OIs. This will be fully analysed in the next chapter.
28. Chapter 5 further analyses the internal character of this axis, where the concept of 'internal OIs' is more precisely defined (via the system of the complete set of firm operations and activities PROSIBB).
29. Many conclusions in this section 4.4 are drawn from Sanidas (2004a).
30. An attempt to combine some OIs together is encountered in other studies. For example, a graphical representation of networks, coordination integration and degree of ownership integration can be found in Langlois and Robertson (1995, p. 124).

5. The process of firm operations: core of the atomic/systemic model of economic growth

5.1 INTRODUCTION

In the previous three chapters, OIs were at the centre of the analysis in order to demonstrate the need and the importance of the link between OIs and economic growth. In Chapter 2, a historical survey of the industrial growth of the USA and Japan revealed that OIs played a substantial role in promoting this growth in the last 130 years or so. In Chapter 3, the first general links between OIs, TIs and economic growth were established from a theoretical aspect. In Chapter 4 these links became more concrete as very specific OIs were scrutinized in both countries, again from a theoretical viewpoint.

More particularly, in the last section of Chapter 4, it was proposed that the 'main' axis of intrafirm operations, represented by the Fordism cum JIT/QC alignment, was the predominant force generating industrial growth (*ceteris paribus* regarding the other axes of the analysis). In this chapter, it remains to justify further why this main axis can be considered as the predominant force. This can take place by further examining the nature of firm activities or operations in order to advance deeper into the core (firm) of industrial evolution. The need for a deeper analysis also comes from two other factors; first, OIs are related to a different process of firm activities from that of TIs and, second, industrial growth is generated by specific firms that lead the overall economy because of the adoption of OIs (*ceteris paribus*).

Thus, in this chapter, two main objectives will be attained. First, in the second section, OIs will be theoretically attached to some concrete processes of activities within the boundaries of firms, so that the deep cause of the growth of firms can be better understood. Second, in the third section the link between the growth of firms and the growth of industrial sectors, as well as the growth of such sectors and the growth of the overall economy, will be established. An atomic/systemic model of economic growth will be suggested in the fourth section as a consequence of the analysis carried out in this and the previous chapters. Consequently, this chapter complements in a valuable way the findings of the previous three chapters.

5.2 THE SYSTEM OF FOUR BIO-SOCIO-ECONOMIC PROCESSES

5.2.1 Introduction

Sanidas (2005b) has introduced a system of holistic processes that fully describe and explain all operations and activities of firms. Here, a summary of this paper will suffice to introduce the reader to all relevant concepts so that finally the role of OIs will be better understood. The underlying overall purpose of this section and chapter is to include OIs – in a more systematic way – in the suggested theory of the firm.[1] Wheatley (1994, p. 78) summarizes some of the intentions of this section: 'To stay viable, open systems maintain a state of non-equilibrium, keeping the system off balance so that it can change and grow … '.

In Chapter 3 and especially Chapter 4 it was concluded that the 'black box' consisted of the following elements: transaction costs, capabilities and competences (a generic concept), organization between humans and equipment, routines, tacit knowledge and learning, teamwork and supervision, strategic and purposeful action, and so on.[2] The black box has also been examined in a different context, such as that of the role of entrepreneurs (Schumpeter, 1934; 1942a/b), or the role of behaviour in economics (Cyert and March, 1963); or that of organizational capital (a type of human capital embodied in organizational relationships) and a holistic model of the firm's socio-economic performance (Tomer, 1987; 1990; 1998b); or that of the links between microeconomic theory and organizatioanl analysis (Leibenstein, 1960; 1966; 1987; this author introduced the concept of x-inefficiency).[3]

From the methodological point of view, the efforts are directed towards being as analytical, as operational and as explicit as possible, thereby avoiding generic concepts or umbrella terms such as that of capabilities. Thus, besides the detailed analysis of many aspects of the firm, the overall result is targeted to be a synthesis of existing theories and concepts into a new framework, although some new elements are also introduced. For example, the standard production function does not include the tangible inputs of organizational movements that entail kinetic costs, as will be seen below, nor does it include psychological, sociological, biological or physical elements. The proposed theory of the firm in this study encompasses the necessary taxonomies of all the above important elements (taxonomies are a fundamental feature of many biological and social sciences), attributes fundamental characteristics to the categories of the taxonomy, uses the theory of systems from sociology to examine the dynamic evolution of firms, proposes ways of measuring the system in relation to production, costs and other socio-economic variables, applies the proposed theory to some major organizational innovations and introduces a taxonomy of innovations. All these features are an attempt to follow some rules of scientific methods and models. Thus, science consists of the interplay of experimentation and theory, and the latter has two separate dimensions, namely, the conceptual aspect and predictive power (Moravcsik, 1988). In social sciences, experimentation is very difficult to achieve and hence measurement of proposed theories with other methods such as econometrics and statistical analysis of surveys becomes a necessity. In brief, this section's aim is to synthesize our existing knowledge[4] by making the concepts more operational and consistent with some theoretical principles.

As an introduction to what will follow, an example of a workplace will illustrate the main points for further analysis. Suppose we have an automobile service that employs six employees and has the necessary buildings, machines and equipment to carry out all required services. Krajewski and Ritzman (1999, p. 114) have the process chart of such a firm where they show the various steps of the job of changing engine oil from the first step of directing the customer into the service bay until the 21st step of receiving payment for the service. In between these two extreme steps we also have other stages such as waiting for customer decision, walking to the storeroom, performing under-car services, inspecting for leaks, parking the car and so on. All these 21 steps or stages involve five types of activities: operation, transport or displacement, inspection, delays and storing. In addition, each one of the steps involves some precise time and distance (ibid.).

The steps just described of changing engine oil express movements of work performed by the six employees and equipment in the above example. These movements necessitate some coordination, teamwork, timing, cooperation, effort and fatigue. They can be called the 'process of movements' (POM). In addition, these employees have some contracts signed with the authorities of the firm, follow some formal and informal rules of behaviour and try to reach some predetermined standards according to the available incentives and trust in the firm. All these extra elements constitute the 'process of contracts' (POC). A third process, called the 'process of strategies' (POS) involves the following extra activities: decisions and mistakes (due to some of these decisions) regarding objectives, plans and risks taken, the momentum of initiatives of action and the survival instinct of the firm. Finally, we have the 'process of wisdom' (POW), which encompasses the experience, knowledge, attitudes, level of training and education of the employees, the culture and innovative spirit of the firm, and the needs for exploration and idiosyncrasies of the firm. For all the pertinent elements of each process see Sanidas (2005b). These four processes will be the objects of analysis of this section and they are called the process(es) of the system of inputs of the black box (PROSIBB).

This firm just described has been through several stages in its life: birth, initial development, healthy growth or just survival, and eventually it will go through the final stage of death. Can this firm's economic growth, with its four processes, as just described, be represented by the typical well-known production functions used in both exogenous or endogenous growth models (usually macroeconomic models)? These functions can, for instance, take the form of (a) Y = F (K, L, N) and (b) of N rate = R (K, L, N), where K is capital, L is labour and N is the stock of knowledge. These two equations can be more mathematically specific and also include intermediate goods, value of the stock market, prices of capital and labour, technology and so on (cf. Marrewijk, 1999). Can these functions represent the kinetic energy by the workers of the automobile service firm example, their attitudes, their plans and so on? In general, can these functions represent the four processes described in the above example? As a brief answer it suffices to say here that six employees and some equipment 'thrown' into the mainstream production function without reference to organized movements, contracts, mistakes and so on will not adequately explain production and growth.

Sub-section 5.2.2 proposes the complete set of the four PROSIBB that embrace the totality of all possible ingredients, operations and capabilities. Sub-section 5.2.3 explores the 'black box' and proposes the relevant opportunity costs. Sub-section 5.2.4 proposes that the four PROSIBB are mutually exclusive and interdependent. Sub-section 5.2.5 proposes that the four PROSIBB constitute an open system with entropy. Sub-section 5.2.6 describes some ways of measuring the contents of the PROSIBB. Finally sub-section 5.2.7 links the PROSIBB to innovations, decision-making process and growth of firms.

5.2.2 Proposition 1: The Complete Set of the Four PROSIBB

As in Sanidas (2005b) it will be shown that the PROSIBB encompass all possible bio-activities[5] of a firm. This author has produced a table (not reproduced here) that indicates all possible elements of all PROSIBB. The following paragraphs summarize his findings.

The POW bestows abilities and memory on the firm, it creates power and potential energy, it expresses experience, culture, intelligence, knowledge, degree of rationality, abilities to research and discoveries. It shows the psychological and emotional state of the employees: for example, imagination and afflatus, illusions, conception and insights. The pertinent opportunity costs are *wisdom costs* (hence the aim would be to minimize negative or insufficient knowledge/wisdom).

The POS bestows initiatives for action on the employees of the firm. It creates a sense for survival and will and provides 'reaction' energy. It expresses ways of winning 'battles', of deciding, of planning for the future, of having vision and policies. It shows the instinctive capacity for attacks, and defence within a framework of uncertainty and risk-taking, hence readiness, emergence and momentum. The pertinent opportunity costs are *strategic costs* (minimize mistakes and inertias).

The POM bestows ways of movement interactions between the tangible inputs of production (TIOP) (for example, machines, people) and generates kinetic energy. It creates a framework for infrastructure and foundations for actual work. It expresses ways of making effort, cooperation, signs of fatigue, teamwork and actual performance. It shows the layout, the kinetic organization of the TIOP, kinetic routines and procedures. The pertinent opportunity costs are *kinetic costs* (minimize waste and misuse). As an example[6] of the POM, we can recall what Womack et al. (1990, p. 28) said:

> Ford soon recognised the problem with moving the worker from assembly stand to assembly stand: Walking, even if only for a yard or two, took time, and jam-ups frequently resulted as faster workers overtook the slower workers in front of them. Ford's stroke of genius in the spring of 1913, at his new Highland Park plant in Detroit, was the introduction of the moving assembly line, which brought the car past the stationary worker. This innovation cut cycle time from 2.3 minutes to 1.19 minutes …

Another example is the evidence indicated by various authors that the U-shaped production line adopted by Japanese firms increases productivity in relation to the L-shaped line adopted by firms not following the JIT/QC system. For instance, Goldman (1993) suggested that the U-shaped line generates reciprocity, cooperative teamwork and immediacy in non-verbal communication and feedback, thus breeding a long-term bonding relationship between workers. All this boosts labour productivity. The layout effect can embrace the whole plant. Blaug (1963, p. 13), for instance, said: 'the mere reorganisation of a plant may be as factor-saving as the introduction of new machines'. The steel plant example (Carnegie's new steel works) analysed by Chandler (1977, ch. 8) is also relevant here, as are many other examples, such as those provided by operations management scholars (for instance, Krajewski and Ritzman, 1999). Another example of layout is the one described by Bornholdt, as quoted by Wilson (1996, p. 28). The importance of the layout factor of the POM is particularly stressed here because any alteration in layout substantially affects the POM and consequently the other PROSIBB.

The POC bestows superstructure of governance, authority, contracts and institutions. It establishes the rules of ownership, conventions, communication and control. It expresses ways of reaching standards, status and hierarchy. It shows how supervision can take place, trust, negotiations and autonomy. The pertinent opportunity costs are *transaction costs* (minimize friction and opportunism).[7] For example, Womack et al. (1990, p. 43) said about the outcome of the Ford production system: 'The result was an ever-growing list of work rules that unquestionably reduced the efficiency of Ford's mass-production factory as workers fought continually for equity and fairness … '.

In brief, the four vital PROSIBB constitute all the primary elements of producing work: we need knowledge and memory (POW) as well as regulations and trust (POC) in order to take initiatives and decide (POS) how much of each tangible input of production[8] is efficient to execute the desired work in an integrated, coordinated and organized way with specific movements of effort (POM) in order to attain the objectives, plans and mission of the firm (POS). If we only have rules, quantities of TIOP, knowledge and decisions, work cannot take place unless the right movements of executed work exist. Thus, the POM (not necessarily the shop floor[9]) is the 'heart' of the firm (Sanidas, 2005b).

Continuing with Sanidas's (2005b) analysis, it becomes evident that the four PROSIBB are the moving force of the activities of firms. A few more remarks will close this section. First, the four PROSIBB when combined with the appropriate quantity and quality of the TIOP produce tangible outputs and non-tangible outputs (such as self-actualization). In turn, both these types of outputs interact with other parts of society such as other firms and non-market activities. Second, it becomes evident that the black box of production which has been ignored for a long time, mainly by economists, now becomes the main issue. Economists have put a heavy emphasis on 'tangible' variables, such as the TIOP and R&D, although the elements of human and organizational capital of the POW, and the POC, have also received a substantial attention. Management scholars have mainly explored elements of the POS, whereas the POM has generally been ignored except by operations management scholars. Overall, the PROSIBB seems to synthesize all existing theories of the firm and also expand on them. Third, as Sanidas (2005b) remarks, the PROSIBB supplements Plato's and Smith's (1776) division of labour. As Heydebrand (1989, p. 326) said: 'A particular division of labour may be both cause and consequence of a particular form of coordination and control … '. Thus, it is possible that the division of labour in a Fordist type of production process is similar to that of the JIT/QC process in terms of detail and precision, but it is more rigid in the former and more flexible and 'blurry' in the latter production system.

Fourth, it is helpful that we can now more precisely define 'capabilities' and 'competences'. Both these concepts are combinations of the various elements of all four PROSIBB.[10] The latter can also be equivalent to the concept of total factor productivity. Fifth, the PROSIBB will lead us to a better definition of technological change than that of Romer's (1990, p. S72): 'technological change is improvement in the instructions for mixing together raw materials … '. Sixth, other theories of the firm can be explained and extended through the use of the PROSIBB. For example, the principal–agency theory (cf. Milgrom and Roberts, 1992; Williamson, 1985) can be explored not only through the elements of the POC (transaction costs, governance, incentives, trust and so on), but also through the POS (reaction energy to the adversities created by the POC), the POW (idiosyncrasies, organizational spirit, illusions and so on) and the POM (the actual movements of work of coordinated efforts).

5.2.3 Proposition 2: The Opportunity Costs of the PROSIBB

There is an interaction between the four PROSIBB and the TIOP. Both types of inputs have opportunity costs. However, since most or all of the elements of the four PROSIBB are not explicitly taken into account in economic analysis, opportunity costs of all inputs are not properly estimated. In addition, the emphasis has been on the explicit costs attached to the TIOP, which is wrong from the theoretical point of view. The relationship between the PROSIBB, the TIOP and the outputs will be explored in this sub-section, again based on Sanidas (2005b). It is important to understand the opportunity costs attached to the PROSIBB elements, because differences in the PROSIBB show differences in firms and their performance, including differences in technological innovations (both OIs and TIs). Overall, the prices of the TIOP ultimately depend on their opportunity costs, which have no corresponding market prices and which are derived from the elements of the four PROSIBB. Furthermore, many of the latter are semi-public goods, hence reinforcing their opportunity costs qualification (see, for example, Ellig, 2001). Another way of explaining the role of these opportunity costs is that of investment streams: a set of TIOP can correspond to different sets of various elements of the PROSIBB. For example, the set $TIOP_n$ can be derived from three different sets of PROSIBB: $PROSIBB_1$, $PROSIBB_2$ and $PROSIBB_3$. Hence, the expenditure for $TIOP_n$ is equivalent to three different streams of investment projects due to the opportunity costs of the three different sets of PROSIBB. It is this possibility of alternatives within the PROSIBB that create substantial differences between firms (Sanidas, 2005b).

Sanidas (2005b) uses the tool of linear programming in order to show the importance of the opportunity costs ‘y’ of the PROSIBB, and argues that if ‘P’ are the explicit pecuniary prices of the TIOP, then we have the following identity: $\Sigma\ (P_j - y_j)\ TIOP_j = \Sigma\ y_j\ PROSIBB_{ij}$. Note that the term $\Sigma\ (P_j - y_j)$ $TIOP_j$ is a summation over the range j of the $TIOP_j$, whereas $\Sigma\ y_j\ PROSIBB_{ij}$ is a summation over the range of all inputs of the PROSIBB. If we let the difference $(\mathbf{P_j} - \mathbf{y_j})$ be called the diamond value of the black box, the higher this diamond value is the higher the value of the firm is for given quantities of the TIOP. The diamond value of the PROSIBB clearly depends on the expression $\Sigma\ y_j$ $PROSIBB_{ij}$, which is the value of the black box, and it can be used to explain the birth, existence, evolution and death of firms. If furthermore we consider the PROSIBB as quantities of production with intrinsic values per unit G_i, we optimize the PROSIBB given the limited amounts of TIOP (whereas in the first case above we optimized outputs of products given the limited amounts of all inputs (TIOP + PROSIBB). When combining the two cases, Sanidas (2005b) arrives at the following important conclusions. First, the opportunity costs (or maximum prices) of the TIOP are higher when the PROSIBB are not included in the optimization process than when they are: $y_j < P_j < y_j^*$, where y_j^* are the maximum prices of TIOP the firm can accept in order to optimize both the outputs and the PROSIBB. Second, the per unit benefits of G_i in the second case are larger than the y_j of the expression $\Sigma\ y_j$ $PROSIBB_{ij}$ of the first case: $G_i > y_j$ and the difference $G_i - y_j$ is another way of appreciating the diamond value as calculated above. Unfortunately, both G_i and y_j are not recorded in any accounting system.[11]

The analysis just described can now be used to elaborate on the standard production function (whether or not of the endogenous growth type) F = A (t) f (K, L, other tangible inputs, knowledge), where A(t) is a technology function (t stands for time), K is capital, L is labour and 'other tangible inputs' are human capital and so on. The production function proposed in this study is the following: Φ = f (TIOP, PROSIBB), where all elements of both TIOP and PROSIBB are in volume terms. In this function Φ, it is not necessary any more to include A(t) as the PROSIBB takes care of it in an explicit and holistic way. Of course, TIOP are measurable whereas it is not obvious yet as to how the PROSIBB can be measured. Sub-section 5.2.6 will make some propositions as to the measurement of the PROSIBB. We can add that although with the present state of the art, the TIOP are more easily or better measurable than the PROSIBB, one could say that even the TIOP are only proxies for what we really want to measure in the complex economic world (for example the capital stock is known for its measurement errors).

More analytically, the equation Φ = f (TIOP, PROSIBB) can be equal to Φ = f (M, K, L, POM, POC, POW, POS).[12] Given that the transaction costs (TCs) are the shadow price of the POC, the kinetic costs (KCs) are the shadow price of the POM, the wisdom costs (WCs) are the shadow price of the POW and the strategic costs (SCs) are the shadow price of the POS, and in a partial equilibrium set-up, the Lagrangian λ is equal to:

$$\lambda = \frac{v}{MP_K} = \frac{w}{MP_L} = \frac{m}{MP_M} = \frac{KCs}{MP_{POM}} = \frac{TCs}{MP_{POC}} = \frac{WCs}{MP_{POW}} = \frac{SCs}{MP_{POS}} \tag{5.1}$$

where MP stands for marginal product, v is the price of capital (K), w is the price of labour (L), m is the price of materials and intermediate goods (M).

This set of equations primarily means that the quantities of K, L and M do not only depend on their relative prices but also on the relative prices (or opportunity costs) of POM, POC, POW and POS. This is only indicative of some possible ramifications in economic theory. The consequences of the foregoing analysis are too numerous to be analysed here, and they would introduce a new set of issues to consider for another agenda of research.

However, it is worth mentioning an important consequence of the production function Φ, namely, the relative weight of the TIOP and the PROSIBB. The *mere* quantities of, for example, K and L of the TIOP very probably contribute very little to the production process, and the regression results of the standard neoclassical production functions and growth are probably based on spurious correlations. The only case that would guarantee that Φ is truly represented exclusively by the TIOP is when the PROSIBB does not change at all. But this is impossible. Even in cases where the firm can be isolated from the outside environment (other firms, the society, the world) there are many ingredients of the PROSIBB that change because we deal with people and people change all the time from within themselves. There are, however, some cases whereby the TIOP are a good proxy of the production process. For instance, the quantity and speed of picking up grapes in a vineyard is well approximated by the number of workers and their tools. Cases like that indicate that generally constant returns to scale are the best cases according to which, changes in TIOP entail predictable changes in output (even though the PROSIBB might still go through some internal changes which do not significantly affect or are affected by the TIOP).

In addition, the well-known positive effect of new investments on economic growth is not denied in this study, in the sense that the PROSIBB associated with new investments is the real moving force that propels economic growth and not the new investments themselves (*per se*). Overall, we need to thoroughly investigate Φ in order to evaluate the relative importance of the PROSIBB and the TIOP. According to the propositions made in this book it is expected that the PROSIBB is the main source of the growth and evolution of firms and hence of sectors and the whole economy, although the TIOP can also be a considerable source of growth through the embodied new technical innovations in intermediate goods and in capital. However, these new TIs are a consequence of the POS and the POW and generally of the PROSIBB (this will be further analysed in section 5.2.7). In addition, with the production function Φ the path dependence and the endogenous nature of the evolutionary process of production is incorporated because of the presence of the PROSIBB. On the contrary the usual production function F is not path dependent, as it does not take into account what happens inside the PROSIBB. Though it is outside the scope of this study, it must be mentioned that the function Φ, according to the propositions made here, is rather well related to the chaos theory, hence to non-linear differential equations that 'feel' more comfortable with the creative and adaptive process of natural phenomena (see Wheatley, 1994, for a relevant philosophical overview).

5.2.4 Proposition 3: The Mutually Exclusive and Interdependent Four PROSIBB

In brief (this sub-section draws the main conclusions from Sanidas, 2005b again), we can say that even if the costs of one particular process are reduced to zero, it is still possible to have positive opportunity costs for the remaining PROSIBB. Although, for example, the POC (superstructure) establishes rules and conventions, in general it does not and cannot determine all movements (POM),[13] thinking (POS) and knowledge (POW). If the POC is ideal (zero opportunity costs), the other three processes cannot necessarily be ideal. To express all this in a different way, we can also say that the firm cannot exist with less than the four PROSIBB; all of them must be present for the firm to start, survive and prosper. Thus, if there are no movements of work (POM) there is no work in the firm; if there are no rules, the firm will degenerate into anarchy; if there is no experience (POW) the firm will not create; and if there are no initiatives for action, the firm will soon die. Furthermore, the existence of each PROSIBB is autonomous. Thus, the POW can self-generate more knowledge and more memory, the POC can self-generate more rules, the POS can self-generate more initiatives and the POM can self-generate more coordinated movements. However, the TIOP cannot self-generate anything (unless there is procreation inside the firm regarding labour). Needless to say, without the TIOP the PROSIBB cannot exist and vice versa.

As Sanidas (2005b) suggests, there are several ways of demonstrating this third proposition. First, the four PROSIBB are at the same time interdependent and mutually exclusive in the same way as a probability[14] tree for dependent events shows. See also Taylor (2002, p. 453) for an example of two processes (or experimental events): the first is to flip a coin, and another process is to choose a coloured ball from a bucket.[15] The joint probabilities of these two dependent events are emphasized here: the flipping of the coin event does not include the picking of a ball from the bucket event (hence the two events are totally different from each other) and yet the outcome of the latter event depends on the outcome of the former event (process). This tree of joint probabilities can thus simulate the relationships between the four PROSIBB.[16] Hence, as each one of the four PROSIBB is linked to a separate aspect of the overall process of production, there is no overlapping between these four processes. Each element of each PROSSIB is unique for that process only, although there is a strong interdependence between all individual elements of all processes. For example, the 'objectives' element of the POS does not overlap with the 'competences' element of the POW although competences may influence the setting of objectives (as strategic outcomes of technical innovations); or, the 'incentives' element of the POC does not overlap with the 'routines' element of the POM although routines may influence the setting of incentives (Sanidas, 2005b).

Second, the parallelism or metaphor (for the use of metaphors see Hatch, 1997) of physical energies are used to further demonstrate the proposition of this sub-section. Thus, we can say that the kinetic energy is represented by the POM, the frictional energy by the POC, the potential energy by the POW and the reaction energy by the POS. These four energies are explained through the law of conservation of energy (Resnick et al., 1992, p. 154): 'in a system in which only conservative forces act, any change in the potential energy must be balanced by an equal and opposite change in the kinetic energy … '. However, if friction or heat acts as a non-conservative force, then friction or heat are transformed into an internal energy, which is lost during the motion that generates the kinetic energy. In a firm the kinetic energy (POM) as an expression of the potential energy (POW) can be somehow wasted due to friction (POC). In addition, it is worth mentioning that kinesis means movement (in both English and Greek) and 'synkinesis'(= with kinesis) means emotion (hence a state due to past experience and memory) in Greek. Thus for a firm, synkinesis is part of the potential energy (POW) that can be transformed into kinesis, the kinetic energy (POM). To complete the energy metaphor, for the kinetic energy to start taking place, we also need reaction energy, which is like an atomic reaction (POS). This reaction energy uses the potential energy of the POW in order to act or react to either external or internal stimuli. If the human input of the firm does not take initiatives to introduce changes, then the firm's 'entropy' or disorder will increase (see next sub-section for more details on entropy). Tomer (1998b) also refers to energies with 'charged electrical energy' coming from intangible soft ingredients (for example, vision) and 'electrical wiring' coming from tangible hard ingredients that provide coordination and control. In general, the potentialities of humans are permeated in the four PROSIBB.

Third, the metaphor of the human being and brain is used as per Sanidas (2005b). Thus, the POS as a decision-making process can be seen as one part of the brain, and the POW as a memory and accumulation of knowledge process as another. This dichotomy of the brain fits well with the principles of cognitive psychology. For example, Hodgkinson, (2003, p. 4) remarks: 'The human information-processing approach to the analysis of skilled performance suggests that the way in which individuals act is driven ultimately by the way in which they interpret their worlds (perception), this in turn being shaped in part by their past experiences and learning … '. Therefore, according to this quote, the verb 'act' is linked with the POS and the nouns 'past experiences', 'perception' and 'learning' are linked to the POW in the present analysis. In addition, Milech and Finucane (1998, p. 292) say: 'Most behavioral decision theories are pitched at the same level of explanation: They describe how information is transformed from the initial statement of a problem to a final judgment or decision … '. Besides the brain (representing the POS and POW), the movements of various parts of the body (legs, arms, head, mouth) represent the POM in the present analysis. However, other parts of the body such as breathing capabilities, hormones, joints, muscle capacity put some limits to these movements, hence fatigue and even diseases take place. These parts represent the POC. Overall, the human being produces output (for example, collecting grapes) by using the whole body (the four PROSIBB) and inputs such as air and food (TIOP). It is evident that air and food cannot produce the output unless the body and mind are used (the four PROSIBB).

5.2.5 Proposition 4: The PROSIBB is an Open Negentropic System

The analysis on the four PROSIBB cannot be complete without this fourth proposition. Again, economics and management can take advantage of concepts developed in other sciences. An open negentropic system as developed and analysed by Buckley (1967) is one of these concepts. Sanidas (2005b) provided a summary of Buckley's (1967) model, and hence only some points will be reminded here. The *process* or *complex adaptive system*[17] is characterized by the elaboration or evolution of organization. The most important element of an organized whole (being more than the sum of it parts) is its *emergence*:[18] 'it is because present behavior is always both emergent from the past and conditioned by possible future results that it is never precisely predictable and always involves an element of spontaneity' (ibid., p. 99).[19] If organizing *routines* do not match the current state of affairs, *work, physical and logical*, must be done to bring the system up to date. This work includes formation, strengthening or dissolution of functional *linkages between various acts* or basic sequences of acts. This complex evolutionary system is also open: 'The important distinction between open and closed systems has often been expressed in terms of 'entropy':[20] closed systems tend to increase in entropy – to 'run down'; open systems are 'negentropic' – tending to decrease in entropy, or to elaborate structure' (Buckley, 1967, p. 51). Also, Hansson (1991) said that any self-organizing system is an open system which exchanges energy, entropy and materials with other self-organizing systems (see also Clark, 1988).

In relation to the PROSIBB, the POM is more plainly related to the characteristics of an open system. At any point of time, a firm has a level of entropy L1 based primarily on the POM. However, it is the POS, which as storage of reaction energy continuously takes initiatives (for example, by removing bottlenecks in specific areas of the production process) in order to resist the tendency of the POM to increase its entropy (for example, due to inadequacies of the POC). The POS together with the POW have the potential to decrease entropy, hence the PROSIBB is an overall open system that becomes negentropic. One of the main conclusions of the present analysis is that the entropic nature of the POM and of all the four PROSIBB is a distinct characteristic of organizational innovations (technical innovations as new products or new processes do not have this trait since TIs depend on the OIs in order to be generated and completed successfully). The mechanisms of *feedback* and *self-regulation* (ibid.)[21] inherent in an open system can be seen through the following sequence of the four PROSIBB: POW → POS → POM → POC → POM → … . As Wheatley (1994, p. 98) remarked: 'In the world of self-organizing structures, everything is open and susceptible to change … Stasis, balance, equilibrium – these are temporary states. What endures is process – dynamic, adaptive, creative … '.

This self-regulation, self-organization of open systems (for example, firms), can now be related to another issue. The four PROSIBB are also linked to other parts of society through the output they produce (products, inventories, defects, self-actualization) and through the PROSIBB 'society' and PROSIBB 'individual' (see Sanidas, 2004c). One of these parts of the society is the shareholders for firms enlisted in the stock exchange market. How are firms, shareholders, and economy interacting? As Bulmash (2002, p. 99) suggests, the economy and the stock market (hence shareholders) are like 'bio-systems that interact with each other in a symbiotic relationship in which they feed each other'. According to this author, the stock market flourishes in a continuous process of convergence, reaction and adaptation to the host economy by seeking self-preservation and growth, and evades dangers by migrating to less threatening 'pastures'. Thus, for at least the public firms, the shareholders are the crucial link that makes firms being self-regulating open systems. In the case where some shareholders are also working in the firm (as managers or employees), these particular shareholders are playing another crucial role in determining the future of the firm. In this case the POW 'individual' (Sanidas, 2004c) will especially influence their decision-making process (hence the POS 'individual') and fortify or weaken the open system of firms (PROSIBB 'firm').

An important consequence for this study of the negentropic nature of the system of the four PROSIBB is a reduction in waste. This waste includes inventories, defects in products, strikes, non-cooperation between employees, corruption and so on. If the firm becomes more efficient by lowering entropy, by introducing OIs that are negentropic (thus by 'producing' a more efficient set of PROSIBB as already explained) there is indeed a reduction of waste.

This important consequence is used as a theoretical support for the choice of the proxy used in Sanidas (2004b) to measure the positive impact of the JIT/QC system on American manufacturing sectoral growth. This will also be shown in Chapter 7. The decrease in entropy and hence waste can also be understood in the context of firms being 'social constructions'. In this context, the decision-makers create the environment they respond to (Hatch, 1997), and the underlying aim (not always conscious) of these decision-makers is also to decrease entropy. The degree of this decrease depends on the interplay and dynamic evolution of the four PROSIBB and the ensuing TIOP, and hence this evolution generates a multitude of various firms in the market at any point of time. As Altman (2002, p. 284) says, 'The fact that market forces do not displace the traditional work cultures in no way demonstrates their economic efficiency … '. However, the numerous combinations of the elements of the four PROSIBB cannot predict the outcome of the performance of firms unless we place it within particular historical, societal and environmental contexts. This leads us to the extension of the system of the four PROSIBB on to a similar PROSIBB for society and for individuals (see Sanidas, 2004c). This agrees with the contingency theory in management and business and it also means that there has to be a balance (not an equilibrium) between the PROSIBB of all three levels (firm, society, individual).

Important elements of this complex system have been used by some scholars to explain and explore concrete cases. Fujimoto (1999), in his detailed examination of Toyota's manufacturing system, called it the 'multi-path system emergence' with an ability of *preparedness*:

> I don't reduce the historical process that formed Toyota-style manufacturing routines to randomness. Instead I describe it as the interplay of both intended and unintended consequences for the people who created the system. I call this highly irregular historical process *multi-path system emergence*, in which decision makers often don't know beforehand which path will lead to successful outcome-deliberate planning, environmental imperatives, intuition, imitation, or luck … (ibid., p. 8)

This system can also be considered as a transmitting and creating information system comprising three layers of organizational capabilities: first, 'routinized manufacturing capability' (taking place in a steady state of repetitive production, development and transactions); second, 'routinized learning capability' (affecting the pace of continuous or repetitive performance improvements, and being recovered from system disruptions or deterioration). An example of this layer is the application of total quality control. Third, 'evolutionary learning capability': 'A nonroutine ability that affects creation of the above routine capabilities themselves through irregular processes of multi-path system emergence' (ibid., p. 17). This third capability is also linked with the organizational culture of 'preparedness': a company such as Toyota is well prepared for reinterpreting, refining and institutionalizing established routines. Also 'it must convert both the intended and the unintended consequences of its actions, the lucky breaks and the well-laid plans, the temporary successes and the failures, into long-term competitive routines' (ibid., p. 23). Fujimoto (1999) has analysed the JIT/QC system as a specific case of system emergence, and hence this study fully supports this author regarding the evolutionary and dynamic development of firm capabilities through the introduction of the four PROSIBB.[22]

5.2.6 Measurement of the Four PROSIBB

Measurement issues are (and should be) an important part of validating theoretical models (see for example, McCloskey, 2000). The following few paragraphs are only a brief outline of the possibilities of measuring the PROSIBB. According to a stepwise procedure, in order to calculate the net benefits for the POM we must measure *the changes* in the TIOP and the impact on productivity by changing the POM but without changing the POC, POW and POS (which is possible if we do not change contracts, training, strategies and so on). Second, in order to calculate the net benefits for the POC we must measure the changes in the TIOP and the impact on productivity by altering the POC but without changing the POW, the POS and the POM; and so on for the remaining two PROSIBB. Finally, if we hold constant the POC, POM, POW and the POS, then we can measure changes in productivity caused by changes in the TIOP itself.[23] All this is possible because the elements of the four PROSIBB are not overlapping. With another experiment, we would be able to measure the changes in net benefits and productivity when two or three of the PROSIBB are altered at the same time. In this way, we could indicate the synchronic effect of the relevant PROSIBB. Overall, with this method we can only measure the marginal effect of each PROSIBB, or a sub-group of the four PROSIBB, or all the four PROSIBB based on changes in the TIOP, thus measuring labour or capital or total factor productivities.

A similar procedure is suggested by Baily and Gersbach (1995, p. 318)[24] in order to measure the impact of organizational changes on productivity:

> Suppose we want to examine the cause of the productivity difference between two plants that use similar machinery and equipment to produce the same products but that use different product designs and organise their workplaces differently. Then techniques such as reverse engineering and process analysis are used to estimate the amount by which labour and material inputs could be reduced if the products were redesigned so that the manufacturing design is comparable to best practice,

> but where the organisation of functions and tasks remains unchanged except for those changes required by the change in manufacturing design. This procedure then allows an estimate of the partial effect of product design on productivity …

Such procedures, namely, reverse engineering and process analysis, are common in practice, especially recently (as Baily and Gersbach, 1995, show). Process re-engineering,[25] in general, is now a standard way of reorganizing work (mainly the POM in the present analysis) and of estimating the relevant costs (Krajewski and Ritzman, 1999, pp. 108–16). Process re-engineering usually cuts costs dramatically, especially by reducing employment, as Krajewski and Ritzman (1999, pp. 88, 108 and so on) show in their examples (this confirms Edquist et al., 2001, who demonstrated in their analysis that organizational process innovations are labour-saving if the compensating effects are not taken into account; see also Chapter 7). In one of these examples (Krajewski and Ritzman, 1999, p. 104), a 'process chart' shows the work movements related to an emergency room admission, and hence it demonstrates the movements and procedures (for instance, it takes three minutes for the nurse to inspect injury, it takes two minutes to walk to the pharmacy, and so on) involved that generate the POM for a services firm.[26]

Also, it is possible to use the various characteristics of the four PROSIBB in order to quantify the impact of an OI, such as the JIT/QC or the M-form of firm governance, on a firm. This is feasible if, for instance, we have a sample of cross-section data of the traits of a PROSIBB for firms producing the same product. Thus, in this case we can use the well-known method of questionnaires and surveys in order to arrive at a detailed exploration of the significance of PROSIBB. The surveys will contain questions on each specific element of each PROSIBB (for example, regarding the impact of the POW, the elements of leadership style, attitudes, beliefs and so on should be the object of longitudinal surveys). Furthermore, the use of proxies in econometrics is well known regarding the quantification of economic phenomena, and once these proxies are 'discovered' and are well justified analytically, they can be used in empirical work to measure the PROSIBB. For example, for the POC we could easily use variables of conflicts in the firm, for example, strikes, or staff complaints; for the POS we could use the number of mistakes or wrong decisions by staff members and their consequences; for the POW we could use R&D patents, attitude traits, level of motivation, training programmes and so on; for the POM we could use effectiveness of feedback through the *kanban* system, and so on. Of course, there are some elements that are easier to measure, such as the effects of layout (see Krajewski and Ritzman, 1999).

In addition, regarding the Φ production function (see sub-section 5.2.3), many proxy variables for the PROSIBB can be divided by the TIOP variables, for example, physical effort per team or per worker and so on, in order to take into account the TIOP. The results of surveys for the 'softer' components can also be included in the estimation of Φ. Moreover, in Chapter 7 the measurement of the PROSIBB as a totality will be attempted in which a well-established proxy for the JIT/QC system (an OI) is used to measure the productivity changes of American manufacturing sectors. Finally, as a very specific example, namely, the physical effort, it is also worth mentioning an attempt to measure a very important component of the POM, that is, the ergonomics part. The Toyota Verification of Assembly Line (TVAL) is an indicator that measures the workload of each assembly job based on a formula developed by Toyota (Fujimoto, 1999, p. 235). This formula is a function of task duration time, work posture and weight of parts/tools, and its aim is to improve physiological aspects of employee satisfaction, so that assembly jobs are less demanding. By using this indicator, 'assembly process planners could identify physically demanding jobs in an objective manner, prioritise the workstations to be improved, and concentrate efforts for improvements (for example, automation, power assist devices, work design changes) on the workstations with high TVAL scores' (ibid.).

5.2.7 Links between the PROSIBB and Growth, Innovations and OIs

The combination of the four PROSIBB is a necessary and sufficient condition for the existence of individual firms, and their dynamic evolution through time and space. Any system of the four PROSIBB is equivalent to a firm. However, systems continually evolve. This proposition is similar to the one proposed by Kaldor (1934), 70 years ago.

> The most satisfactory definition of a firm is that of a 'productive combination possessing a given unit of coordinating ability' … Firms whose coordinating ability changes, while preserving their legal identity, would not remain the same firms; but then all the theoretically relevant characteristics of a firm change with changes in coordinating ability. It might as well be treated, therefore, as a different firm … (Kaldor, 1934, pp. 69, 70).

Note that Kaldor's 'coordinating ability' very probably encompasses all PROSIBB. Overall, it seems that this analysis allows for a perpetual creation of the firm, thus confirming its dynamic nature and Schumpeter's (1942) continuous 'creative destruction'.

Any change in any of the four PROSIBB would generate an increase (or decrease) in the firm's productivity, because of changes in the underlying opportunity costs attached to each PROSIBB and the differences between the prices of the TIOP and these opportunity costs (the diamond value of the black box, as mentioned earlier). The evolution of firms depends on the evolutionary interplay between the four PROSIBB. In this respect the following schema will help to grasp a probable sequence of interdependence between the various PROSIBB: POS → POW → POM → TIOP → POC → POS → … and so on. Of course, it is possible that once we are at a particular PROSIBB the sequence is reversed in order to get feedback, for example, POM → POW → POS → TIOP → POC → and so on. Initiatives are taken by the POS (which is not necessarily represented by the upper management, but also by the participation of many other employees) in order to change the POM on which the whole production system depends. These initiatives can spring from the two types of technological innovations, namely the organizational innovations and technical innovations (or other types of innovations, see further below).

Regarding the TIs, a distinction must be made as to whether a TI is produced in the firm (for example, as a new product) or a TI (produced by another firm) is incorporated in the stock of capital and intermediate goods. In the first case of a TI produced in the firm, the knowledge and wisdom of the firm is increased (for example, through R&D), hence the POW is affected, and if this TI is an intermediate good or a type of equipment that increases the capital stock, then the TIOP is affected as well. Consequently, TIs are exclusively generated by the POW and/or incorporated or embodied in the TIOP. In brief the following schemas summarize the sequence of the decision tree:

- (POM and TIOP) → POS → POW → POS → (TI to be produced: final good) → POW → POM → (POS, POW, TIOP, POC, POM)
- (POM and TIOP) → POS → POW → POS → (TI to be produced: intermediate good or equipment) → (POW and TIOP) → POM → (POS, POW, TIOP, POC, POM)
- (POM and TIOP) → POS → POW → POS → (TI to be adopted, hence bought) → TIOP → POM → (POS, POW, TIOP, POC, POM)

In each of the above cases the requirements of the POM and TIOP initially provide feedback to the POS, which then initiates a set of decisions through all of the PROSIBB.[27] If we take an OI,[28] for example, the JIT/QC system, then, we have the following schematic interactions between all PROSIBB: (POM and TIOP) → POS → POW → POS → JIT/QC → POM → (POS, POW, TIOP, POC, POM).[29]

First the POM and TIOP provide feedback to the POS and POW, as to the possibilities of improving the POM. Second, the POS decides to implement JIT/QC. Third, the latter will primarily affect the POM. Fourth, the suggested changes in the POM will affect the other PROSIBB. This sequence of interactions can also be represented by a decision tree as suggested in a previous sub-section.

Overall, innovations that emanate from each one of the four PROSIBB are different types of technological innovations. Thus, 'kinetic innovations' are generated by elements of the POM exclusively because this process is the only one to include aspects of organizational energy, such as procedures, routines, timing, layout and so on, which all involve kinetic energy. Also, according to this analysis, when the firm produces TIs, the POW is affected (wisdom is enhanced) and hence these TIs can also be called 'wisdom innovations'. There are, of course, many other wisdom innovations. For example, the quality control circles[30] can generate much more knowledge and many more competences. In the same way, innovations that emanate from the POS ought to be called 'strategic' or 'decision' innovations, and those innovations stemming from the POC are 'contract' innovations. For example, if we consider the OI of vertical integration, the decision to vertically integrate is primarily a strategic decision, because it examines the relative position of the firm vis-à-vis the other firms, in other words the firm's competitive advantage (Grant, 1998; Thomson and Strickland 2001). If we consider the OI of M-form of governance of a firm, the decision to implement this form directly impinges on the POC and indirectly on the POM.[31] In this case, we have an example of a contract innovation. Hence, we have the following schemas that represent the decision innovation of vertical integration, and the contract innovation of the M-form respectively. Note that these schemas also follow conclusions such as those by Chandler (1962; 1977) who argues that structure (POC) follows strategy (POS).

- (POM and TIOP) → POS → POW → POS → (decision innovation to vertically integrate) → POS → POM → (POS, POW, TIOP, POC, POM)
- (POM and TIOP) → POS → POW → POS → (contract innovation to introduce the M-form) → POC → POM → (POS, POW, TIOP, POC, POM)

This section closes with a further examination of the relation between the four PROSIBB and OIs. For example, the OI of vertical integration cum mass production is a mixture of the four PROSIBB as follows. If a firm decides to further vertically integrate parts of the value chain, the POM will primarily be affected, as the new parts will have to be integrated into the existing POM. Overall, the evolution of vertical integration depends on how each process (PROSIBB) changes according to the particular nature of incorporated OIs or TIs. Thus, for example, as products become more and more sophisticated the embodied technology (TI) primarily affects the POW (accumulated knowledge and memory), which strengthens the focal character of specialized firms and hence encourages outsourcing. In the same way, as the use of the JIT/QC system in more and more firms in the economy takes place, this system principally impinges on the POM, which again strengthens the focal character of specialized firms and hence encourages outsourcing. Thus, the gradual reinforcement of the POM and POW, and not so much the POC as the transaction costs theorists would have suggested, is the main reason behind the recent tendencies for vertical disintegration. However, the POC will be substantially reconsidered mainly as a consequence of the changes in the POM. If we take another OI, for example the introduction of the M-form of governance, the various PROSIBB are affected in similar ways (vertical integration and the M-form are of course related). Thus, the POC is affected from the rules and contracts point of view directly, and indirectly based on changes of the POM, as the latter is heavily and directly influenced because the actual effort of work in terms of teamwork and other tasks is different under the M-form than under other forms of governance. The four basic processes suggested in this study can be eventually used to empirically test the importance of the various types of firm governance, by using the unique characteristics of each PROSIBB.

If we consider the JIT/QC system, it is worth showing the POM as the single-card *kanban* procedure (see Krajewski and Ritzman, 1999, p. 743, for a relevant chart), where it becomes clear that the POM is about movements and kinetic energy or costs, as already explained. It also becomes clear that there are several ways of arranging these movements in the same space in order to produce a given amount of output, hence there is always room for improvements in the organization of the kinetic energy, and hence there is always room for increasing productivity. These improvements can occur with or without changes in the quantities of the physical factors of production, that is the TIOP (or even with or without changes in the other PROSIBB). The JIT/QC system is of course much more than the *kanban* procedure, as explained in several places in the present study (recall for example, Harrison, 1992) and hence all the PROSIBB are involved. Krajewski and Ritzman (1999) argue, more cooperation and trust is needed for the successful implementation of JIT/QC. Accordingly the POM and POC are affected. In addition, the reward systems and labour classifications need to be altered, thus the POC is involved again. The workers might feel more stressed and pushed to work harder, thus the POM and POW are affected this time. The layouts also need to be altered, thus the POM is involved again. 'A survey of 68 firms using JIT systems indicated that the single most important factor in successful implementation is changing product flows and layout to a cellular design' (ibid., p. 752).

Furthermore, it is important to emphasize that the POM is usually put into place first and then all other processes follow, but after the initial interactions between the POS and the POW have made their own impact. Substantial evidence on the leading role of the POM (for instance, via JIT/QC) is provided by Womack and Jones (1996) and Mintzberg and Quinn (1991) who explored many firms and their successful attempts to radically change their firm operations and capabilities. Another way of showing the precedence of the POM over the other processes (again after the initial 'debate' between the POS and the POW has reached appropriate decisions) is the relationship between the POM and the POC. For instance, the ambiguity of transactions and contracts is generated by organizational ambiguities. Thus, if the POM is not clearly demarcated (teamwork, routines, layout, coordination and so on) contracts with employees cannot be clearly defined either.[32] A detailed analysis of all these issues is, however, outside the scope of this study.

5.3 LEADING FIRMS AND SECTORS, AND DIFFUSION OF OIs

In the second section of this chapter the nucleus of industrial growth was explored. That is, the specific operations which take place inside firms. Thus, the focus of analysis became gradually more narrow from the general historical review in Chapter 2, to the set up of a general theoretical framework in Chapter 3, to a more precise relationship between OIs and industrial growth in Chapter 4, and finally to the core of firms, namely the four PROSIBB. However, in order to complete the analysis, the focus has to become less narrow again so that the role of OIs in industrial growth and hence economic growth can be fully explored. In other words, how can growth on a sectoral basis occur if initially growth occurs only in one firm or several firms? Thus, there is a need to explore the role and significance of leading firms and sectors, and this will be the object of this section.

5.3.1 Leading Firms

In the previous section it was established that four primary processes or energies constitute the basic components of the functioning of a firm, and that OIs are affected by the four PROSIBB. Hence the firm is also affected accordingly. In general, leading firms owe their existence to the outstanding performance of one (such as the POM) or more of the four PROSIBB, because it is normal to expect that there are substantial differences between the PROSIBB of various firms. Furthermore, once we know how a firm expands in terms of its fundamental energy forces, namely the four PROSIBB, we can explore how some firms grow faster and larger to become leaders in their corresponding industries. Subsequently, the leading firms can drive some industries to become leading sectors in the economy and, finally, the latter moves forward into sustained and even higher rates of growth.

The process of creative destruction, as initially developed by Schumpeter (1942a) and subsequently taken up by many other scholars, is a good starting point for this analysis. The following passage (ibid., p. 83) summarizes the concept of this destruction.

> The opening up of new markets, foreign or domestic, and the organisational development from the craft shop and factory to such concerns as U.S. Steel illustrate the same process of industrial mutation – if I may use that biological term – that incessantly revolutionises the economic structure from within, incessantly destroying the old one, incessantly creating a new one. This process of Creative Destruction is the essential fact about capitalism ...

Thus, this process continuously creates and destroys. It takes place every time there is something new in the market, such as a new product, a new technical method, a new method of production, an OI, and so on. In addition, this leads to the conclusion by Schumpeter that it is monopolistic or generally large firms and not perfectly competitive firms that generate economic growth, because the former have the means and the power to create and destroy (ibid., ch. 8).

Based on these Schumpeterian premises, it is possible to expand the analysis further. Let us assume that in the first place an industry is strongly competitive, in the sense that there are many firms supplying an almost identical or similar product and that entry into the industry is quite easy. One of the existing firms, or a newly created firm that just entered the industry, discovers a new technique (for example, a new OI) that reduces the cost of its product quite substantially in relation to the good already sold in the market. This innovation will transform the innovating firm into a dominant one within the industry, because this company now has a comparative advantage vis-à-vis the other companies due to its lower cost of production, the new features of the product that wield a better quality, and so on. In other words the innovating firm will experience increasing returns to scale, will grow faster and soon become a leading firm in the industry.

In this way this innovative firm will destroy the calm waters of the existing firms and will create a storm, which will eventually generate a new situation in the industry with rather predictable consequences. Namely, the innovative firm will become the leader. This leadership has the tendency to lead to a monopolistic situation in the industry, unless other firms quickly imitate the leading firm or become even more innovative. In this case the tendency would be the formation of an oligopoly or a monopolistic competition in the market. Thus, the existence of a non-perfectly competitive environment in the economy, which is the case in reality, is equivalent to saying that there are some stronger firms and some weaker firms (in the sense of leadership as in the example above). This entails that some firms survive and perhaps lead, whereas other firms vanish according to the principles of natural selection (in this respect, evolutionary economics is relevant here, cf. Boulding, 1981, or Nelson and Winter, 1982). This selection takes place because there are differences, sometimes substantial, between firms in terms of the four PROSIBB, that is, the POM, the POW and so on.

The above example shows how a leading or dominant firm is born. Economic history is full of other examples; some of these have already been noted in Chapter 2. Some of these examples, whereby innovating firms became leading firms within their industries, still operate in the global markets as leading companies. These instances are only a small sample of what actually happened in economic history not only in the USA and Japan but also in many other countries. The results are shown in Table 5.1. These leading firms in Table 5.1, and many others, became leaders because of their innovations in many areas, such as new technical or organizational features. For example, Sony became one of the leaders in Japan and the world by developing new products such as the transistor radio, the Trinitron colour television, the Walkman, the compact disc player, the floppy disc, and others

Table 5.1 Examples of major leading firms

Leading firm	Founding date	Industry
Boeing (USA)	1910s	Transport
Bridgestone (Japan)	1931	Tyres
Coca-Cola (USA)	1886	Food
Dell (USA)	1984	Computers
Eastman Kodak (USA)	1884	Instruments
Ford Motor (USA)	1903	Transport
General Electric (USA)	1892	Electrical machinery
Gillette (USA)	1901	Fabricated metals
Goodyear Tire and Rubber (USA)	1898	Rubber, tyres
Intel (USA)	1968	Electronics
IBM (USA)	1914	Machinery
International Paper (USA)	1898	Pulp and paper
Johnson and Johnson (USA)	1885	Chemicals
Fuji (Japan)	1934	Film
Matsushita (Japan)	1918	Electrical
Microsoft (USA)	1975	Computer software
Mitsubishi (Japan)	1918 (corporation)	Machinery, cars, etc.
Sony (Japan)	1945	Electrical/electronic
Toyota (Japan)	1930s	Transport

Source: These data are based on Chandler (1977, 1990), Harris Corporation (1996), Abegglen and Stalk (1985, p. 168), and Internet sites for most of the Japanese firms (for the founding dates).

(Beamish, 1999). All these new products show the importance of TIs. In addition, it is well known that Sony has excellent product quality, which was achieved through good OIs such as the JIT/QC system.[33] Other dominant companies developed very effective POM, POW, POC and POS, which allowed them to evolve into sustainable and expanding growth. Thus, Ford Motor Company developed the semi-automatic Fordist assembly line (which is mainly an element of the POM)[34] and became the world leader in the vehicle sector. Toyota developed the JIT/QC system and became the leader of the automobile industry in Japan. General Electric became one of the first fully multidivisional companies (POM and POC) and with their technical expertise (POW) became the world leader in electrical machinery.

For more concrete historical evidence (besides Chandler's monumental work, for which see below) reference is now made to Knoedler (1990) who examined three leading companies in three leading industries at the turn of the 20th century. Here is his conclusion about one of them:

> In 1903, DuPont, now the largest explosives producer in the United States, turned to a more efficient strategy to dominate the industry. The key to this new strategy was DuPont's organisational innovation: the bureaucratic, hierarchically organised firm, headed by an executive department whose primary job was to conduct long-range and strategic corporate planning. The executive department quickly moved to rationalise production in DuPont's large number of acquired plants and subsidiaries, and as a consequence, lowered its unit costs to levels that competitors could not meet … (ibid., p. 399)

General Electric and AT&T achieved the same result in their respective industries, as DuPont did, despite the Sherman Antitrust Act.

It is worth noting that a leading firm in a given sector can be succeeded by another leading firm, which in turn might survive being on top for only a limited time, and so on. A good recent example of the succession of leading firms is in the industry of personal computers, that is, from IBM to Compaq and now to Dell. Of course, this presupposes that the initial leading company did not become a monopoly, thus stifling all competition, though the tendency for monopolistic power is always present in many situations (in our example, IBM was a virtual monopolist for some time).

The analysis of leading firms cannot be satisfactory unless reference is also made to Chandler's work, since this author has provided monumental evidence about their importance. His theoretical views are summarized in the following points (Chandler, 1992):

- The first firms to invest in manufacturing, marketing and management exploited economies of scope and scale, and consequently quickly dominated their industries.

- In the new and transformed capital-intensive, oligopolistic industries, the first movers and challengers competed for market share in national and international markets. Such competition kept oligopolies from becoming monopolies.
- The cumulative learning honed by such oligopolistic competition generated organizational capabilities that became powerful barriers to new entrants. These capabilities and the resulting retained earnings became the basis for their continued growth.
- The first movers can be either new firms or established ones.

5.3.2 Leading Sectors and Diffusion of OIs

The importance of leading firms and leading sectors in determining economic growth is not new. According to Thomas (1972) this importance was already a hot topic in the 1950s, mainly with the work of Perroux (1955), Hirschman (1958) and Blaug (Rostow's contribution will be separately analysed further below). Perroux's 'poles of growth' are based on the effect of the lower costs achieved by the propulsive and key firms and industries, and the diffusion of growth to other firms and industries through the working of the input–output table. The effect of lower costs is easier within vertically integrated industries through lower input prices and higher demand for lower-priced outputs. Thomas's (1972) analysis tells us that the external economies generated by the poles of growth acquire their significance from the assumptions of *dynamic disequilibrium* (formulated by Chenery, Scitovsky and Rosenstein-Rodan amongst others), which states the existence of an interdependence between investments of one firm or industry and the profitability of another. Furthermore, as Perroux indicates, poles of growth eventually can become poles of stagnation, thus the Schumpeterian thesis about creative destruction is pertinent here.

Thomas (1972) emphasizes the fact that leading firms and industries have scale economies and innovations as the major sources of internal economies that generate economic growth. Indivisibilities, increased production, specialization, more efficient methods of organizing production and the learning effect influence economies of scale. In addition,

> We find in growth pole literature the suggestion that the lead firms and industries that supposedly accrue substantial scale economies also greatly benefit from the economies resulting from process, product, and organisational innovations. However, we have little quantitative knowledge of the economies that accrue as a result of investments in research and development by the firm or industry … (ibid., p. 64)

This was published in 1972. Since then, a huge literature on the role of R&D (an element of the POW and the TIOP) has demonstrated its importance in economic growth through the actions of leading firms and sectors (cf. Rostow, 1990, or endogenous growth models, the evolutionary type of analysis and, to some extent, neoclassical writers). This, however, took place to the detriment of the role of organizational innovations.

A final major issue will close Thomas's (1972) review of growth poles, which is the issue of diffusion. Based on some other studies he cites, he says that it frequently takes some 10 to 20 years for major new techniques to be adopted by all major firms in an industry. Certainly this is true for OIs such as the JIT/QC system, which took at least 10 years to spread out to other vehicle firms after Toyota introduced it in the 1960s. This diffusion is related to the spreading out of innovations from the leading firm(s) to other firms within a given industry. It is worth noting that the medium of networks greatly facilitates the diffusion process. This is one of the reasons (as already emphasized in Chapter 2) of the prevalence of networks in the Japanese economy, that is, the focal firm is part of a network or several networks so that innovations can be easily spread out from one firm to another.

With the advent of leading companies, a tendency for oligopolistic or monopolistic industrial structures started forming in the USA and other industrializing countries such as Germany and Japan about a hundred years ago. In Chapter 2 a brief account of this tendency has already been provided. In turn, these new industrial structures led to the appearance of leading industries. For example, the oligopoly of electrical machinery companies led this sector to become one of the leading sectors at the beginning of the 20th century in the USA. After World War II, IBM and Intel became the leaders of their respective oligopolistic industries, which in turn became the leading sectors of the American economy at least in the last 20 years.

Rostow (1990), in his chapter 20 entitled 'What don't we know about economic growth', emphasized the importance of technology (unfortunately he only refers to embodied technology), which has been relatively neglected since the adoption in economics of the marginal analysis from the 1870s. Thus, he puts forward eight propositions, one of which is particularly related to leading sectors. He states, 'Schumpeter appeared to have in mind in his *Theory of Development* a pioneering innovational breakthrough in a single major sector, followed by a strong bandwagon effect bringing into that sector a large number of new firms, with entrepreneurs of lesser breed, to exploit the demonstrated profitability of the innovation' (p. 458). Rostow then mentions the time it took for some major industries in Britain to reign as leading sectors, which were about 40 to 80 years as he calculated them according to his own method. For example, cotton textiles were a leading sector from the 1780s to the 1840s in that country. Also, Rostow, after extensively quoting Kuznets's sectoral analysis according to which the American economy owed its growth from the latter part of the 19th century up to 1930 to successive waves of expansion by leading sectors[35] concludes:

> In short, a modern economy is not driven forward by some sort of productivity factor operating incrementally and evenly across the board. It is driven forward by the complex direct and indirect structural impact of a limited number of rapidly expanding leading sectors within which new technologies are being efficiently absorbed and diffused. And it is this process of technological absorption that substantially generates, directly and indirectly, the economy's flow of investment via the plowback of profits for plant and equipment, enlarged public revenues for infrastructure, and enlarged private incomes for residential housing … (Rostow, 1990, p. 469)

Rostow's conclusion regarding the 'rapidly expanding leading sectors' is also the conclusion in the present theoretical explorations, and it will be freely used to justify the analysis in the forthcoming chapters where empirical evidence for the role of key industries in promoting economic growth in the USA and Japan is presented. However, it must be stressed again that Rostow was exclusively preoccupied with the impact of TIs on the fate of leading sectors, whereas the analysis presented here is predominantly preoccupied with the impact of OIs. Whether Schumpeter's bandwagon effect mentioned above is stronger in the case of TIs than in the case of OIs, is difficult to say and is outside the scope of this study. However, at least within the boundaries of the same country, OIs will not be easier to transfer from the leading firm to other firms and to other industries than TIs despite the fact that TIs often involve specific assets, patents and stronger secrecy than OIs do. Overall, it seems that the bandwagon effect, plus the more general demonstration effects, plus Hirschman's (1958) externalities (due to linkages) are the main mechanisms of transmission and diffusion of not only the TIs but also the OIs.[36]

In order to demonstrate the importance of leading sectors in terms of timing and succession, reference can be made to a diagram (not shown here) used in Best (1990, p. 190) regarding the Japanese manufacturing sub-sectors' historical evolution. Resources were shifted from unskilled-labour intensive sectors through medium capital- and raw-material-intensive or labour-intensive industries to knowledge-intensive sectors. A similar diagram could be used to demonstrate the links between OIs, leading sectors and dynamic changes in the economy through time. Thus, the knowledge-intensive industries in Japan are those that mostly introduced significant OIs (this will be further shown in Chapter 6).

5.3.3 Networks and Diffusion of OIs

It is not only the leading firms and leading sectors which generate economic growth. It must be stressed that a strong impetus for the substantial impact of leading firms and sectors on the whole economy is the imitation process for survival of the non-leading firms. This imitation is a powerful tool of the learning progression of any modern society and, in particular, in the business world. A very good example is the strong competition that exists in the American computer industry. For the last seven years or so most of the leaders in this industry made considerable efforts to imitate the amazing growth of Dell Computer Corporation so that they could maintain their leading position, and unless they succeeded in doing so they would have been relegated to the followers' position in the market (Thomson and Gamble, 2001). Needless to say that the imitation process is not easy to carry out as the case of Dell (and many others) indicate: Dell is now the world leader in personal computers.

The imitation process can be facilitated by some other factors in the economy. Thus, the role played by a third driving force on the micro economy, namely that of networks of firms (which were extensively analysed in the previous chapter), reinforce this imitation process. A network is a medium of business that heavily influences firms and industries as well as markets. It is a nexus of firms and industries that facilitates communication, transactions, diffusion of innovations, and cooperation. Consequently, within networks, the boundaries of a firm become blurred.[37] In a schematic way, Figure 5.1 shows two networks that transact with each other.

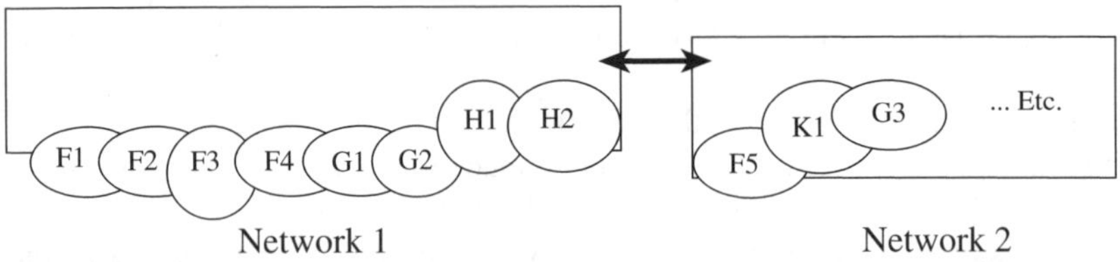

Figure 5.1 Firms, networks, industries and markets

The firms F1 to F5 belong to the industry F in Figure 5.1, the firms G1 and G3 belong to the industry G, and the firms H1 and H2 belong to the industry H. Network 1 contains eight firms of these three industries; however, these industries contain more firms than those belonging to network 1. As for network 2, it also contains firms from other industries. Firm F3 is the leading firm of industry F; the firm G3 is the leading firm of industry G, and so on. The market forces are stronger between networks than between firms within the same network. However, competition is very strong at every level of the network or industry.

In addition, particular mention must be made regarding the formation of clusters, industrial districts and other similar entities that include networks, industries and firms. The importance of these entities was analysed in the previous chapter. Here, it should be repeated that leading firms and industries are almost always closely linked with clusters or similar organizational entities, thus making the role of these firms and industries even more important in the process of economic growth.

5.3.4 Globalization and Diffusion of OIs

The diffusion of OIs throughout the economy of a nation is generated, as we saw in the previous sub-sections, by the leading firms in the first place and, subsequently, by the leading sectors, as well as by networks. However, the diffusion of OIs from one country to another also needs to be clarified. Certainly, leading firms can, through their overseas operations, transfer their innovations to other countries; these firms become the so-called transnational corporations (TNCs). Also, the diffusion of OIs can take place through other channels such as the publication of well-known books and scholarly activities in various universities; a good example of this type of diffusion was briefly described in Chapter 4 with the dissemination of Taylor's writings and its impact on organizational behaviour within the Japanese economy.

Some prevalent theories of why some firms establish production plants in other countries are briefly described in Dicken (1998, ch. 6). Thus, for example, a TNC prefers to establish a plant in a different country from its origin because it has a specific advantage in terms of assets such as knowledge, organization and human skills. This is exactly what took place with some Japanese carmakers such as Toyota and Honda; they had the specific advantage of the revolutionary OI of JIT/QC, which they transferred to their overseas plants and created a competitive edge against the local car manufacturers in the USA and other countries (Dicken, 1998, p. 339, lists 12 Japanese automobile plants in the USA, established in that country from 1982 to 1989). In addition, the specific asset advantages the TNCs possess are safely internalized by establishing overseas plants, thus markets are also internalized and their specific asset advantages are protected from being imitated by other firms for a longer time. However, eventually, the process of imitation is unavoidable and local firms gradually copy the specific knowledge of the TNC. A similar description of the diffusion process through globalization is provided by Gersbach and Baily (1996), who emphasize the important role of the 'productivity leaders' (transplants) in this process and provide some relative measurement of the relationship between a globalization index and relative productivity (cf. graphs 11 and 12 in their article).

The imitation process is further analysed and quantified in Chapter 7 of this study in which it is verified that the Japanese system of JIT/QC has been successfully transplanted in the USA during the last 20 years or so. It is interesting to note here, that changes in appropriate OIs in the framework of a globally competitive environment can have a significant impact on economies and, indeed, on any economy such as the American one. Lazonick predicted this in 1990, as this extract testifies:

> During the 1990s, new ventures in the institutional transformation of the shop floor must confront outmoded labour–management relations and work practices as well as conflicting ideologies of the 'rights' of management and labour that are deeply embedded in the American past. At the same time movements toward cooperation and commitment must confront the value-creating power of the new competition … (Lazonick, 1990, p. 298)

Thus far, it seems that the leading TNC firms, through their specific advantageous assets (for example, OIs) prefer to establish overseas plants in order to become global in operations, and increase or conserve their market share and profits, but, at the same time, by transplanting some of these assets local firms will eventually imitate the TNCs in terms of these transferred specific assets. According to this process some convergence will eventually take place in most cases, whereby the leading TNC firms will be successfully imitated by local leading firms. However, convergence of TIs or OIs is not universal. As Dicken (1998, pp. 198–9) observes:

> It is virtually impossible to transfer the whole package of firm advantages and practices to a different national environment … Recent detailed empirical research … provides strong evidence to support the view that nationally-based differences between TNCs tend to persist contrary to the notion of a convergence towards a standard global model of firm structure and behaviour …

In this respect, Dicken (1998, p. 198) provides a list of differences between American, German and Japanese TNCs in terms of corporate governance and financing, R&D, and direct investment and intrafirm trade. Some of these differences have already been mentioned in the detailed analysis of the characteristics of American and Japanese firms in previous chapters.

There are other reasons why leading TNC firms establish plants overseas, including lower labour and location costs, tax incentives and so on. However, these reasons are not covered here as they are less directly related to OIs. Furthermore, the process of globalization, and thus diffusion of some OIs, is more pronounced in some sectors than in others. For instance, the automobile industry becomes gradually more concentrated and more interconnected at the same time as it becomes more global (Dicken, 1998, ch. 10) and, consequently, the diffusion of various organizational practices becomes more intense.

It must be stressed that the diffusion of OIs through globalization tendencies can be further explored if we take into account the analysis on a PROSIBB basis. Thus, the characteristics of the POM differ from country to country for firms that produce similar products, such as cars, and hence a firm in one country that has a comparative advantage in its POM can potentially exploit its advantage by locating its operations in a different country. This is what happened with the Japanese car manufacturers that heavily invested in the USA and consequently took advantage of their superior JIT/QC system (in Chapter 7 this point will be tested empirically within a more general framework of all manufacturing industries). Similar examples can be found for other PROSIBB.

Finally, the diffusion of OIs through TNCs can also be seen through their networking capabilities. Thus, within the framework of the global economy the TNCs have been more and more involved in networking with more and more countries, other TNCs and smaller firms. The traditional 'arm's length transactions' of the international economy are increasingly complemented by the other types of linkages, namely joint ventures, marketing agreements and especially the subcontracting system. This more complex and denser network of economic relations instigated by the TNCs, and especially the larger ones, creates more value-added in the world economy (Dicken, 1986, p. 184). Out of the above four types of linkages perhaps the most important is the one relating larger firms with smaller ones, that is, mainly the subcontracting and sourcing systems which started becoming international in extent only from the mid-1960s (Dicken, 1986, p. 188). Subcontracting firms have been described as performing 'shock-absorbing' functions for large firms, which is pertinent to TNCs given their international or transnational exposure.

5.4 AN ATOMIC/SYSTEMIC MODEL OF ECONOMIC GROWTH

In this section a model of economic growth will be presented, based on the findings of this chapter. This model (non-mathematical but conceptual) is microeconomic-founded and therefore contrasts with models developed by the mainstream economics (see for example Barro and Sala-i-Martin, 1995, or Jones, 1998, or Rostow, 1990, for comprehensive reviews). Furthermore, the proposed model here is, of course, closely related to OIs and other innovations as described in previous sub-sections.

To explore the atomic model (atomic in the sense that it goes deeper into the core of the economic process), Figure 5.2[38] summarizes the key proposi-

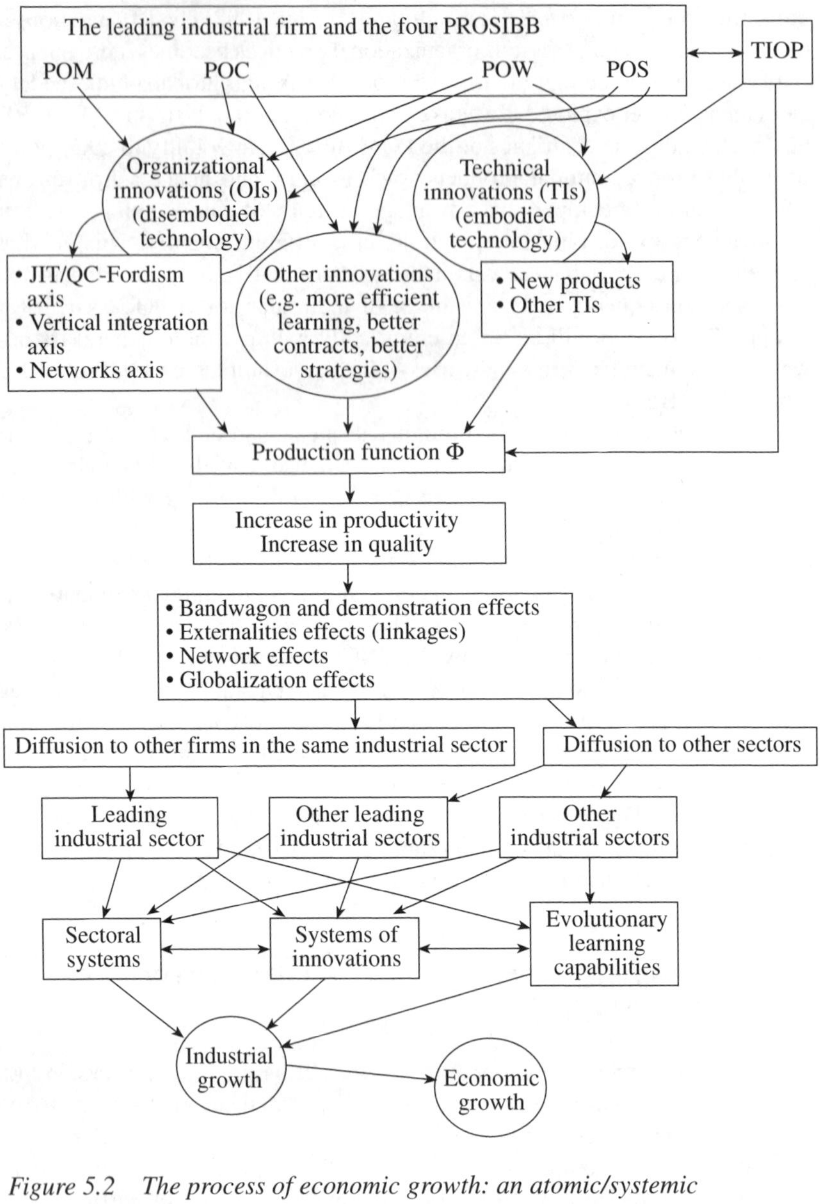

Figure 5.2 The process of economic growth: an atomic/systemic proposition

tions. This model is also systemic in the sense that it functions like a system (as already described). Thus, in Figure 5.2 it can be seen that the core, or atom, of industrial growth (and hence overall economic growth through the various linkages between the broad sectors of services, agriculture, mining and manufacturing) is what takes place inside the firm, and especially the leading firm, which is continually in the process of creative destruction. Within this firm, various types of OIs, TIs and other types of innovations emanate from the four PROSIBB. These innovations are the moving forces of industrial growth. Thus, kinetic innovations originate from the POM, the TIs from the POW and the TIOP, and many other types of innovations through the POC, POW and POS (for example, more efficient learning within the POW, better contracts within the POC, better strategies within the POS and so on) in order to increase productivity and hence growth. Regarding OIs, three specific interdependent groups were synthesized in Chapter 4, namely, the JIT/QC main axis, the vertical integration axis and the networks axis. These OIs are a dynamic synthesis of the four PROSIBB plus the TIOP.

OIs based on the interplay between the four PROSIBB generate productivity increases and product quality increases that make the innovating firm a leader in its own industrial sector, which eventually becomes a leading sector in the economy, especially if other firms adopt the OIs of the leading firm. Thus, the full 'cycle' of economic growth starts from the leading firm and propagates to other firms within the same industry. During this propagation there is a continuous interplay between changes in the four PROSIBB within firms, which in turn continuously alters the position of competitive advantage of the various firms within the same industry. Thus, some companies become weaker, even die, whereas others become stronger and even dominate. There is an incessant 'creative destruction'[39] to use Schumpeter's term.[40] As already seen in section 5.3, leading firms generate leading sectors through the bandwagon, demonstration, externalities, networks and globalization effects.

A leading industrial sector, together with other leading industrial sectors plus other sectors that take advantage of the leading sectors (on the importance of established sectors in relation to leading ones, cf. Robertson et al., 2001), propel the broad overall industrial sector of the economy and eventually the whole economy. The impact of leading sectors on the whole economy, though not analytically covered in this volume (although it will be descriptively covered in Chapter 6 and empirically in Chapter 7), can be easily understood by the following example. Based on data available regarding the German automobile sector, the impact of such a leading sector in similar economies such those of the USA and Japan is accordingly huge: according to the VDA (the Association of the German Automotive Industry) (2000), 'In 1998 the gross product dependent on the development, manufacture, sale and use of motor vehicles represented almost a fifth of the national product … '.

The importance of the auto industry in both countries, the USA and Japan, is thus paramount, and this conclusion helps us to appreciate the strong competition between the auto sectors of these two countries in the last 50 years. In particular, it is worth noting that as the JIT/QC was the system that propelled the Japanese auto industry in its way to dominate the world market,[41] the same system, JIT/QC, was also the moving force behind the revival of the American auto sector in the last 20 years (Chapter 7 provides evidence for this argument).

Leading firms and leading sectors *per se* are not sufficient to make the proposed model really dynamic. To complete the description of the atomic model see Figure 5.3. Some additions to the system model are now necessary. Edquist (1997), Edquist et al. (2001) use the concept of system to describe the systems of innovations (SI) approach for studying innovation. Edquist (1997) enumerates nine characteristics: first, innovations and learning are at the centre of focus, thus they are treated as endogenous to the system variables. Second, SI are more than systems of R&D, and include not only economic but also institutional, organizational, social and political factors. Third, a historical perspective is natural in SI, thus emphasizing path depend-

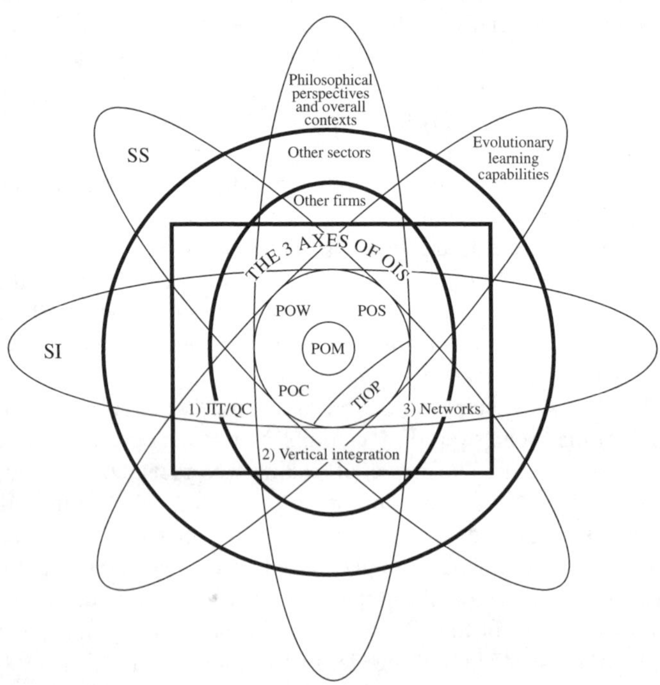

Figure 5.3 The complete atomic/systemic model

ence and feedback. Fourth, differences between SI are stressed and focused upon, and hence there is no optimality (equilibrium) attached to the evolution of SI. Fifth, there is interdependence and non-linearity between the elements (firms, customers, demand, and so on) of SI, thus there is *emergence*[42] in the systems. Sixth, SI include both 'technological' (for example, new products) and organizational innovations.[43] Seventh, institutions play a central role in SI. Eighth, SI are conceptually diffuse and pluralistic. Ninth, SI are characterized by conceptual frameworks rather than formal theories. Overall, contrary to mainstream economics, 'the SI approach incorporates both explicit principles of change and a view of innovation as a collective learning and selection process, inherited from evolutionary economics' (Edquist et al., 2001, p. 4). The close relationship between SI and evolutionary theories was also recently analysed by Saviotti (1997) who argues that, 'systems theory and non-equilibrium thermodynamics predict structure formation, qualitative change, indeterminacy, irreversibility, path dependence, and multistability, properties that are commonly displayed by national SIs' (p. 182).

Malerba (2002) proposed the concept of 'sectoral systems of innovation and production' (SSIP) based on the evolutionary theory and the system approach. These SSIP are multidimensional, integrated and dynamic systems of sectors, thus having a knowledge base, technologies, inputs and demand, and generating new and established products for specific uses. The agents that carry out market and non-market interactions for the creation, production, and sale of these products are, 'individuals and organisations of various levels of aggregation, with specific learning processes, competencies, organisational structure, beliefs, objectives and behaviors' (ibid., p. 248). These interactions take place through communication, exchange, cooperation, competition and command, and they are shaped by institutions. In addition, the co-evolution of all these elements changes and transforms SSIP according to four theoretical traditions. First, sectors change over time via laws of motion, dynamics, *emergence* and transformation. Second, sectoral boundaries change over time through mutual links and interdependencies. Third, the SI approach (see the paragraph above) is adopted. Fourth, evolutionary elements are included: processes of creation, replication and selection with emergent properties and of a metastable nature. Also note that in the whole process of SSIP interactions among firms and non-firm organizations are included and geographical boundaries are considered as well.

These brief additions to the already described systems approach to innovations and production make the atomic/systemic model of economic growth presented in this section a comprehensive, dynamic, evolutionary and innovative model.[44] At the core of this model is the sub-system of the four PROSIBB of firms. These PROSIBB (which are in turn four sub-systems) include all possible and probable operations of firms, such as the well-known generic transactions, competences, capabilities, technologies or, more precisely, the POM, POC, POW and POS. As a consequence of the POM, its interactions with the other PROSIBB and with the other elements of the whole system (which includes the SI, the sectoral systems [SS],[45] and the emergent evolutionary learning capabilities) specific OIs, TIs, and other types of innovations are generated that propel the growth of firms, sectors and the whole national and global economy.[46] Figure 5.3 shows this model.

The exact way of how all the actors of this model are interlinked together would necessitate much more research to be determined. For example, the SI 'has not had so much to say about how innovations translate into economic growth. It is more a theory of innovation than a theory of growth' (Gregersen and Johnson, 2000). It is one of the intentions of this study to pinpoint more precisely how innovations and, in particular, OIs affect economic growth through the introduction of the four PROSIBB. The main message of the latter is that unless the researcher goes to the core of the firm (the four PROSIBB) the economic process cannot be complete. For instance, Reati (1998, p. 93) proposes a descriptive model that links several variables together in order to explain how a process[47] technological revolution affects demand and output. Thus, employment in the sector is the result of two factors acting in opposite directions: first, the rate of increase in productivity in the sector producing the final commodity i, which reduces employment; and, second, the rate of growth of demand for commodity i, which increases employment – the compensation effect. The net result depends primarily on the price elasticity of demand and to a much lesser extent the income elasticity. However, how the rate of change in productivity of the sector involved is linked with the technological revolution is not explained. If we introduce the four PROSIBB, it is possible to determine more precisely which PROSIBB affects firm growth and hence sector growth. For example, as it is briefly indicated in this chapter and Chapter 7, the JIT/QC process (an OI) primarily affects the POM and hence productivity increases through the POM.

A final remark will supplement this section. It concerns the micro basis of national economies. This study argues that national economies have their growth-generating core in the micro sector and in particular in the firm and its OIs. The former USSR is a good example to recall; Chandler (1992, p. 84) summarizes it:

> On the other hand, the economies that followed the Soviet model – relying on central planning agencies to coordinate production and distribution and to allocate resources for the future – prevented managers in units of production and distribution from learning how to coordinate effectively. These managers never developed hands-on organization knowledge about current facilities, available supplies and market demand. The failure to develop such capabilities has been central to the disintegration of these centrally planned economies …

Twelve years after this article was written, one can add that the USSR, and most of the other former Soviet type of countries, have also shown this failure in their very poor economic performance during these last 12 years or so (cf. GDP growth rates, World Bank Group, 2001).

5.5 CONCLUSIONS

The purpose of this chapter has three dimensions. The first is to lay the foundations for a comprehensive theory of the firm that analyses and explains the role of OIs in the evolution and growth of firms. Once this role is understood, economic growth of many firms and sectors and the whole economy is also analysed in the light of leading firms and sectors and the process of diffusion of OIs, and economic growth throughout the economy. Then, a holistic 'atomic/systemic' model of economic growth within a more general framework of systems of innovations, sectoral systems and evolutionary learning capabilities is suggested that synthesizes all the main points of this chapter and previous chapters. Figures 5.2 and 5.3 summarize a large part of Chapter 5.

More precisely, this study has opened two important boxes in economic analysis. The black box is 'revealed' and a diamond of valuable inputs for the production process is described and analysed. Four propositions have established the nature of this diamond: a complete open system of all possible fundamental ingredients of the black box, which are categorized in four mutually exclusive but interdependent processes (the PROSIBB). The latter can assist everybody in the firm to ask, for example (many other similar questions are analysed in the chapter), four fundamental questions: (1) are we moving (as human bodies) around at work in the best organized way possible (POM)? (2) Are we supporting these movements with the right contracts and rules (POC)? (3) Are we creating the necessary knowledge to acquire power and feed our needs (POW)? (4) Are we taking the right decisions given what happens outside and inside our firm in order to survive and perhaps dominate (POS)? The simultaneous answer to these questions determines the quantities of the tangible inputs of production. Hence the PROSIBB (POW, POS, POM, POC) interact with and determine the TIOP. At the same time they interact with other firms, other parts of the society and non-market activities. The four PROSIBB contain inputs that are closely related to economic, psychological, sociological and human factors. The numerous possibilities that exist within the black box as represented by the PROSIBB and its diamond shape and traits, make sure that there are substantial differences between firms and their evolution.

Increases in productivity, and economies of scale, are primarily due to the dynamic evolution of the four PROSIBB and their interplay with the TIOP, other firms, and the society. It is shown that when the inputs of the four PROSIBB are taken into account the difference between the actual market prices of the TIOP and their true opportunity costs constitute the diamond value of the black box which is not properly considered by mainstream economics. This diamond value depends on all innovations, hence not only technical ones, which emanate from all the inputs of the PROSIBB. Organizational innovations, which are a synthesis of various combinations of the four PROSIBB, become a central issue of the growth of firms and, hence, economic growth. Overall, different types of innovations, namely, three types of TIs, OIs (such as the JIT/QC and the M-form of firm structure), kinetic, strategic, contract and wisdom innovations are used as examples to describe both innovations and decision-making. This taxonomy of innovations is in evident contrast with the term of technological innovations usually used in many studies.

To put it differently (and perhaps bluntly), the mere 'throwing' into the production function of the quantities of the TIOP, plus some other generic inputs such as human capital and knowledge, plus a vague term for technology and expecting to produce cars and computers and other thousands of products seems to be an amazing artifice in traditional economic theory. All these products cannot be produced unless there is coordinated kinetic energy, organized movements (walking, lifting, talking, and so on), acts, actions, activities, 'kinesis' and 'synkinesis' (= emotion), vision, perception, trust, appropriate rules, mistakes, motivation, and all the other unique and fundamental inputs of the PROSIBB. Above all, the four PROSIBB constitute an organized open system with emergence, state of readiness and entropy. If the firm (and its production function) does not contain the appropriate elements of organized movements (POM), if the firm does not possess the appropriate state of readiness and emergence (POS), if the firm does contain too much tension (POC) and if the firm does not contain the information needed (POW), the entropy (for example, waste) of the open system of the firm (or the country and so on) will increase, hence more disorder will ensue. In addition, one of the erroneous conclusions of the existing economic analysis is that the opportunity costs of the TIOP are much higher than they actually are because the inputs of the PROSIBB are not taken into account.

Chapter 5 concludes the series of four chapters whose main aim was to provide theoretical evidence that OIs, or disembodied technology, have their own place of merit in the exploration of the very complex phenomenon of industrial and economic growth. The next two chapters will provide some empirical evidence of this phenomenon and of many other issues that have been raised so far regarding the importance of OIs in shaping and developing industries. In particular, in Chapter 6 a general discussion of the importance of leading manufacturing sectors (and firms) in the USA and Japan in terms of OIs will take place; thus Chapter 6 is a necessary link to the previous four chapters and especially the fifth as it emphasises the connection between OIs, leading sectors and industrial growth. In Chapter 7, the axis of JIT/QC of OIs will be further explored by quantitatively linking this axis to manufacturing sectoral growth in the USA.

NOTES

1. This suggested theory and the included OIs are one of the main preoccupations of this book.
2. Some relevant references are: Williamson (1997), Dietrich (1993a), Dietrich (1994), Foss and Knudsen (1996), Dosi and Malerba (1996a), Foss (1997a; 1997b), Ricketts (2002), Foss and Loasby (1998), Langlois (1992), Kogut and Zander (1992), Khalil (1996), Demsetz (1988) and Pitelis (1998).
3. Hattwick (1999), Yang and Lester (1995), Stern (1990) and Morgan (1998) express similar ideas. Tomer (1998b, p. 328) suggests that we also need 'soft' elements such as, 'passion, inspiration, esprit de corps, enthusiasm, vigor, zest, vision, strongly held values, deep commitment, spirituality, and highly ethical orientation … '.
4. Various ways of analysing the firm can also be found in Eliasson (1996, p. 99 and so on).
5. The prefix 'bio' has the more general meaning of anything related to life, such as life activities, energies, physical states and so on.
6. Another example of the POM is the activities related to the JIT process (see previous chapters) and to the 'cell manufacturing' by Dell (Thomson and Gamble, 2001).
7. For more details of all elements of the four PROSIBB, see Sanidas (2005b, table 1).
8. The TIOP are, for example, labour, machines, space, intermediate goods and tools.
9. The 'shop floor' can also include all types of administrative and managerial offices.
10. The present analysis does not propose to synthesize all 'capabilities' mentioned in the literature. For example, the engine is a core competence or a strong capability of the firm Honda. In the present analysis only the black box is considered (the engine is an output or an intermediate input). Moreover, capabilities can be expressed in various ways. For instance, Hagstrom and Hedlund (1998) suggested a three-dimensional model of the internal structure of the firm comprising position (similar to the POC), knowledge (similar to the POW) and action (similar to the POS). However, the POM is still absent in their analysis. Comparable partial syntheses are also present in Coriat and Dosi (1998) or Nelson (1996) and others.
11. A new accounting system that measures all elements of the four PROSIBB would certainly introduce a completely new way of conducting business.
12. This function Φ now complements the analysis started in Chapter 3.
13. Even if the POC establishes *all* the rules describing every movement of people and machines (hence the POM) in the firm, the existence of the POM would still constitute a non-overlapping entity.
14. Given the uncertainty and bounded rationality that typifies the decision-making process of firm operations, the probabilistic view of the four PROSIBB is very plausible in reality.
15. Many other relevant examples can be used in this context.
16. Bayesian analysis can also be included in this analysis in order to revise probabilities due to feedback mechanisms.
17. The other two types of systems are the equilibrium model that reaches equilibrium but

loses organization, and the organismic or homeostatic model that maintains a high level of organization against ever-present tendencies to reduce it (Buckley, 1967, p. 40).

18. The concept of emergence dates back to 1877 according to Sawyer's (1999) example of improvisational theatre.
19. Emergent systems are also compatible with the contingency view of management.
20. An advocate of the importance of entropy in economic theory is Georgescu-Roegen (1971). His pioneering analysis is still pertinent today because of the development of evolutionary economics. For instance, Boulding (1981), used the concepts of open system and self-organization as well as entropy in his analysis (Dopfer, 1994).
21. Recently, another feature has been added to complex adaptive systems, that of 'punctuated equilibrium' which expresses discontinuous changes (Romanelli and Tushman, 1994; Wollin, 1999). This concept of punctuated equilibrium has paralleled a more thorough exploration of complex evolving systems (Wollin, 1999).
22. Fujimoto's 'evolutionary learning capability' is a dynamic interplay primarily between the POW and the POS (and consequently between all four PROSIBB).
23. However it must be emphasized that the quantities of the TIOP are a consequence of interactions between the four PROSIBB amongst each other and between the PROSIBB and the TIOP (hence the explicit prices of the TIOP matter as well, see analysis above).
24. Also see Gersbach and Baily (1996).
25. 'Reengineering is the fundamental rethinking and radical redesign of processes to improve performance dramatically in terms of cost, quality, service, and speed' (Krajewski and Ritzman, 1999, p. 108).
26. A service firm is chosen on purpose as an example so that it demonstrates that the POM and all the PROSIBB are applicable not only to the manufacturing sector but to any sector of the economy.
27. A good example of how TIs affect the division of labour, and how in turn some other elements such as power relations and transaction costs affect TIs, is provided by Robertson and Alston (1992).
28. For other examples of OIs, see previous chapters.
29. This schema and the others presented here are only indicative of what can take place in reality.
30. The organization of QC circles belongs to the POM; however, the concept of such circles and their outcome in terms of knowledge is a wisdom innovation.
31. Whether or not, actually, the top management decides to take into account the consequences on the POM if a change in the governance form is implemented is debatable. Here, it is suggested that these consequences are taken into account. In any case, the conclusions remain the same.
32. This does not mean that all tasks should be meticulously defined. It is possible to have multi-task workers, as is the case for the JIT/QC system.
33. The same TI can lead to bad or good quality depending on the OIs.
34. The appropriate and relevant TIs of course supported the POM and the four PROSIBB.
35. For this see also Street (1988).
36. A good recent summary of the literature regarding leading sectors from the 1950s is provided by Robertson et al. (2001).
37. Nonetheless, there is always a boundary of the firm that allows it to organize itself. The firm is an evolving open system that must always improve its organizing abilities in order to survive (as already analysed).
38. Figure 5.2 emphasizes the role of industrial growth generating economic growth. However, the role of services growth can be as important.
39. Eliasson's (1994) pertinent construction of what he calls the 'experimentally organised economy', which considers firms as 'experimental learning machines', captures this process of continuous creative destruction to some extent. Thus, the accumulation of firm-based knowledge ('organizational learning') is the source of business competence and the driving force behind macroeconomic growth. In this set-up, business failure becomes a standard cost of such growth and the constant competitive struggle between firms, through innovations, failures and successes moves the macro economy. However, this model has to be

adapted to include the elements of the four PROSIBB and the role of leading firms and sectors.

40. Malerba (2002) summarizes the differences between Schumpeter Mark I and Schumpeter Mark II innovations. Mark I is more related to 'creative destruction' and is characterized by technological opportunities, low appropriability and low cumulativeness at the firm level, whereas Mark II is more related to 'creative accumulation' and is characterized by high appropriability and high cumulativeness. This distincion is more appropriate to the nature of TIs, but a more thorough study on the links between OIs and TIs could probably include this distinction for analysis purposes (see also Chapter 3).
41. See Kawahara (1997), for a detailed account of this 50-year-old intense competition of the leading auto sector in the USA and Japan, including the impact of the JIT/QC system.
42. Thus, the 'emergence' trait also appears in this analysis according to the concept of system of innovations.
43. See Chapter 3 for a more detailed account of Edquist's (1997; Edquist et al., 2001) contribution to the distinction between TIs and OIs.
44. This model excludes a simple version of economic growth as it is often described by traditional economics.
45. Sectoral systems are defined to include SS like the SSIP suggested by Malerba (2002).
46. Clark (2000), in his holistic approach to organizations, examines their competition 'between contexts', thus making the analysis very general (for example, Ford would have failed if he had started his model T from England, ibid., p. 201).
47. Reati's (1998) analysis is similar to that of Edquist et al. (2001). Also, note that these explanations depend on the interaction between process innovation and product innovation: the process innovation introduces better product quality thus enhancing product innovation.

6. OIs and manufacturing sectors' growth in the USA and Japan

6.1 INTRODUCTION

In the previous four chapters, the important role of OIs in promoting industrial growth in the USA and Japan has been extensively discussed from a theoretical point of view. In this chapter, a general discussion of manufacturing sector growth[1] in these two countries will provide additional support for this view. Emphasis will be given to the role of leading firms and sectors in terms of OIs. This chapter is primarily descriptive, and discusses sectoral performance in terms of OIs and at times in terms of OIs and TIs together. Also, some basic quantitative evidence will be provided; however, a more rigorous empirical exploration is conducted in the next chapter.

Two periods will be analysed separately; first, from the end of the 19th century up to World War II (section 6.2), and from the 1950s up to the end of the 20th century (section 6.3). For the former period and for the USA, growth rates of real output and total factor productivity (TFP) for various critical sub-periods will be calculated, and appropriate conclusions will be drawn regarding the impact of OIs on manufacturing sub-sectors. For Japan, due to the lack of substantive data, a simpler analysis will be conducted by examining the time series of some basic manufacturing sectors' real production. For the second period,[2] a graphical representation of the three-digit industrial sectors of the two countries will be shown, with the aim of comparing the growth in real output and TFP between sectors and between countries; at times this comparison will be assisted with basic cointegration analysis, which will facilitate the drawing of relevant conclusions regarding OIs. Most emphasis will be given to the American economy. In this second section the recent prolonged Japanese recession will be examined in the light of OIs and leading sectors. Also, a parallel examination of TIs and OIs in some key industries in the two countries will be provided in section 6.3. Finally, section 6.4 provides a summary of the major conclusions from this chapter.

6.2 THE PERIOD FROM THE END OF THE 19TH CENTURY TO WORLD WAR II

6.2.1 USA

Kuznets (1971) analysed the structural changes of all industries in the USA in terms of shares in total output and growth for the period from 1880 to 1948. Chandler says in this respect: 'Kuznets's data support the assertion that the industries spearheading American economic growth were those dominated by a small number of large managerial enterprises' (Chandler, 1990, p. 226). In this section, the data used by Kuznets is scrutinized in order to link sectoral growth with OIs.

Two variables will be examined, namely, the growth in the real output of manufacturing sectors and the corresponding TFPs (the definitions and discussion of which are provided in Chapter 7 where they will be used extensively). Although the data for this period are not as comprehensive as that of the last 50 years, available records from official American sources are used to construct these two variables on a two-digit SIC equivalent sub-sector basis. The results can be deemed to be quite accurate for the purpose of this analysis, which is to detect differences in each one of the two variables (real output and TFP) between the sub-periods 1899–1914, 1915–29 and 1930–37 and relate these differences to various OIs. Unfortunately, no relevant data are available for the computation of TFPs before 1899. Table 6.1 summarizes the results analysed in this section. Upon examination of Table 6.1, some conclusions can be drawn. First, and most important, although the rate of average annual growth in real output was almost as high during 1915 to 1929 as it was during 1899 to 1914, the TFP average annual growth was substantially much higher during 1915 to 1929. For the period 1930 to 1937, as might be expected, both real output and TFP declined equally because of the consequences of the Great Crash. A pronounced slowdown of industrial productivity during the period 1890 to 1913 was also observed by David (1990, p. 356). Second, sectoral growth in terms of real output and TFP (referring mainly to the period 1915–29) can be grouped into several categories as shown in Table 6.2.

The results in Table 6.2 depict the qualitative analysis carried out so far in this study. Manufacturing sectors, such as transportation, and machinery (both electrical and other) led the American supremacy in industrial development during the second half of the Second Industrial Revolution (1915–29).

What have been the reasons for the differences between the period 1899–1914 and 1915–29, and for the differences between sectors in terms of OIs? The most important reasons can be summarized as follows. First, during 1915–29, the expansion took place through an expansion in firm size. This

Table 6.1 Sectoral productivity growth in the USA from 1899 to 1937

	Empl/t (000)	Real output (in 1929 $)	Output deflator	Annual ave. TFP (%) growth	% Real output growth	Number establ/s	Empl/t per establ/s	Empl/t (000)	Real output (in 1929 $)	Output deflator	Annual ave. TFP (%) growth	% Real output growth	Number establ/s	Empl/t per establ/s
				Food							Tobacco			
1899	476	6041	0.43			42 000	11	140	472	0.56			14 959	9
1914	760	11 050	1.48			51 502	15	196	770	0.64	–0.12	4.21	13 951	14
1929	872	13 730	1.00	1.60	3.03	55 325	16	126	1284	1.00	0.20	4.45	1888	67
1937	1049	16 862	0.86	0.31	2.85	48 763	22	98	1484	0.88	0.20	1.95	852	115
				Textiles							Apparel			
1899	716	2076	0.50			5930	121	364	1203	0.52			12 619	29
1914	1013	3500	0.61	–0.15	4.57	6756	150	618	2289	0.60	0.07	6.02	18 015	34
1929	1190	6662	1.00	0.76	6.02	7415	160	681	3866	1.00	0.74	4.59	22 470	30
1937	1209	7231	0.73	0.81	1.07	6096	198	779	3055	0.98	–1.67	–2.62	16 389	48
				Wood							Furniture			
1899	602	2037	0.35			32 456	19	111.2	1277	0.35			2614	43
1914	718	1940	0.53	–1.26	–0.32	37 949	19	169	985	0.74	–0.57	–1.52	4844	35
1929	651	1583	1.00	0.22	–1.23	20 928	31	248	1392	1.00	1.68	2.75	5491	45
1937	473	1142	1.06	–2.58	–3.48	11 747	40	225.4	1256	0.90	–0.39	–1.22	4469	50
				Paper							Printing			
1899	100	332	0.60			1895	53	244	965	0.42			24 363	10
1914	182	794	0.65	–0.24	9.28	2344	78	406	1805	0.52	–0.06	5.80	34 241	12
1929	258	1761	1.00	1.31	8.12	2973	87	566	3122	1.00	0.63	4.86	27 364	21
1937	301	1834	1.03	0.21	0.52	3084	98	555	2871	0.94	–0.94	–1.00	22 674	24
				Chemicals							Petroleum			
1899	170	829	0.63			7669	22	31	173	0.72			308	101
1914	269	1713	0.74	–0.28	7.11	10 698	25	60	429	0.92	–0.04	9.87	591	102
1929	382	4254	1.00	1.18	9.89	9327	41	133.8	4737	1.00	0.22	66.95	922	145
1937	377	4911	0.89	0.79	1.93	8337	45	138.6	6150	0.85	–0.28	3.73	934	148

			Rubber								Leather			
1899	39	37	2.70			301	130	265	1435	0.41			5785	46
1914	89	174	1.73	1.75	24.68	342	260	341	1959	0.56	–0.10	2.43	6798	50
1929	172	1102	1.00	2.61	35.56	525	328	351	1747	1.00	0.61	–0.72	4285	82
1937	150	1089	1.01	–0.84	–0.15	578	260	362	1510	0.94	–0.45	–1.70	3249	111
			Clay, glass etc.								Primary and fabricated metals			
1899	243	505	0.54			11 571	21	591	2209	0.84			14 949	40
1914	405	1203	0.51	0.33	9.21	14 793	27	1107	5362	0.66	0.05	9.52	25 205	44
1929	372	1655	1.00	1.18	2.50	8788	42	1549	10 641	1.00	1.25	6.56	13 785	112
1937	331	1622	1.00	–0.09	–0.25	6114	54	1635	9895	0.98	–0.62	–0.47	11 933	137
			Non–electrical machinery								Electrical			
1899	456	1468	0.52			8995	51	49	190	0.49			592	83
1914	774	2266	0.56	–0.35	3.62	15 792	49	156	627	0.58	0.39	15.33	1048	149
1929	769	4250	1.00	2.05	5.84	8529	90	421	1735	1.00	2.25	11.78	1861	226
1937	795	4393	1.00	–0.18	0.42	7327	109	374	1801	0.91	0.69	0.48	1597	234
			Transportation								Miscellaneous			
1899	70	312	0.54			3404	21	138	434	0.50			14 123	10
1914	331	1000	0.90	–0.21	14.70	4151	80	293	842	0.72	–0.11	6.27	21 504	14
1929	631	5365	1.00	2.04	29.10	2246	281	290	2334	1.00	1.68	11.81	12 545	23
1937	680	5894	0.91	–0.17	1.23	1958	347	288	1687	0.92	0.08	–3.47	10 690	27
			Total man/g											
1899	4850	21 984	0.50			204 754	24							
1914	7513	36 434	0.64	–0.35	4.38	268 436	28							
1929	9660	71 220	1.00	0.94	6.37	206 663	47							
1937	9786	74 687	0.90	–0.16	0.61	166 794	59							

Notes:

1. TFP, as estimated, includes as inputs the number of employees, materials or intermediate goods, and stock of capital (see Chapter 7 for the exact formulae and other information on TFP).
2. The output deflator used in the calculation of the TFP is derived from the series of nominal and real output.
3. The materials series used in the calculation of the TFP is derived from the series of nominal output and value added.
4. Some numbers for some years are only estimates; this does not affect the overall picture of Table 6.1.

Source: Creamer et al. (1960) for the data of nominal output and real output (1929 prices) (Table A-10), and total capital stock in 1929 prices (Table A-8); US Bureau of the Census (1975) for the data on number of employees, payroll, value-added and establishments (Series P 58-67).

Table 6.2 *Sectoral growth in the USA (1899–1937, based on the sub-periods of Table 6.1)*

Category	Very high growth in output (above 12%) and very high growth in TFP (around 2%)	High growth in output (above 6%) and high growth in TFP (around 1.2%)	Average (or low) growth in output (about 2%) and Average (or low) growth in TFP (around 0.70%)	Very high growth in output and low or average growth in TFP or vice versa.
Sectors	Transportation Electrical machinery Rubber	Paper Chemicals Clay, glass etc. Primary and fabricated metals Miscellaneous	Tobacco Textiles Apparel Wood (lowest) Printing Leather	Food (high TFP) Furniture (high TFP) Petroleum (low TFP) Non-electrical machinery (very high TFP)

Source: Based on Table 6.1.

can be testified by reference to several factors, such as vertical integration, integration of mass production and distribution, and a continuous evolutionary technological improvement (as opposed to the revolutionary technological improvement which took place in the last three decades of the 19th century). Statistical support for this expansion through integration can be seen in Table 6.1, with reference to the ratio of the number of employed people per establishment. This ratio increased dramatically during 1915–29. Note, that although these developments (vertical integration and integration of mass production and distribution) started at the end of the 19th century and continued during 1899–1914, it was not until 1915–29 that these changes spread out into all layers of the economy and not just in the largest companies (for the latter, details can be found in Chandler, 1962; 1977; 1990).

Second, the initial steps of rationalization in industrial production, better known as scientific management or Taylorism, started during 1890 to 1910 and did not really become applicable until later during 1915–29 and beyond, especially with the advent of Fordism in the transportation sector. Tugwell (1927) and, more recently, Nyland (1989) in their detailed accounts confirm the conclusion that applications of Taylorism (for example, through Fordism, cf. Nyland, 1989, p. 141 and generally ch. 4) became increasingly prevalent in the American economy during World War I and in the 1920s. In addition, Tugwell (1927, p. 126) argued that:

> So far as specific cases go, it would appear that scientific management has contributed enormously to increased productivity. Judged by the test of measuring material outputs before and after the introduction of the system, the cases cited above, and others, show a definite correlation between the introduction of the system and increased productivity …

Tugwell (1927, in particular ch. 4) gives a detailed account of many industrial cases that benefited from various applications of scientific management (such as time-and-motion studies). He suggests that productivity increases that took place in American manufacturing between 1914 and the mid-1920s were due to a large extent to the advent of applications of Taylor's principles.[3] As can be clearly seen from Tables 6.1 and 6.2, the sectors which grew most rapidly both in terms of real output and TFP were those for which scientific management, Taylorism and Fordism became most applicable, that is, mainly the transportation, electrical machinery, primary and fabricated metals, and non-electrical machinery sectors (Chandler, 1977; Tugwell, 1927). It is worth noting for the latter that although its real output did not grow very fast, its TFP did.

Third, and as a corollary to the second point, since TFP seems to be so small before 1915, one could conclude that almost all growth in real output took place because of corresponding increases in capital, labour and materials (hence TIs-related technology). Furthermore, OIs during 1880–1914 'allowed' the increases in these inputs to take place because they set the right organizational and managerial environment for such growth of inputs and output. Specific reference is made here to OIs such as vertical integration, or integration of mass consumption and mass distribution, centralization of the management process, replacement of owners by salaried managers, as well as the replacement of the domestic system by the factory system. Most likely the domestic system almost completely disappeared by the end of the 19th century (Chandler, 1977), but the new system still had to work itself out for another 15 years before it also contributed to some extent to the sharp increase in TFP.

In order to further support the above comments, Figures 6.1 and 6.2 summarize the TFP and real output annual growth rates for the three sub-periods examined. TFPs and real output did not grow in a parallel way in all sectors; for instance, though the petroleum industry's real output grew by about 65 per cent annually from 1914 to 1929, TFP only grew by 0.22 per cent annually during the same period. On the contrary, although the electrical industry's real output grew by about 12 per cent per annum during 1914–29, TFP grew by 2.25 per cent per annum during the same period. These differences are due to OIs, TIs and other types of innovations.[4] In the above examples, the petroleum industry grew so fast primarily because of capital investment (hence high TIs and a low TFP),[5] whereas the electrical industry had a much higher TFP growth mainly because pioneering electrical companies such as GE had set such an internal organization that new products were mass produced by following Tayloristic and Fordist methods. Similar comments apply to other industries such as that of transportation (for example, cf. Chandler, 1977).

As quantitative support of the conclusions in this sub-section, it is worth quoting the analysis carried out by Kogut (1996). This author emphasized the importance of the Taylorist rationalization process that took place in the American economy during the period 1915 to 1935: 'Taylorist ideas had spread remarkably wide by the 1920s, from manufacturing to the organization of the offices and hospitals. By 1935, 34 per cent of all American manufacturing firms with more than 250 employees reported the use of time and motion studies.' (ibid., p. 267). Using these time and motion studies he statistically tested the determinants of the stock of American foreign direct investment in 1930 and provided evidence that it was not size economies, nor R&D, nor marketing, but the degree of standardization of the work that was the significant determinant. Furthermore, 'the multidivisional structure was, itself, an expression of the application of rationalisation to the newly evolved corporate hierarchy, complete with new accounting principles' (ibid., p. 268).

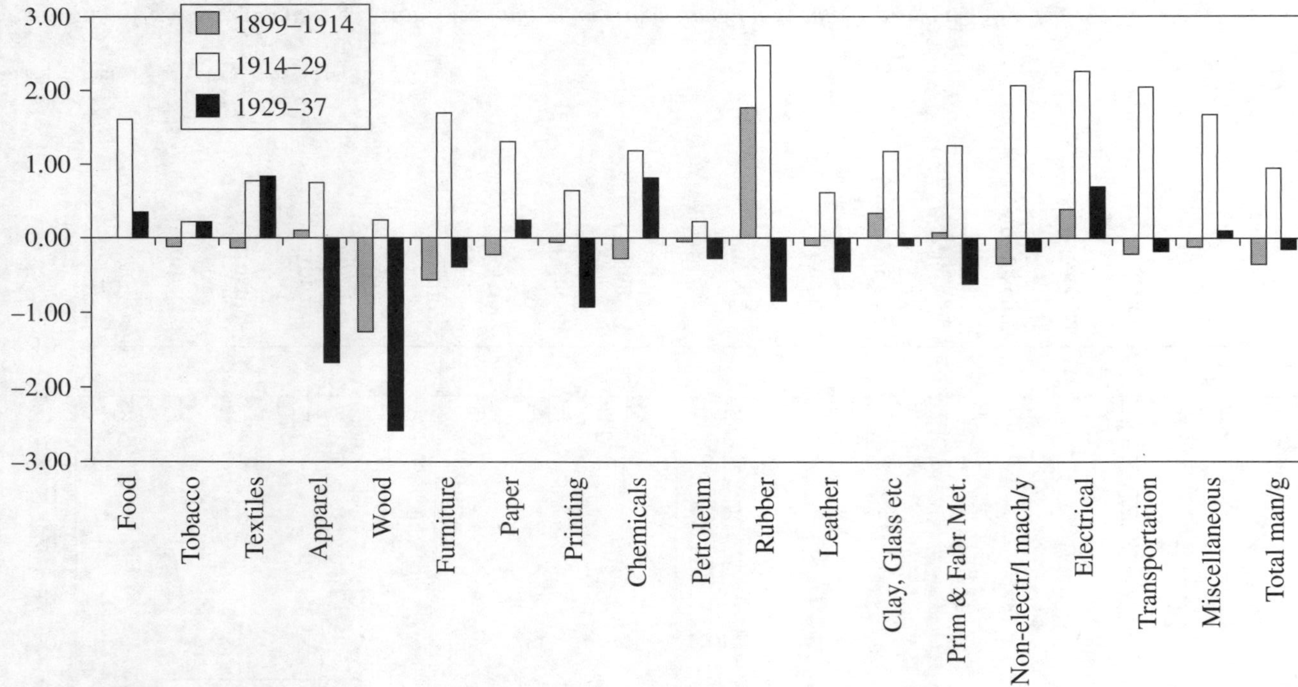

Source: Author's calculations based on Table 6.1.

Figure 6.1 USA, TFP of manufacturing sectors, average annual percentage growth, 1899–1937

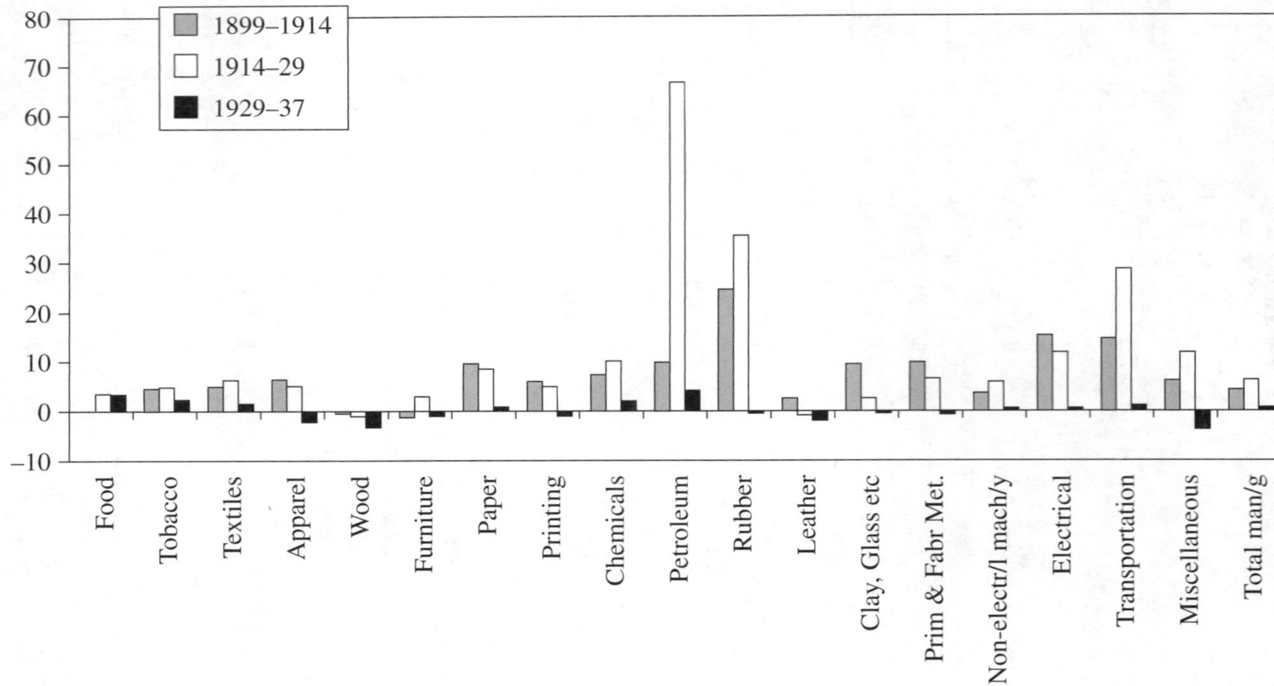

Source: Author's calculations based on Table 6.1.

Figure 6.2 USA, manufacturing sectors, real output, average annual percentage growth, 1899–1937

6.2.2 Japan

In Japan, there are two main reasons why a detailed analysis similar to that for the USA is not possible or as important. First, a detailed analysis is not possible because of lack of basic data since only eight manufacturing sectors (see Figure 6.3) have appropriate records, and capital stock is not included in them. Second, a detailed analysis is not important because, for most of the period examined, the textiles industry consistently constituted about a third of total manufacturing. Figure 6.3 shows the data for real output of these eight sectors. It is obvious from this graph that the index of total production has followed very closely that of textiles. Also it can be seen that in the 1930s the sub-sectors of machinery, chemicals and metals have accelerated their growth. To see this picture more clearly, two more figures are shown.

Figure 6.4 shows the five-year moving average of growth rates of the sub-sectors textiles, food, forest products, stone and miscellaneous, whereas Figure 6.5 shows the five-year moving average of growth rates of the sub-sectors textiles, chemicals, metals, machinery and miscellaneous. It becomes clearer in these graphs that textiles became a strong leading sector in the last 15 years or so of the 19th century. The sub-sector machinery followed its own path both at the beginning of the period examined and from World War I to the 1920s. This latter part of the path has already been explained in Chapter 2 mainly through the fact that the demand generated by the combating nations dissipated after World War I. Another important observation is that both the machinery and metals industries grew together and had some exceptional rates of growth in some sub-periods such as those at the end of the 19th century and again at the beginning of the 20th century.

In order to see more clearly the periods where there was some important divergence of growth rates between sub-sectors, cointegration analysis was used and two cointegrating vectors were calculated to detect these periods. Figure 6.6 shows the results, whereby it becomes obvious that the main divergence and structural changes took place first in the 1930s and to a lesser extent in the period of World War I, the 1920s and at the turn of the 20th century.

These brief remarks can now be related to OIs in Japan during the period examined. According to the analysis conducted in Chapter 2, the focal factories took off on a large scale in the 1930s. Also, during and just after World War I, the Japanese industrial system, and in particular the machinery sub-sector, was reorganized and many new firms were created, mainly as a response to foreign demand. These OI-related structural shifts coincided with some constitutional changes as well, which always play a preponderant role in the business sector in Japan. For example, the government in the 1880s initiated large-scale privatization, and in the 1930s it encouraged industrial growth for

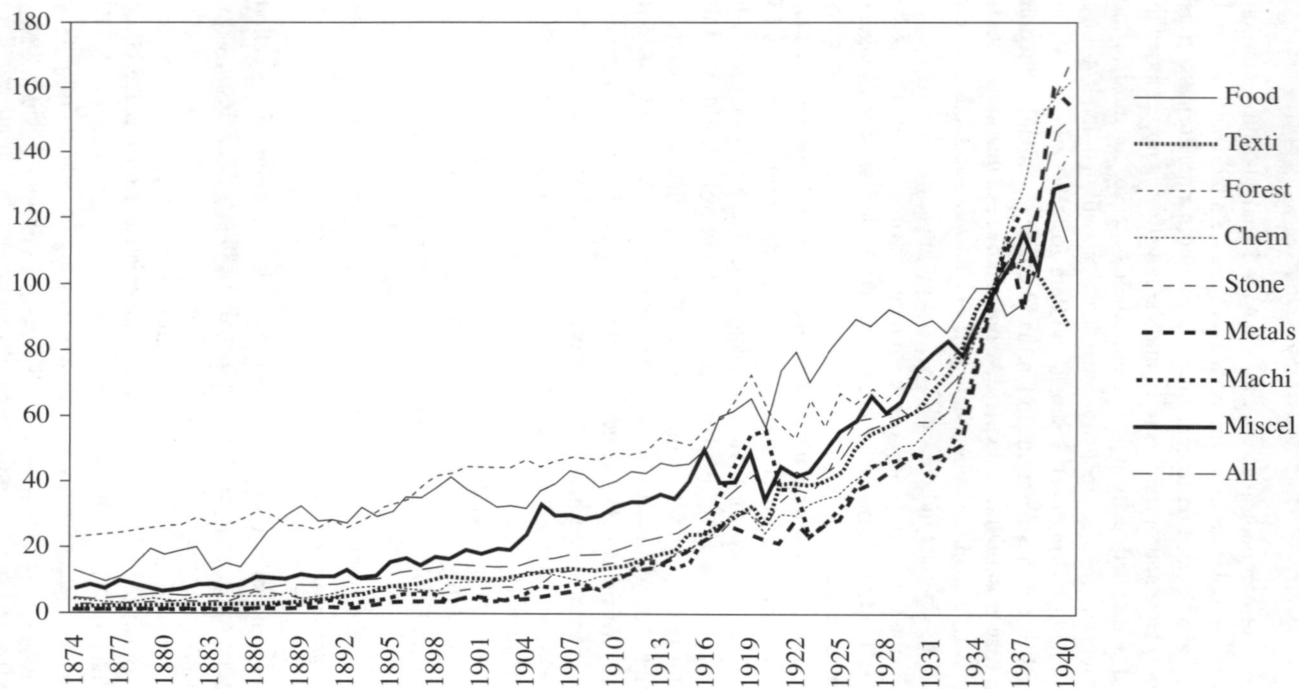

Source: Based on data provided in Klein and Ohkawa (1968).

Figure 6.3 Japan, real output indexes for eight industries

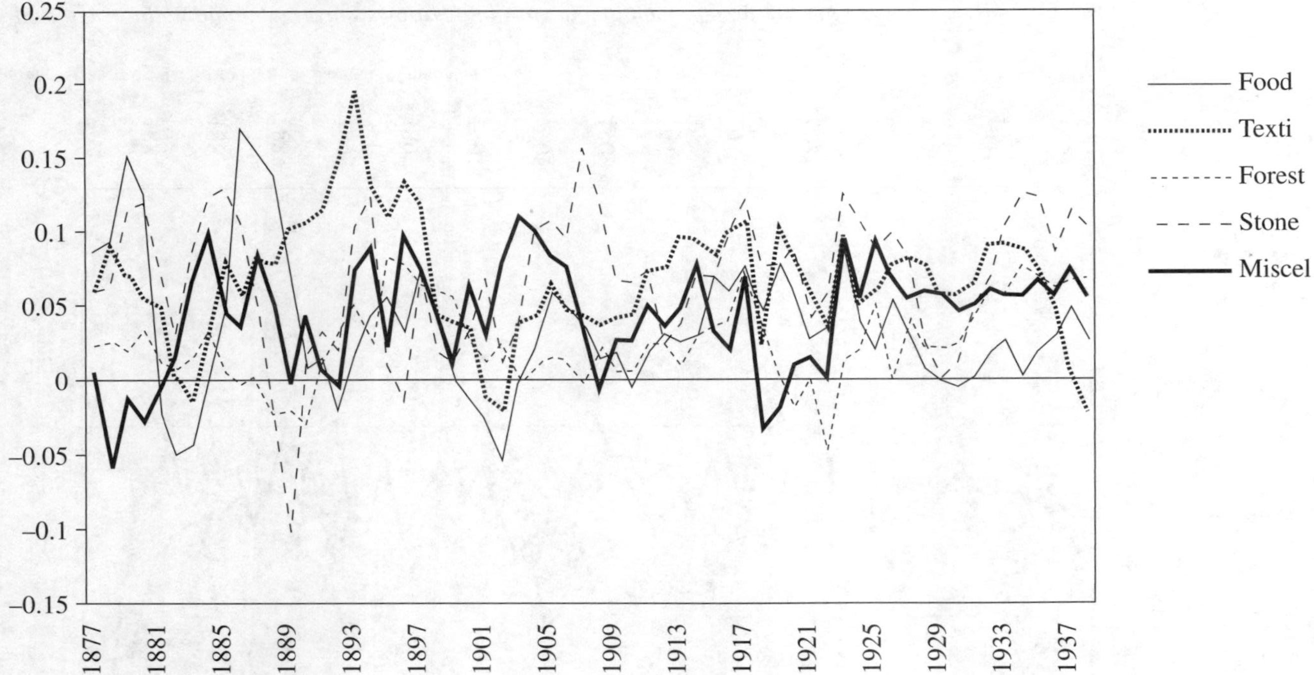

Source: Based on data provided in Klein and Ohkawa (1968).

Figure 6.4 Japan, manufacturing sectors, five-year moving average of growth rates

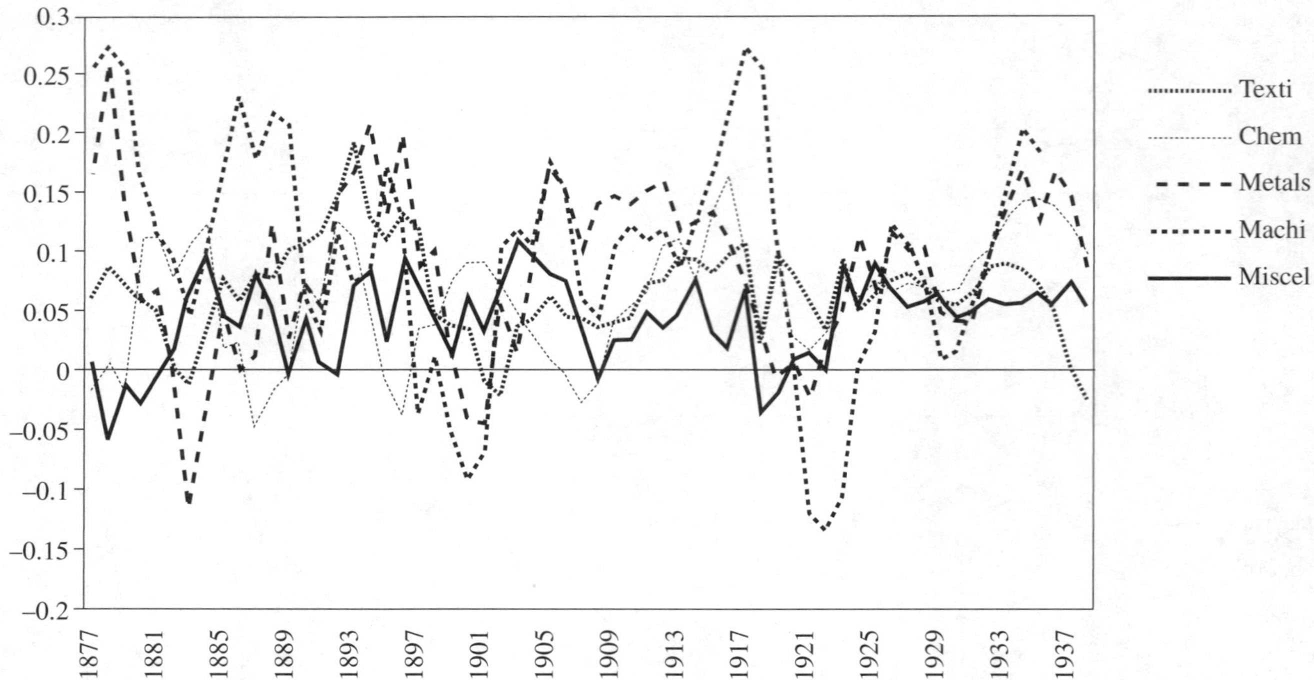

Source: Based on data provided in Klein and Ohkawa (1968).

Figure 6.5 Japan, manufacturing sectors, five-year moving average of growth rates

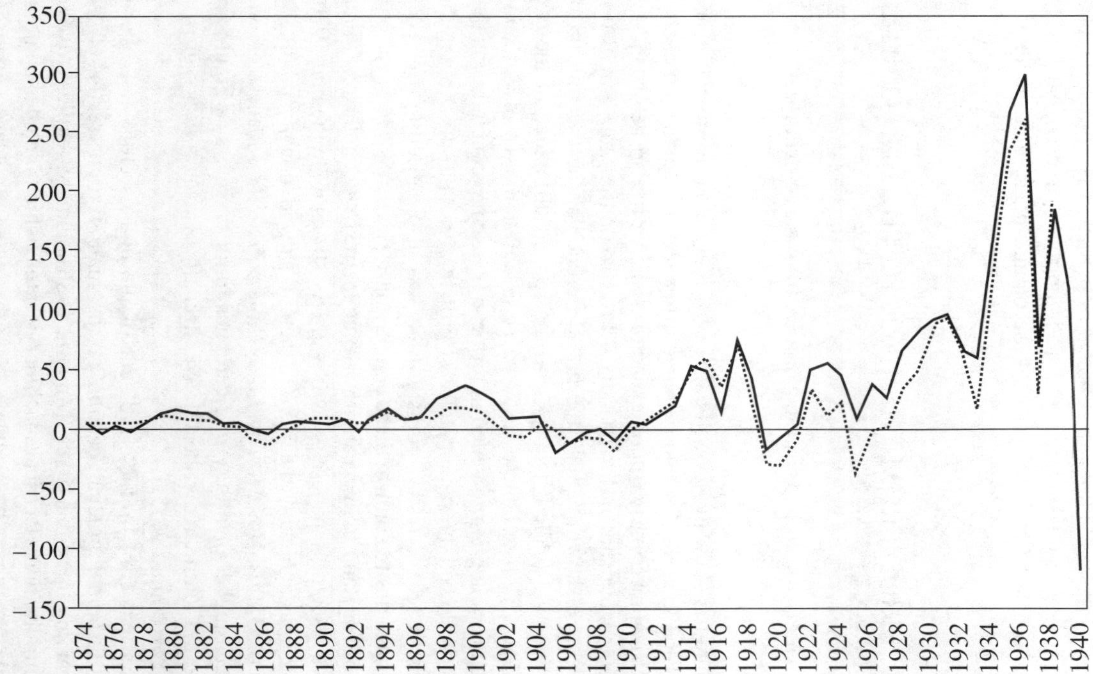

Note: The y-axis represents index numbers.

Figure 6.6 Japan, eight industries, 1874–1940, two cointegrating vectors

military reasons. Furthermore, the textiles sector played an overwhelmingly important role for a long period in Japan because it continuously reorganized itself to become decentralized and flexible to accommodate global changes. Overall, as Japan was still an agricultural country for most of the period under consideration, (and although it becomes perhaps easy to overemphasize the importance of OIs), OIs played an important role especially in the leading sectors of textiles initially and those of machinery and metals at a later stage.

6.3 THE PERIOD FROM THE 1960s TO THE END OF THE 20TH CENTURY

6.3.1 The American and Japanese Manufacturing Sectoral Real Output

Figures 6.7 and 6.8 show the growth patterns of real output for the USA and Japan from 1963 to 1998, for two manufacturing sectors – industrial chemicals and electrical machinery[6] (another sector is shown further below and the remaining sectors are shown in Appendix 6.1). Also on the same graphs two more variables are shown, namely the first difference of the two indexes of industrial production (called 'jadus'), and the second difference of the two indexes (called 'djadus'). These last two series offer a way of detecting stationarity and cointegration, which in turn can be used to conclude whether a given sector followed the same pattern of growth or not through time in the two countries. A formal test of cointegration was also carried out for each pair of sectoral series to confirm these conclusions.[7]

From Figures 6.7 and 6.8 we can draw some interesting conclusions. First, for almost every sector, Japanese growth was very strong up to the mid-1970s (the catching-up stage with the USA), then it slowed down, and finally decreased in the 1990s. However, over the whole period, the Japanese growth rates have been higher than the American ones in most sectors as Figure 6.9 shows. Table 6.3 summarizes the relative strength of each sector's growth between the two countries.

For the first category in Table 6.3, out of the eight sectors, it is worth noting the substantial gradual decline of the main leading sector of the Japanese economy up to the 1930s (Minami, 1986), namely, textiles. Also, it is worth noting the non-electrical machinery sector (which includes computers) where the Americans have always been the leaders (however, note that the higher American growth in this sector is not substantial). In the second category, it is worth mentioning the equal strength of industrial chemicals and rubber. In the third category out of the 14 sectors almost all heavy

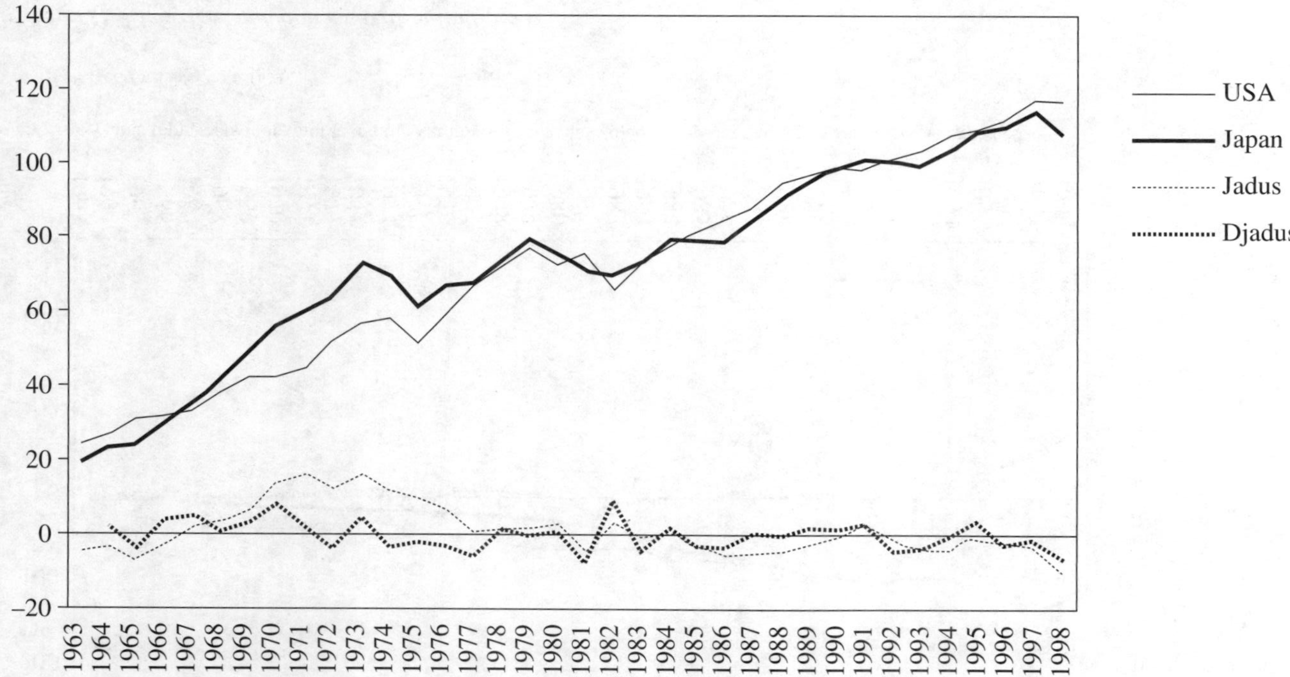

Note: The y-axis represents index numbers; the variables 'jadus' and 'djadus' are fluctuating around zero, as expected.

Source: Based on UNIDO (2000).

Figure 6.7 *Industrial chemicals, real output index*

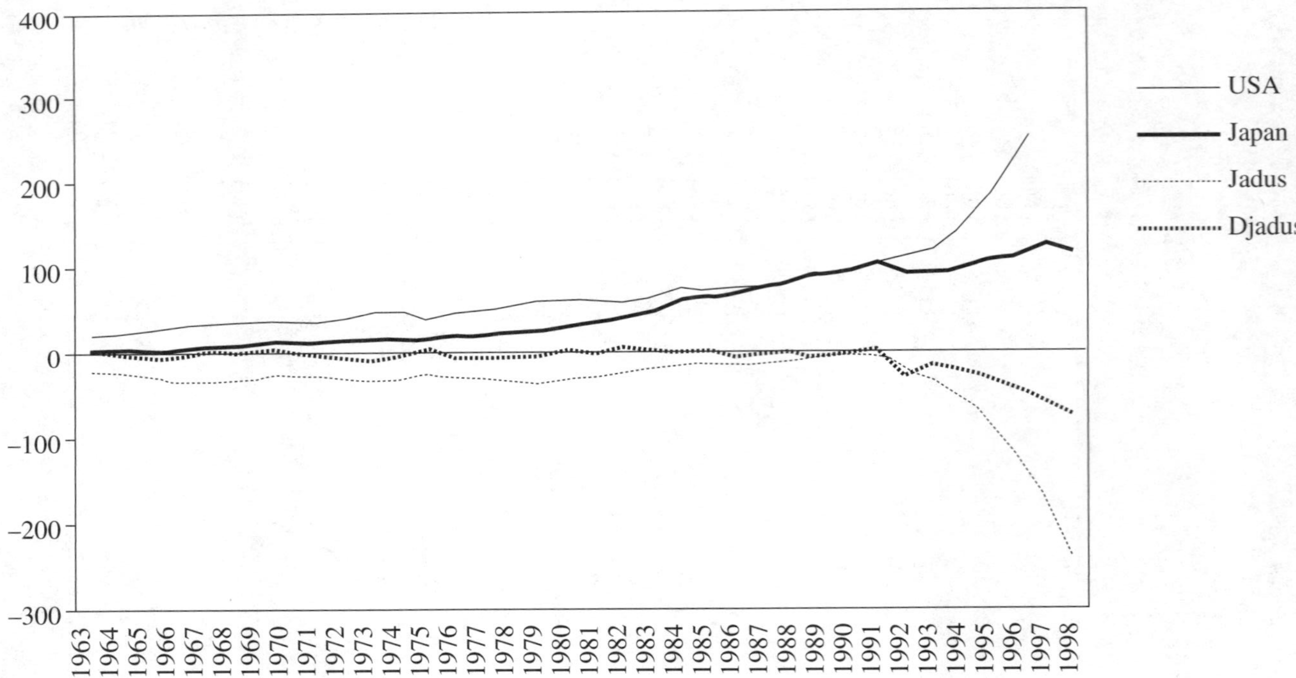

Note: The y-axis represents index numbers; the variables 'jadus' and 'djadus' are fluctuating around zero, as expected.

Source: Based on UNIDO (2000).

Figure 6.8 Electrical machinery, real output index

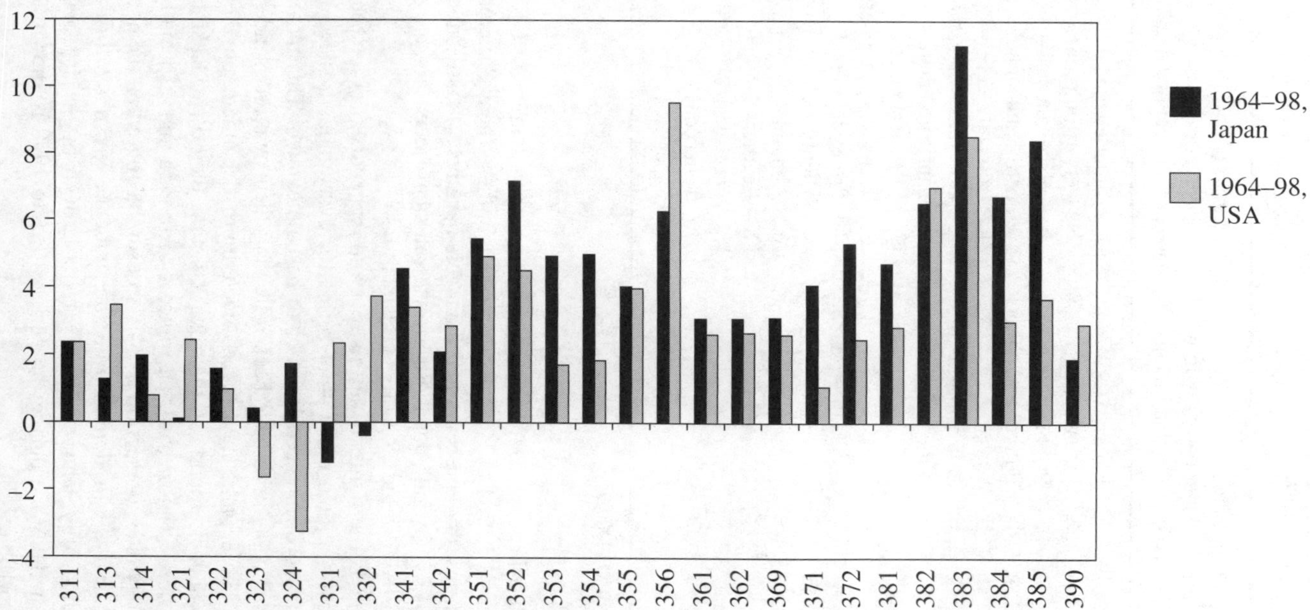

Note: The code numbers of each sector are indicated on the x-axis. See Appendix 6.2 for the corresponding names.

Source: UNIDO (2000).

Figure 6.9 Real output, annual percentage growth rates, for the USA and Japan, 1964–98

Table 6.3 Growth rates in real output from 1964 to 1998, industry differences

Category (1) American growth was higher than that of Japan	Category (2) American growth was close to that of Japan	Category (3) American growth was lower than that of Japan
1. Beverages	1. Food	1. Other chemicals
2. Textiles	2. Industrial chemicals	2. Petroleum refineries
3. Wood	3. Rubber	3. Miscellaneous petroleum and coal
4. Furniture	4. Pottery and china	4. Iron and steel
5. Printing	5. Glass	5. Non-ferrous metals
6. Plastic	6. Other non-metal minerals	6. Fabricated metals
7. Non-electrical machinery		7. Electrical machinery
8. Others		8. Transport equipment
		9. Professional and scientific equipment
		10. Apparel
		11. Leather
		12. Footwear
		13. Tobacco
		14. Paper

Source: Based on UNIDO (2000).

manufacturing industries have been more dynamic in Japan than in the USA during the whole period 1964 to 1998. It is precisely in these 'heavy' sectors that the Japanese firms, entrepreneurs and managers have been mostly innovative in terms of OIs. Overall, Japanese growth rates have been higher, or almost as high, in 20 out of the 28 sectors considered.

Furthermore, and as the figures (see this text and Appendix 6.1) for each sector show, the two national real output series for every industry need in most cases to be twice differenced in order to be cointegrated. This means that the two national economies have been following their own independent paths of growth. These paths have been much more dependent on their underlying historical background and their own OIs than their own TIs, since Japan has always been in most cases an imitator of foreign and, especially, American technology (see sub-section 6.3.4). An example, based on the non-ferrous metals industry, will demonstrate the evidence given by formal cointegration methodology. Thus a steadily faster growth of the Japanese sector than the American one during the period 1963 to 1998 is shown, as Figures 6.10a and 6.10b exhibit.

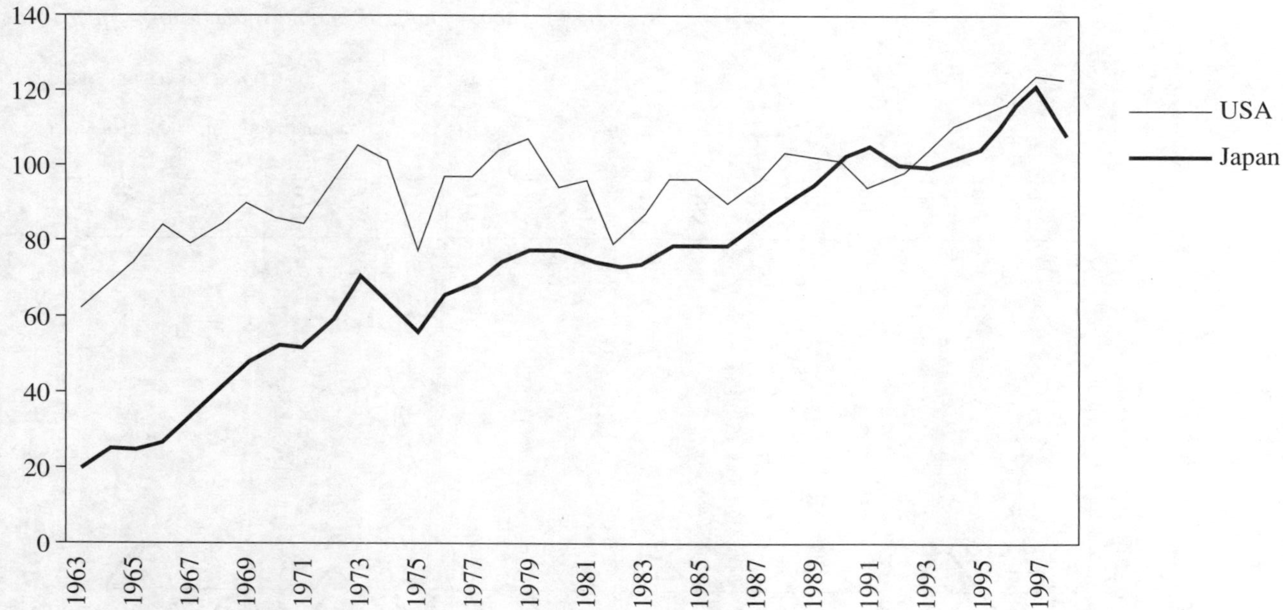

Note: The y-axis represents index numbers.

Source: Based on UNIDO (2000).

Figure 6.10a Non-ferrous metals, indices of industrial production, 1963–98

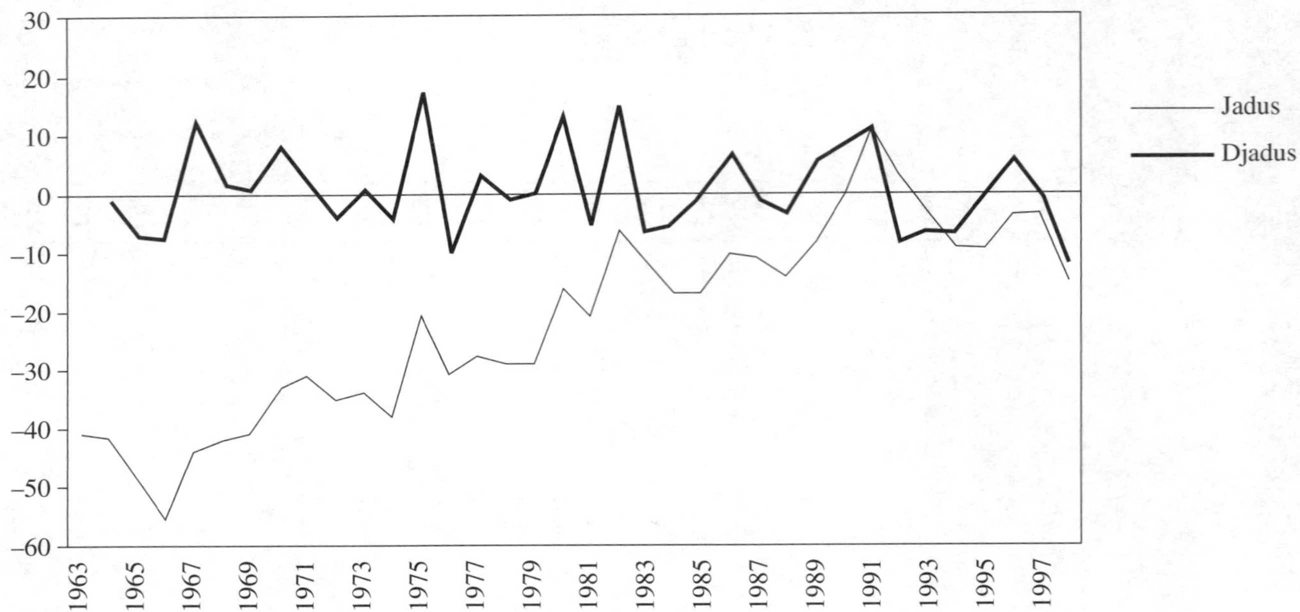

Note: The y-axis represents index numbers.

Source: Based on UNIDO (2000).

Figure 6.10b Non-ferrous metals, first and second differences, 1963–98

From Figure 6.10a we would expect no cointegration as the two national series grow at different speeds. Effectively, all the appropriate tests, such as the one based on the maximum eigenvalue of the stochastic matrix, confirm the non-cointegration of the two indexes of real output for non-ferrous metals. From Figure 6.10b we can conclude that the first difference between the Japanese and the American series of output ('jadus') is not enough to produce a stationary series; however, when 'jadus' is differenced once more ('djadus') then we have a stationary series; the unit root tests conducted for the four time series (the two original ones plus jadus plus djadus) confirm these conclusions.

The period 1964 to 1998 will now be split into three sub-periods so that more appropriate conclusions can be drawn. These sub-periods are between 10 and 13 years long, and each one of them includes a major depression: the period 1964–76 includes the effect of the first oil price shock, the period 1977–86 includes the effect of the second oil price shock and the period 1987–98 includes the 1990–91 depression. Figures 6.11 and 6.12 show the changes in growth between the three sub-periods considered.

From Figures 6.11 and 6.12 the following comments can be made. For the USA the period 1964 to 1976 saw the highest growth rates in almost all sectors, whereas the other two periods saw lower growth rates. However, there are some outstanding exceptions to this pattern. For the 1987 to 1998 period two industries were leading the American economy, namely, non-electrical machinery (mainly the computers component) and electrical machinery (mainly semiconductors). These two sectors have been experiencing the highest growth rates (at least double the others) amongst all 28 industries during 1987–98, and higher growth rates than in the other two sub-periods. Also, during 1987–98, iron and steel has been strongly reviving, and rubber continued its steady high growth. It will be further shown in this and the next chapter that these currently leading sectors of the American economy have been, more than any other sector, following and imitating Japanese OIs.[8]

For Japan it is remarkable how the two oil price shocks considerably slowed down almost all sectors during the second period 1977 to 1986. The three exceptions to this pattern were the two machinery sectors, especially electrical machinery (in which the Japanese semiconductor sub-sector was booming during that period) and the professional and precision instruments sector (in which the photographic equipment and the watches sub-sectors established themselves in the world). During the third sub-period, 1987 to 1998, all sectors experienced a deep plunge, with the least affected sectors being the electrical machinery, chemicals, non-ferrous metals, petroleum, paper and transport. Some explanations will be provided in sub-section 6.3.3 regarding Japanese manufacturing industries' protracted recession during the 1990s.

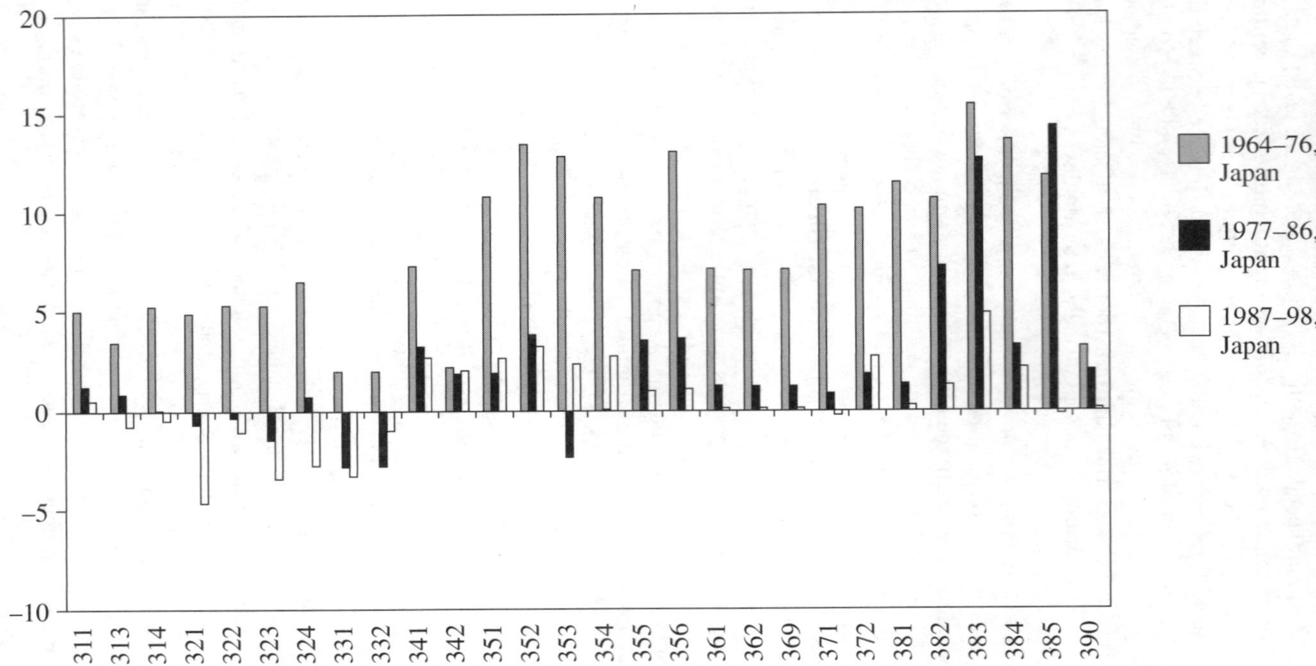

Source: Based on UNIDO (2000).

Figure 6.11 Japan, real output, annual percentage growth rates, for three sub-periods

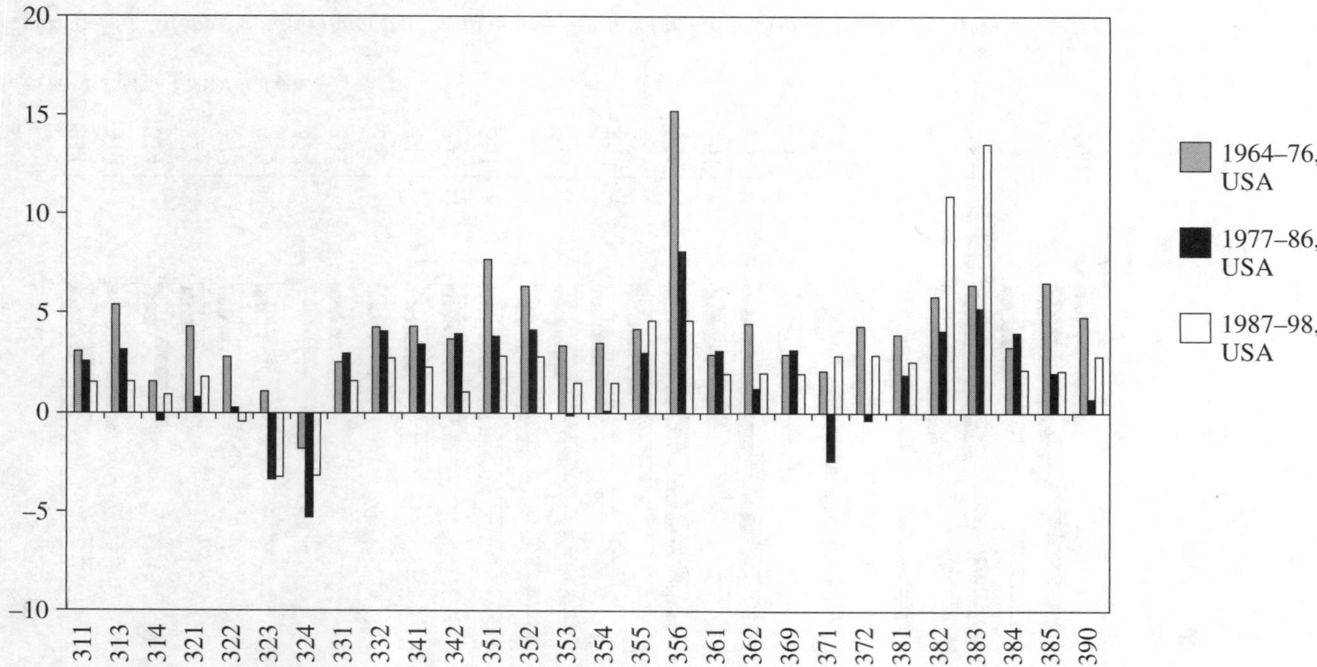

Source: Based on UNIDO (2000).

Figure 6.12 USA, real output, annual percentage growth rates, for three sub-periods

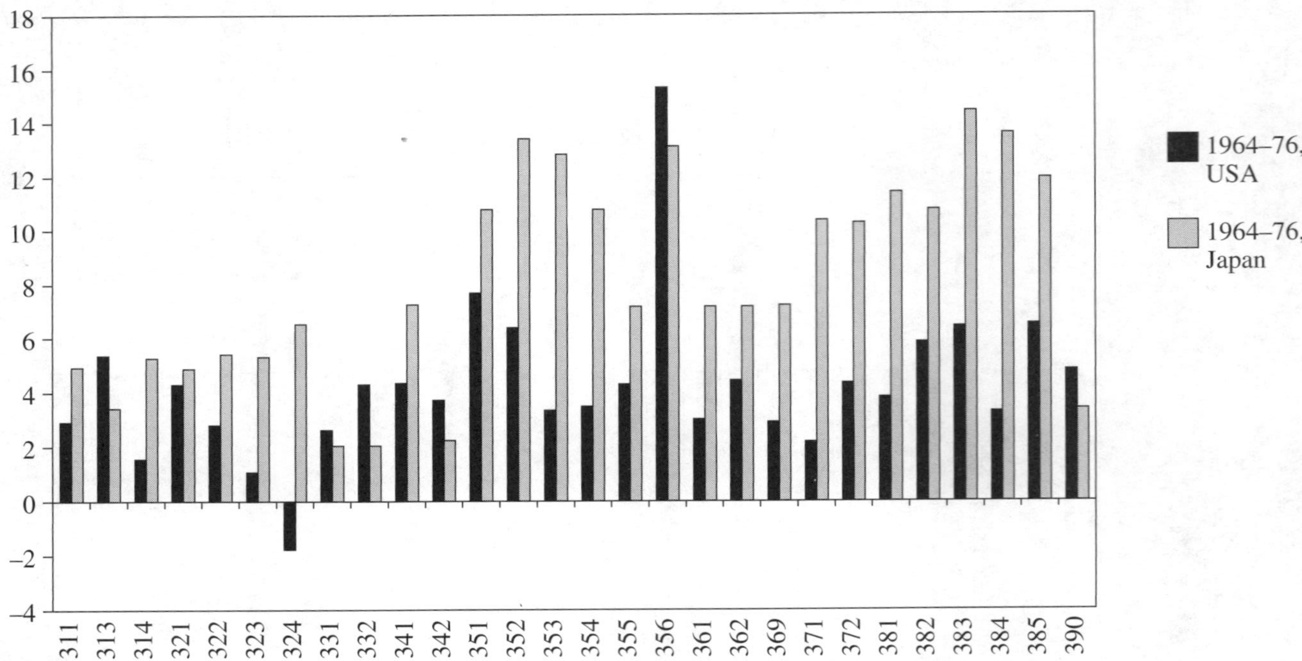

Source: Based on UNIDO (2000).

Figure 6.13a Real output, annual percentage growth rates for the USA and Japan, for 1964–76

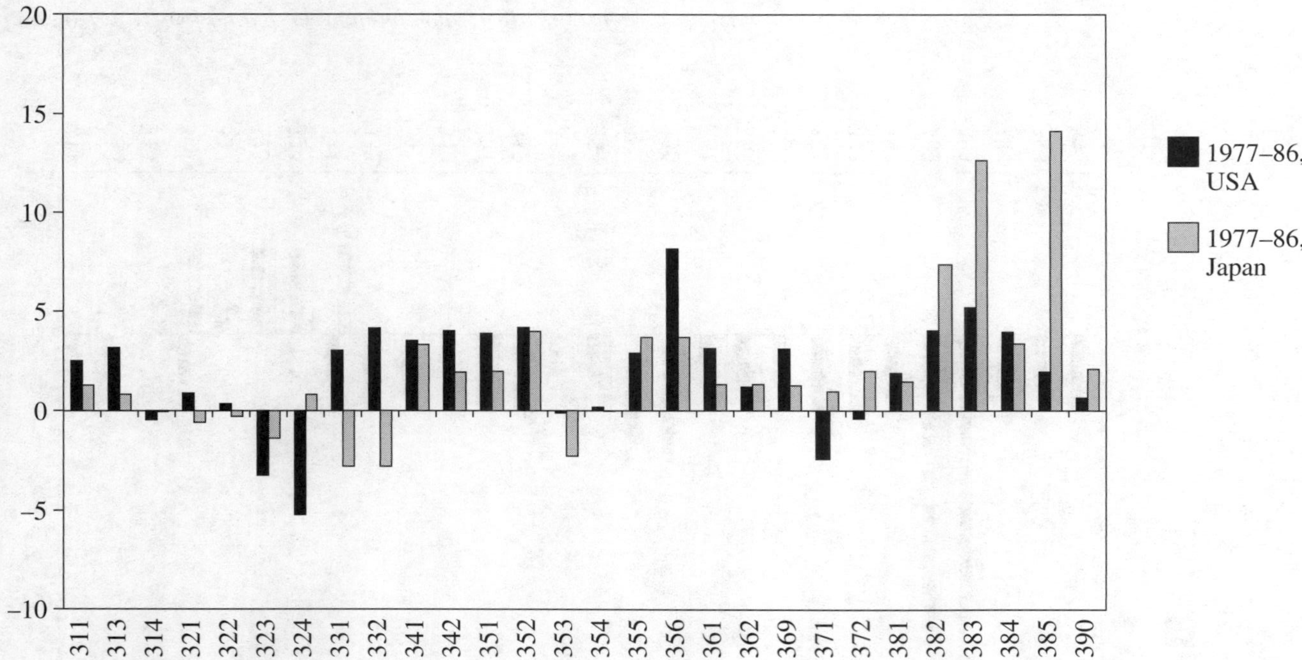

Source: Based on UNIDO (2000).

Figure 6.13b Real output, annual percentage growth rates for the USA and Japan, 1977–86

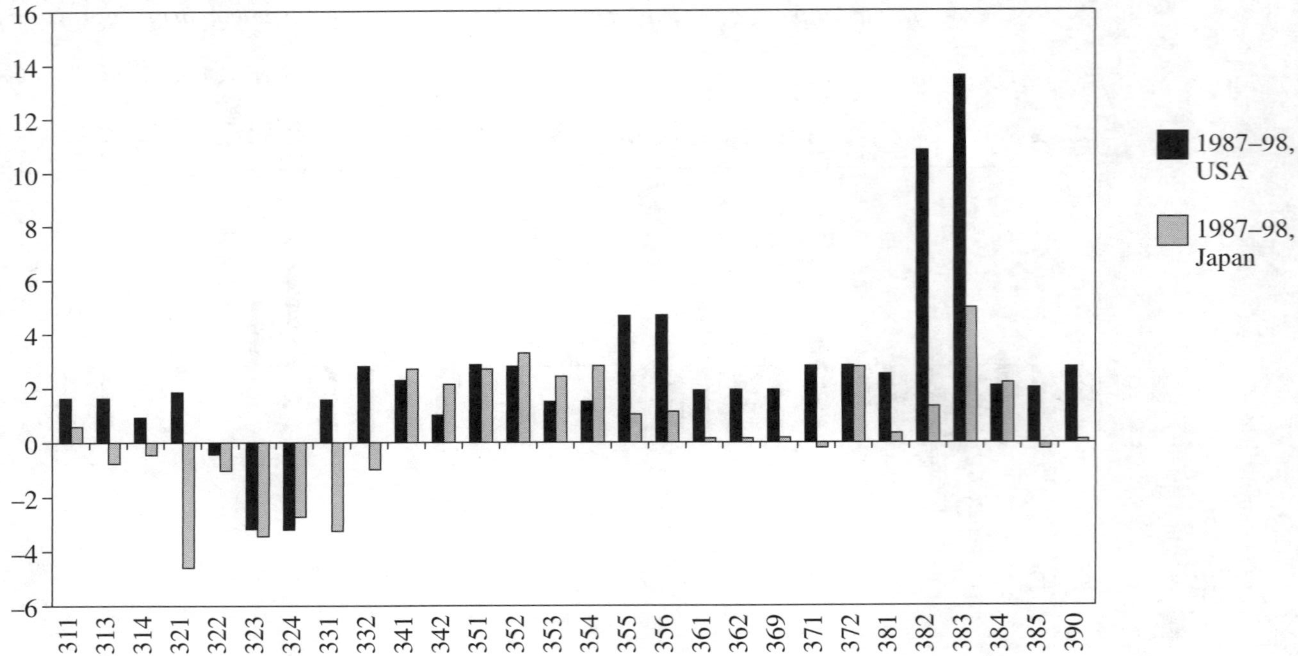

Source: Based on UNIDO (2000).

Figure 6.13c Real output, annual percentage growth rates for the USA and Japan, 1987–98

In addition, for each sub-period, the growth rates are compared for each sector between the two countries (see Figures 6.13a, 6.13b and 6.13c). Some remarks are necessary. First, during the period 1964 to 1976, only a limited number of industries showed a higher growth rate in the USA than in Japan (beverages, wood, furniture, printing, plastic and 'others'). Second, during both periods 1977 to 1986 and 1987 to 1998 the Americans did perform better than the Japanese in most industries. However, an interesting reversal took place between these two periods. Whereas, during 1977 to 1986 the Japanese grew faster in the three leading sectors of non-electrical machinery, electrical machinery, and professional and precision instruments, the situation was reversed during the period 1987 to 1998. This is an important finding and will be further explored in the next chapter where it will be shown that when the Americans imitated the Japanese in terms of OIs they started performing like the Japanese (which means they grew very fast).[9]

Second, there were some sectors (chemicals, petrol, non-ferrous metals and paper) for which the Japanese growth was still positive in the 1990s. It is especially remarkable for the chemical industries to have been growing strongly and continually during the whole period of 1963 to 1998, despite some criticisms. For example, Arora et al. (1999, p. 245) observed,

> The availability of imported technology and the general backward state of the chemical industry itself enabled users – firms in downstream sectors – to play a more prominent role in the chemical industry … The *keiretsu* structure therefore exacerbated the tendency, caused by import protection, toward production at scales that were too small to be economic …

However, it was precisely the Japanese characteristics of the chemical industry organization – *keiretsu* structure, many producers, high degree of product customization and so on – that was the reason for the strong continuous growth of that industry. Furthermore, as Figure 6.7 shows, both the American and Japanese chemical sectors grew in the same or parallel way after the mid-1970s, despite substantial differences in OI traits between the two countries.[10] This confirms one of the key assertions of this study that OIs are conducive to higher economic growth according to specific country contexts, and hence what is good for one country may not be for another in terms of OIs or other types of innovations.

6.3.2 TFP and Real Output in the USA and Japan from 1960 to 1985

Figure 6.14 shows the average annual rate of change in TFP for Japan and the USA for 28 two-digit sectors from 1960 to 1980, as calculated by two expert teams, namely, first by Jorgenson et al. (1995), and second by Kuroda et al. (1996). The methodology used by both teams was quite similar and

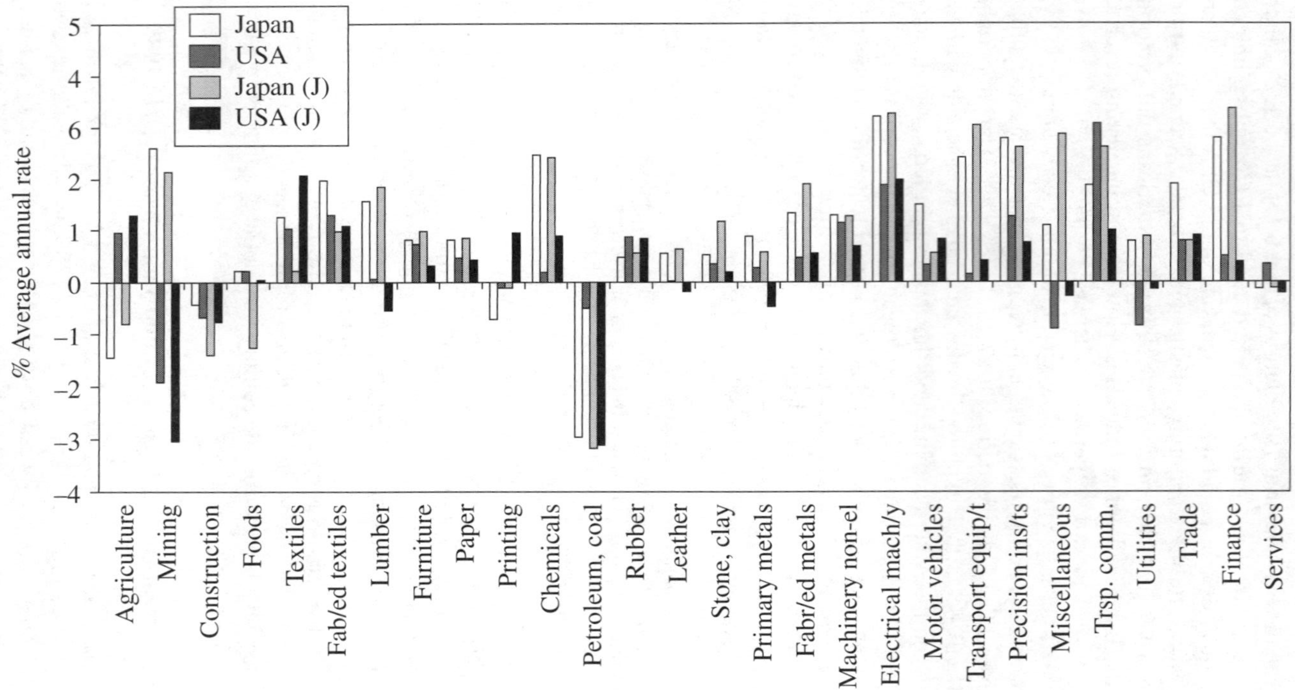

Source: Kuroda et al. (1996): Japan (J) (third bar) and USA (J) (fourth bar) refer to Jorgenson et al. (1995).

Figure 6.14 TFP annual percentage changes 1960–80 for the USA and Japan

based on Jorgenson (1990). Figure 6.14 shows that the differences between the two sources are insignificant. Other similar work (for example, Denny et al., 1992) confirms these results. The main conclusion out of this comparison is the higher TFP in Japan than in the USA in almost all industries, with the clear exceptions being the agriculture, construction, printing, petroleum, rubber and services sectors. Hence, the overwhelming majority of the manufacturing two-digit sectors exhibited higher TFP growth rates for Japan than for the USA during that period.

Another conclusion from Figure 6.14 is that both countries have been experiencing higher growth rates in TFP in almost the same sectors, namely, the electrical machinery, precision instruments, non-electrical machinery, transport and communication, and fabricated metals. Japan's growth has been more pronounced in the chemicals, fabricated metals, lumber, leather, transport equipment, miscellaneous manufacturing and finance sectors.

Figure 6.15 concentrates on TFP growth rates from 1960 to 1970 and from 1960 to 1985, as calculated by Kuroda et al. (1996). As expected, TFP was much higher in general for the former period in Japan, because the 1960–85 period included two major depressions due to oil price shocks. On the contrary, these depressions did not affect the USA to the same extent (which is not a surprising result for this country on account of its large natural resources). Furthermore, the comments made for Figure 6.14 are also valid for Figure 6.15. Figure 6.16 shows the rates of growth of real output for Japan and the USA for the period 1960–85. These rates are consistent with the rates of TFP. The Japanese growth rates are much higher for both real output and TFP.

Why did the Japanese economy achieve such a high and sustained economic growth during 1960 to 1985, which was much higher than that experienced by the USA during the same period? What role did OIs play in this development? The leading manufacturing sectors in Japan have been mainly electrical machinery, motor vehicles, precision equipment, miscellaneous manufacturing and, to a lesser extent, fabricated metals, chemicals and fabricated textiles. In all these principal industries the *quality* of the product[11] has been the principal force through which Japanese firms quickly penetrated foreign markets such as the American one; these markets have always been open to free competition from abroad (this openness started after World War II, contrary to Japan which remained more protective of its infant industries until quite recently).[12] However, this statement about quality begs the question of why Japanese firms achieved such a reputation for quality in their products after World War II. Chapters 4, 5 and 7 extensively answer this question (summarized in the next paragraph). Before World War II the quality of Japanese products was absent (for confirmation of this historical fact see Best, 1990, Odagiri and Goto, 1996, and Juran, 1995).

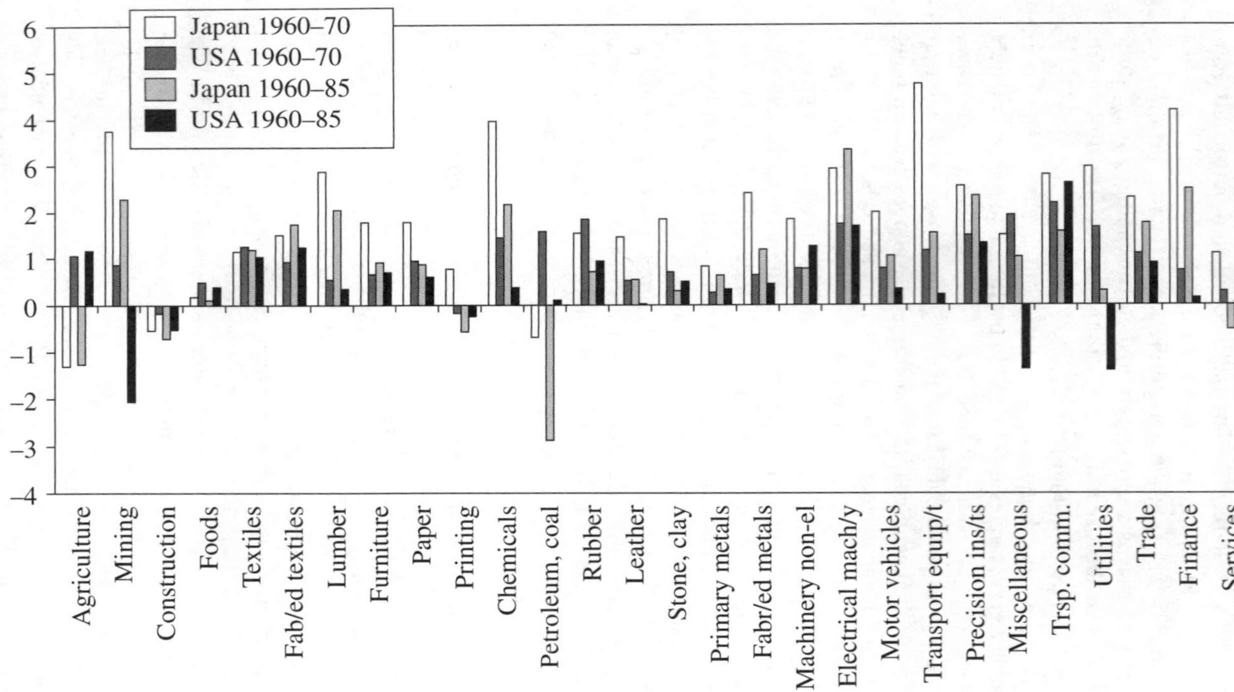

Note: 'Japan 1960–85' and 'USA 1960–85' are shown on the third and fourth bars of the figure.

Source: Kuroda et al. (1996).

Figure 6.15 *TFP annual percentage changes, two-digit sectors, 1960–70 and 1960–85, for the USA and Japan*

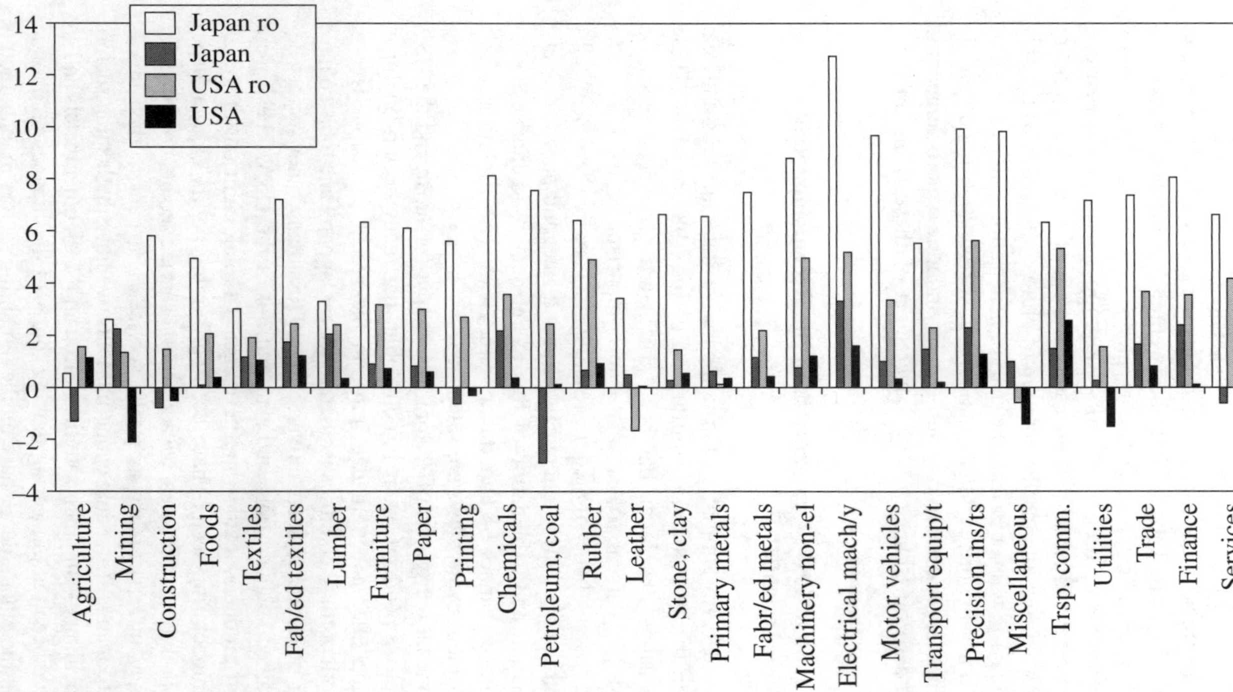

Note: In the legend, 'ro' means real output.

Source: Kuroda et al. (1996).

Figure 6.16 Real output and TFP annual percentage changes, 1960–85, USA and Japan

In previous chapters, some evidence was provided about the historical predispositions that are necessary during a certain period for a particular new structure to be formed in a subsequent period. This new structure may represent a noticeable break in economic terms. This is exactly what happened to Japan after World War II. The preconditions were the existence of a cooperative focal firm specializing in its core competences, with a strong emphasis on human development, strong networks and cooperation, plus a patriotic government. The new structure was the establishment of a practical spirit for *quality products* through the implementation of philosophies such as the JIT/QC system (hence low inventories), the market penetration criterion in business, government support, a large number of SMEs, a strong domestic competition for many manufacturing sub-sectors under new more democratic institutions, and so on (other chapters cover most of these points in more detail).

6.3.3 From 1986 to 1997: Explanations of the Recent Japanese Economic Downfall

Figures 6.17 and 6.18 summarize the comparative data for TFP and real output between the two countries during the period 1987 to 1998. These figures clearly show that Japan has been going through a deep recession during the last 12 years or so, especially in comparison to the USA. The manufacturing sectors that exhibited the least reversal are non-ferrous metals, transport, electrical machinery, chemicals, iron and steel and paper products. A full examination and explanation of this recent decline in Japanese economic performance cannot be undertaken here. However, in relation to this book some partial relevant views are suggested.

If we examine all the graphs of sectoral real output for Japan and the USA since 1963 (see Appendix 6.1), it is noticeable that many sectors in Japan stopped growing after a certain date prior to the recent recession. Hence, although for both of these countries leather, footwear, apparel and tobacco have been declining or remained stagnant since the 1970s, many other sectors declined or remained stagnant only in Japan and not in the USA. These were furniture, wood products, textiles and beverages. In total, eight out of the 28 three-digit manufacturing sectors have been performing poorly or even negatively. If, on top of these industries, we add several service sectors notorious for their inefficiency in Japan (such as public services and banks), then this is suggestive that Japan is still a dual economy. Thus, we have the leading sectors – transport, machinery – and their satellites – iron and steel, other metals and so on – which have been the pioneering moving forces of the Japanese economic miracle, and on the other side we have all the remaining sectors in manufacturing and services which lag behind in performance and effectiveness.

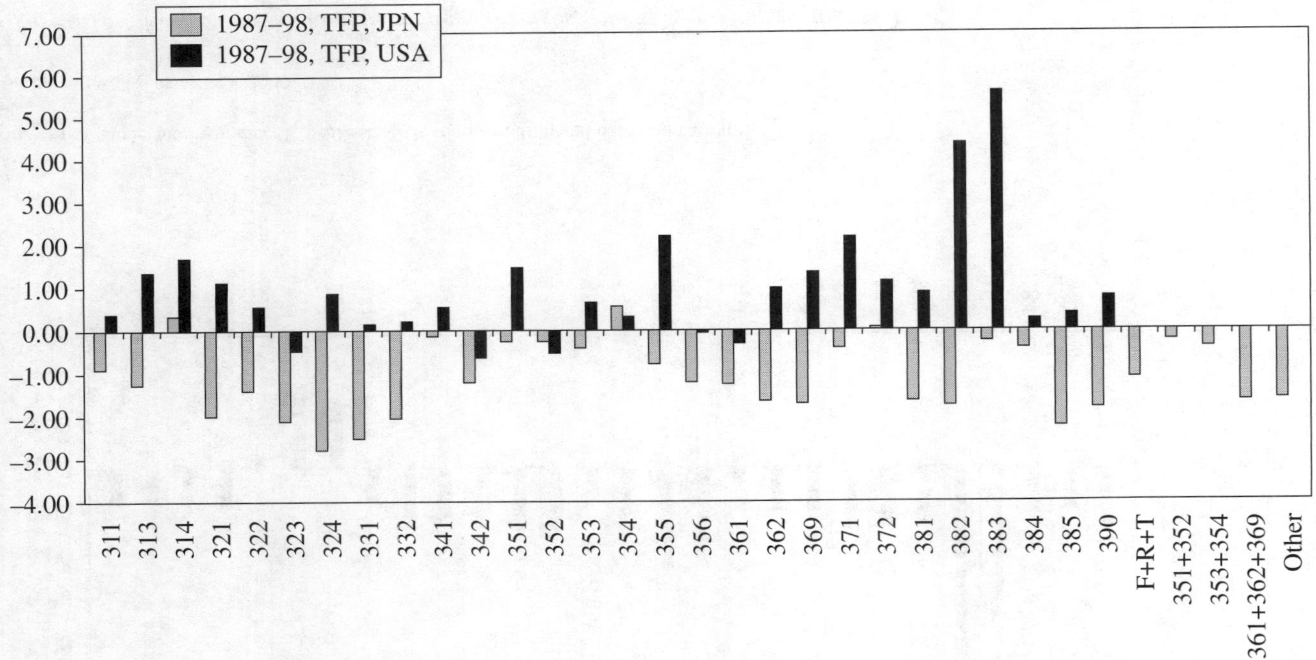

Note: F: finance and insurance; R: real estate; T: transportation and communication.

Source: Based on UNIDO, 2000 for both countries; the capital stock data were obtained from NBER (2000) for the USA, and from the Economic Planning Agency (EPA) for Japan.

Figure 6.17 TFP annual percentage changes, 1987–98, in Japan and the USA

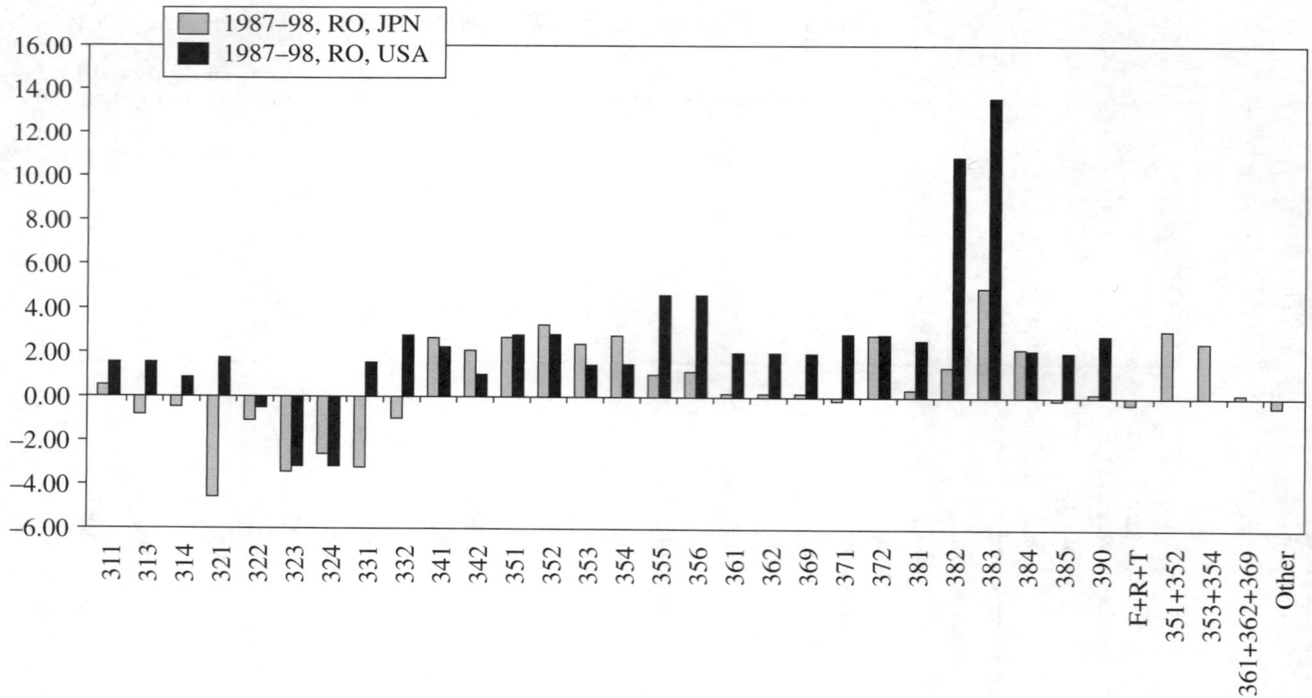

Note: F: finance and insurance; R: real estate; T: transportation and communication.

Source: Based on UNIDO (2000).

Figure 6.18 *Real output, annual percentage changes, 1987–98, for Japan and the USA*

Many scholars have made similar comments. Porter and Takeuchi (1999) have emphasized the microeconomic nature of recent Japanese failures. First, they stress, 'the consensus over Japan's past success has come overwhelmingly from the robust growth of a relatively small number of industries' (ibid., p. 67) such as semi-conductors, machine tools, steel and vehicles. In addition, Japan's exports are dominated by a relatively small number of industries in 'automotives, consumer electronics, office machines and production machinery. In huge areas of the economy there are few if any successful exporters, including chemicals, packaged goods, services and health care' (ibid., p. 72). Second, the same authors in their extensive study (ibid., p. 78) remark:

> Many of Japan's failures can also be traced to fragmented, inefficient, and anachronistic domestic sectors such as retailing, wholesaling, logistics, financial services, health care, energy, trucking, telecommunications, housing, and agriculture. By design, government policies have created two Japans: one composed of highly productive export industries, the other containing domestic sectors … The inefficient Japan drives up business costs across the board, weakening the competitiveness of the export industries …

From the OIs point of view the same writers Porter and Takeuchi (1999) recognize the importance of JIT/QM in Japan's economic growth (p. 71):

> The model of Japanese corporate success centres on the notion that a company can achieve both high quality and low cost by employing – and continuously improving – fundamentally better managerial practices. The idea is that companies compete by relentlessly staying at the frontier of best practice. This model is not an abstract theory but stems from extraordinary advances made by Japanese companies after the introduction of now well-known managerial practices, such as total quality management (TQM), lean production, and close supplier relationships …

However, still from the OIs point of view, the same authors pinpoint some relative weaknesses in Japan's organizational structures. These are various activities such as planning and control, finance, logistics, distribution, order processing, customer information, after-sales service, information technology, the Internet, marketing and office operations. Consequently, Porter and Tekeuchi (1999, p. 81) suggest that the 'companies must move from an exclusively egalitarian, seniority-driven model to one where doing things differently is rewarded in compensation, advancement, and opportunities for entrepreneurship'. In the book version of the just cited article, Porter et al. (2000, p. 189) recommended that Japan must move beyond competition just based on quality products to competing on strategy and innovation that result in 'true profitability'. If we refer to the analysis of the previous chapter, we can add that the POS is not yet well developed in Japan, contrary to the POM for the both the firm and society levels (see also Sanidas, 2004c).

According to Mroczkowski and Hanaoka (1998), the following changes have already been taking place in Japan:

1. Performance-based evaluation and rewards: by 1995, 75 per cent of Japanese companies administered pay by competency and merit (and hence not by seniority), whereas in 1987 the relevant percentage was 54 per cent, and in 1978 it was 42 per cent. These figures are supported by other surveys as well.
2. Evidence of change in social values: Japanese employees' attitudes are moving away from loyalty to the company and towards identification with their profession.
3. Manipulation of working time: overall hours worked (including overtime) are being reduced in larger companies. Also the use of flextime systems is increasing. However, both these elements are not true for SMEs.
4. Early retirement: this means a job transfer to a subsidiary or affiliated company under less attractive conditions. Again, SMEs lag behind the larger firms in implementing this policy (in 1995, 17 per cent as against 40 per cent respectively).
5. Lifetime employment and transfers: the *koyochosei* (employment adjustment) is now replaced by the *shukko* system (either temporary or permanent transfer to other companies).[13]
6. Employment of women and foreigners: between 1992 and 1995 the proportion of women in management positions has doubled, although is still low by international standards. Although the attitude of employers has changed rapidly, 40 per cent of Japanese companies employ foreigners whose status is still inferior to that of Japanese workers.

The same authors (ibid., 1998) predict that probably by 2010 the Japanese management system, at least in terms of human resources, will be like that in the West. This prediction is based on surveys they conducted with Japanese company managers, academics and management consultants. However, it must be added that it seems too difficult to change a history of more than a 100 years within a time frame of 10 to 20 years. Japan will emerge out of this new crisis (like many other crises during the last two centuries) with a new organizational outlook, which will be a marriage between the old features and some new ones, and hence it will not be just another Western country. In this respect it is important to refer to Fujimoto's (1999, p. 6) analysis: 'Japanese-ness no longer seems such a magic bullet, and it's now clear that the *organisational capabilities* of individual companies – especially Toyota –

make a real difference' (emphasis added). Thus, for Toyota, these capabilities have been taking place during the 1990s, together with some developments in TIs, as summarized by Fujimoto (1999) in his analysis of Toyota's new assembly systems.[14] In addition, for Toyota, there is a strong force of what Fujimoto calls 'the organisational preparedness' that constantly associates the outcomes of the system emergence (emergence to find solutions) with opportunities for capability-building[15] (ibid., p. 273).

The analysis conducted so far in this sub-section has shown that Japan still has much to accomplish in order to have a more stable economy. It is often forgotten that this country's enormous progress since World War II does not necessarily mean that problems of underdevelopment no longer exist. The Japanese dual economic nature will eventually disappear, but until then many changes in terms of OIs must take place. This is consistent with one of the key points of this study, namely, that OIs which are good for one country's economic growth are not necessarily good for another's, and that OIs which are good in one historical stage of economic development are not necessarily good for another historical stage. This point is again consistent with the contingency theory of management.

Several other authors have taken a similar stance. It is worth noting that 20 years ago, Abernathy et al. (1983, p. 84), expressed the same ideas and predicted a Japanese downfall:

> The modern Japanese system of production is not some manufacturing Nirvana, free of all tensions and problems that beset such systems elsewhere. It is the result of a set of deliberate choices and trade-offs and is appropriate not to every economic context imaginable but, rather, to the specific context of postwar Japan. If that country's social or political stability becomes problematical, if the work force ages too greatly, if expectations about living standards rise too quickly, if key industries cannot sustain their rate of growth, if – in short – changing conditions give the lie to the assumptions on which much of that production system rests, it will inevitably show the strain …

This extract hints at the contingency theory of management, and the 'changing conditions' they mention have actually taken place during the last decade in Japan. In brief, there are no leading or key sectors any more in that country (as demonstrated by the sectoral analysis above), the population is ageing rapidly, the younger generation has higher expectations about living standards without the willingness to do the 'dirty' jobs any more, and social and political stability is no longer so strong in Japan (for example see EAAU, 1997).

The importance of leading sectors in the present economic situation in Japan has become a key issue in economic circles of that country, according to Ito (1996, p. 236): 'Unless there is some structural reform, new industries will not emerge and the slow growth will continue. The appropriate kind of reform and the new leading industries that might emerge are debated intensively in Japan … '. This author examined several sectors which might benefit from structural reform in terms of promoting competition: airlines, telecommunications and broadcasting, financial services, distribution, agriculture and land use. The same author, like many others, also analysed the importance of the share and land price bubble and its collapse in the late 1990s in explaining the long recession in Japan. Finally, it is worth noting the role of the globalization tendencies present in many Japanese firms and their consequences in terms of growth in Japan and in terms of changes in OIs such as the subcontracting process. Ito (1996, p. 220) quotes some figures of employment that are very revealing of the globalization tendencies: the number of employees in Japanese Asian and North American subsidiaries substantially increased between 1990 and 1994, whereas employment in Japanese manufacturing fell in a comparable way during the same period.

Regarding the subcontracting system, there is gathering evidence that it has undergone substantial change to accommodate the very weak demand for consumer products. Turner (1994) observed that Japanese companies were subcontracting more labour-intensive work overseas, particularly within Asia; also, he remarked that the *keiretsu* system was beginning to loosen up. A similar view is expressed by Sugiura (2002) who talks about the meltdown of the automobile *keiretsu*, the metamorphosis of industrial agglomerations and the weakening of the entrepreneurial spirit. However, we cannot as yet generalize; for instance Lincoln et al. (1998, p. 242) concluded, 'While some prominent *keiretsu* partnerships are indeed loosening, elsewhere the form is alive and well … '.

Other radical changes are taking place in the last few years: for example, the competition with imports is intensified (Lux, 1997); also, Japan finally 'goes web crazy' (Rohwer, 2000). This new direction will certainly alter several economic structures as this last author comments, 'Japan is what you might call a middleman economy, and if there is anything the Internet is great at, it's killing off middlemen. Whether it's banking, retailing, or health care, the Internet will lower transaction costs, reduce the number of workers, and streamline communications' (ibid., p. 116).

All these changes and perhaps many others,[16] such as firing of employees (a rare phenomenon before this prolonged recession), are indicative that many new OIs will eventually emerge and probably new leading sectors will come forward. Thus, Lux (1997, p. 38) suggests: 'By combining the best of Western management with the best of Japan and the dedication of Japanese employees, companies that will be able to accomplish this transformation process successfully will become awesome competitors again … '. Kono and Clegg (2001) also suggest a new hybrid Japanese model of management and production that will retain many existing features and adopt new ones such as horizontal alliances and more flexible employment. Finally, the dilemma about whether the 'Western' or 'Japanese' overall system of production is superior is only a rhetorical question. As emphasized several times so far, what is 'good' in one country or in one period (for instance, regarding OIs or TIs) is not *necessarily* 'good' in another country or in another period.

6.3.4 The Role of TIs in the USA and Japan since World War II

In this sub-section a comparison of TIs for some key sectors[17] in the two countries will be outlined (although the analysis is far from exhaustive). This comparison is more an account of what happened in Japan than in the USA, but since Japan has been mainly an imitator of technologies the comparison is rather implicit. As will be seen, both nations adopted similar technologies (TIs related) almost everywhere, despite some lags and leads of a short duration, mainly because Japanese firms substantially copied American firms. This process of technical copying was accompanied by adaptations and appropriate changes to fit local circumstances; however, the Japanese also introduced new products mainly in the electronics industry (see below) and in personal items such as crystal quartz watches and automatic cameras (cf. Kono and Clegg, 2001, p. 206). Thus, the conclusions of this comparison further support the argument that differences between the rates of growth of American and Japanese industries are due to differences in OIs and not TIs, since TIs have been basically similar in the two countries in most cases.

Before briefly analysing some major industries in terms of TIs (and incidentally of OIs) it is worth mentioning six important aspects of the Japanese way of importing foreign technology (Odagiri and Goto, 1996, pp. 39–40):

1. Imported machinery and equipment assisted many industries in a critical way to improve product quality and productivity. Domestic machinery manufacturing then tried to reverse-engineer, by copying the imported capital until eventually they completely replaced it.
2. Japanese firms eagerly sought technological agreements primarily with the USA and also with European firms.
3. Consultants, mostly Americans, were hired to help to modernize the production processes.
4. The purchase of blueprints was also common.
5. Japanese companies often sent their engineers abroad to seek promising technologies.
6. Japan restricted foreign direct investment (FDI) until the gradual liberalization in the late 1960s and the early 1970s. However, even today, FDI still remains at a relatively low level.

The *iron and steel industries* in Japan, like the others described in this subsection (and generally like all industries), were initially protected by government measures until they took off and, at the same time, they were very competitive in their structure as oligopolies. This competition was one of the main reasons why some leading steel firms in Japan were pioneers in introducing new imported technologies and subsequently substantially improving on them. Following Odagiri and Goto's (1996) account, some examples will illustrate the industry.

At the start of the 1950s, Kawasaki Steel decided to build a new large plant despite the fact that two-thirds of the existing furnaces were then idle in Japan. The real novelty of this plant was related not to TIs but to OIs in terms of location and layout. Whereas it was common to build an ironworks near a coal mine or an iron-ore mine, Kawasaki's entrepreneurial spirit (through its president) built the new plant near the huge market of Tokyo and it developed a layout such that movements of materials and half-made products were minimized. This OI was soon followed by other steel producers in a very competitive oligopolistic market.

However, it was also TIs that made Japanese firms the best in the world during the second half of the 20th century. In the first place, the new technology of a basic oxygen furnace was imported from Austria, but it was soon found that there were major problems in operating such furnaces. These problems were resolved with some TIs introduced by the Japanese firms and in particular by the Yawata Company, which invented an oxygen converter gas recovery system that was soon adopted worldwide. Second, another technology known as continuous casting, that was originally developed in Switzerland, was imported by the Kawasaki firm who built a new plant with this technology in 1967, pushed again to take such a risky decision by the intense prevailing oligopolistic competition. By 1980 the process of continuous casting had been adopted by 60 per cent of Japanese plants but only 20 per cent of American plants (Odagiri and Goto, 1996, p. 152). As these two authors remarked:

> As a result of this and other innovations, Japan's productivity increase has outpaced that of other countries with companies starting to export technology and know-how in plant construction and operation to many countries, including both developing countries, such as Brazil, and developed countries, such as Italy and the USA. In 1974 receipt of royalties exceeded payments for the first time among Japanese industries … (ibid.)

These 'other innovations' referred to in the above quote were not only TIs but also, and perhaps mainly, OIs, as these two writers emphasized. The OIs included the active participation of workers in management and technical matters, so that firms attained company-wide involvement in productivity improvement. Florida and Kenney (1992) analysed some of these OIs in Japan and in Japanese FDI in the USA; thus, the firm NKK, for example, pioneered the use of QC circles in the steel industry; the same authors have also stressed differences between the American and Japanese TIs and OIs, for instance 'The US steel industry was the paradigmatic case of "Taylorist" scientific management ... In contrast, the Japanese steel industry developed a system of production organization and labour-management relations that harnessed workers' intellectual as well as physical capabilities' (ibid., p. 150).

Cars were first developed by the Europeans at the end of the 19th century, but it was the Americans, with Ford's mass production, that popularized car consumption in the 1910s and 1920s. The Japanese made many attempts to establish their own vehicle industry from the start of the 20th century (Odagiri and Goto, 1996), but it was only in the 1930s that they finally succeeded with Nissan and Toyota. Again, the technology (TIs) was mostly imported or, through reverse engineering, adapted and probably improved. Odagiri and Goto, (1996) provide us with some details. Nissan, in their Yokohama plant before World War II, bought the whole production equipment including jigs and tools as well as technology from the Graham-Paige Company in Detroit, and shipped them to Japan (this American company was the fourteenth largest auto producer in the USA but was planning to liquidate because of financial difficulties). Nissan repeated technology importation after World War II when it formed a tie-up with British Austin.

For Toyota the story is similar to Nissan's, but is more creative in terms of both TIs and OIs. One of Toyota's engineers stayed in the USA at Ford and other plants, and upon his return to Japan he started to disassemble a Chevrolet engine and copy it in an experimental plant built within Toyoda[18] Automatic Loom. Eventually, Toyota decided to use many of Chevrolet's and Ford's parts, and to buy many materials and components from outside suppliers. This was the origin of this company's supplier system. By 1938 Toyota had an R&D division, thus putting an official emphasis on its determination to create its own car technology. This took place much later after World War II with the development of the Corolla model and so on.

The impossibility of foreign firms exporting to Japan, or setting up production facilities within Japan, led some of them to sell their technology (TIs) to Japanese firms. Thus, Rootes (UK, producer of Hillman, later acquired by Chrysler and then Peugeot), Renault (France) and Willys-Overland (USA, producer of jeep, later acquired by American Motors and then Chrysler) consented to sell their technology to, respectively, Isuzu, Hino and Mitsubishi. After these tie-ups started in 1952–53, Japanese firms succeeded in complete domestic production within five years (Odagiri and Goto, 1996, p. 196).

These brief reports of some major car producers in Japan clearly show that the technology (TIs) in both the USA and Japan was similar, since the

Japanese mostly imitated Western vehicle producers. Consequently, one can safely conclude that the success of the Japanese car industry did not lie in TIs but in OIs. Odagiri and Goto (1996, p. 202) summed up the situation as follows, 'The strength of the post-war Japanese automobile industry is probably most evident in the fields of production management and human resource management, including training programs and the TQC (total quality control) movement. Toyota's kanban and just-in-time production system and keiretsu supplier system are well-known ... '.

Regarding the *electrical and electronic industries*, the major electrical appliances (for example, clothes dryer, dishwasher, home computer, and so on) were introduced in the post-war era, from the 1950s to the 1980s (Mowery and Rosenberg, 1998). These appliances were mostly Western innovations, but the Japanese firms also contributed, especially after 1970 with new electronic goods. Although colour television was initially developed by the Americans and Europeans in the 1960s the Japanese not only, once more, successfully imitated at the beginning, but they also introduced their own creations in terms of the Trinitron tube by Sony and solid-state colour receiver by Hitachi (Clark, 1987). Several other new products were developed in this industry, such as games machines, high-quality liquid crystal displays (LCDs) and digital versatile discs (DVDs).

Although the Americans invented the transistor in 1947 it was only in 1955 that the Japanese succeeded in making a transistor themselves after some years of struggle, especially in purifying silicon to almost 100 per cent. Odagiri and Goto (1996, p. 166) quoted one of the Japanese 'inventors', Kikuchi saying: 'Some people say that the transistor was just a borrowed technology. I would like to say from my own experience that the transistor is a kind of thing that, if you can copy it, it in itself is a spectacular achievement ... '. Sony's transistor radio tells us a similar story. Although Sony had already developed and marketed tape recorders by 1952, it was not until it made the transistor radio another successful consumer product that this company became a household name and a multinational. At the beginning of production and marketing of this radio the ratio of defects was very high despite the huge efforts by Sony's officer Ibuka to technically improve the transistor quality; at the same time competition with other large domestic producers became intense very quickly. As Odagiri and Goto (1996, p. 168) narrated:

> the only solution was to reduce defects. According to Ibuka, it was the instinct and insistence of a female production worker to scrutinise all the processes to find out the causes of defects. Following her suggestions, the engineers started the cumbersome task of testing every product at every point of the process and came to the conclusion that the use of antimony caused the problem. After several trials, they started using phosphorus instead and the yield rate greatly improved. This innova-

tion helped Sony to solve both of the problems above and put the company in a more advantageous position in its competition against other larger rivals …

Sony's case also shows another aspect of interplay between OIs and TIs; that is, the workers' participation in improving the product and the production process, which is an OI, had a positive impact on TIs.

Integrated circuits (ICs) became a practical device around 1959 with the invention of solid-state circuits by Texas Instruments, and the introduction of planar processing to interconnect circuit elements at Fairchild Semiconductors in the USA. Despite some new methods to bypass the planar processing by Hitachi and Toshiba, semiconductor technology was further advanced in Japan because of the world's first electronic calculator introduced by Sharp in 1964 (Odagiri and Goto, 1996, p. 171) priced at US$1400. By 1969, Sharp had reduced its price to US$300 by using American large-scale ICs (LSI) (produced by Rockwell). Before this last version, Sharp had used Japanese metal oxide semiconductor ICs produced by Mitsubishi, Hitachi and NEC. In 1971 there were at least 20 Japanese and several foreign firms (thus making a total of 33) producing a similar calculator to the one by Sharp. The calculator war eventually ended leaving only two survivors, Sharp and Casio.

However, as Odagiri and Goto (1996, p. 173) remarked: 'the Japanese suppliers had by then established LSI technology and their reliability in quality and delivery helped them to regain a position as the main suppliers of semiconductors to calculator producers … '. For instance, in terms of quality, in a test of about 300 000 memory chips (bought from American and Japanese firms) conducted by Hewlett-Packard in the late 1970s, none of the Japanese lots was rejected on account of failures, whereas the failures for American chips ranged from 0.11 to 0.19 per cent (ibid., p. 274). As extensively analysed elsewhere in this study, the quality control and just-in-time processes were the main weapons that Japanese firms had against American technical superiority in order to establish themselves in world markets. Thus, once more, OIs seem to prove themselves as being very important in promoting industrial growth through leading firms.

Finally, regarding the origins of microprocessors, Intel in the USA was the innovator in 1971 with the 4004 model. Once again the Japanese firms were the followers, although, as Odagiri and Goto (1996, p. 173) remarked, it was also a Japanese engineer who helped Intel to develop the first two microprocessors which triggered the PC revolution. Regarding the semiconductor industry, in Chapter 1 an important extract from Langlois and Steinmueller (1999, pp. 19–78)[19] demonstrated the importance of OIs in both the American and Japanese firms, and how the former firms successfully imitated the latter firms in terms of quality control. Although this industry started in the USA, especially in 'Silicon Valley', the Japanese eventually developed their own technology (TIs) and combined it with OIs such as the JIT and QC. The rivalry between these two countries' firms regarding this industry still continues today.

The *personal computer* industry has been more recently closely related to the Dell Computers Corporation (Dell CC). Dell CC is one of the most successful business stories[20] in the USA in the last 20 years. It is also an excellent example of how an entrepreneur (M. Dell), his managers and his personnel have pioneered the introduction of OIs (and not TIs) in order to become the industry's leader not only in the USA, but also globally. The initial OIs were marketing oriented: Dell's new company in 1984 was able to sell IBM clones at about 40 per cent below the price of an IBM PC; also, Dell sold his computers directly to large customers and eventually to individual customers through the Internet. By late 1997 Dell had become the global industry leader in keeping costs down by achieving what Dell called a 'virtual integrated' firm – 'a stitching together of Dell's business with its supply partners and customers in real time such that all three appeared to be part of the same organisational team' (Thomson and Gamble, 2001, p. C-136).

Overall, Dell's three golden rules have become its OIs cornerstones: (1) disdain inventory, (2) always listen to the customer, and (3) never sell indirect. The first rule will now be closely scrutinized.

The companies which started the PC industry in the 1980s manufactured many of the components themselves, thus being at least partially vertically integrated. However, as the industry grew very rapidly, as TIs were introduced more frequently, the PC manufacturers could not keep pace with being experts on all fronts, thus more and more specialist firms emerged that could mass-produce cheaper and more technologically advanced components. Consequently, vertical disintegration became more and more prevalent in PC producers. Dell CC's first steps consistent with its established rule to sell directly to customers captured this tendency for vertical disintegration. All this entailed two fundamental features of this company: an extensive outsourcing and virtually no in-house stock of finished goods inventories. Dell's build-to-order policy has been working in all directions.

Dell CC established long-term relationships with its best suppliers, such as Intel and Sony, and laid the basis for JIT delivery of suppliers' products to Dell's assembly plants. At the same time Dell CC itself practised JIT, thus yielding major cost advantages and shortening the time it took for Dell to get new generations of its computer models into the marketplace. Thomson and Gamble (2001, p. C-150) quoted Dell himself explaining the economics of minimal inventories.

> If I've got 11 days of inventory and my competitor has 80 and Intel comes out with a new 450-megahertz chip, that means I'm going to get to market 69 days

> sooner. In the computer industry, inventory can be a pretty massive risk because if the cost of materials is going down 50 per cent a year and you have two or three months of inventory versus 11 days, you've got a big cost disadvantage. And you're vulnerable to product transitions, when you can get stuck with obsolete inventory …

The results of this deliberate JIT philosophy are impressive: only a few days of inventory for some components and a few hours for others. In 1995, Dell CC averaged an inventory turn ratio of 32 days; in 1999, the ratio was six days' supply; the long-term goal is to reach a three-day average supply. All these efforts have made Dell CC the low-cost leader of the PC industry, and a high-profit company.

Still regarding the JIT system, it is worth quoting the authors Thomson and Gamble (2001, p. C-147–8) about a change of the operations on the shop floor that generated a huge productivity increase. This quote shows in a very concrete manner a good example of the process of movements and its related kinetic costs, as fully explained in the previous chapter:

> Until 1997, Dell operated its assembly lines in traditional fashion, with each worker performing a single operation. An order form accompanied each metal chassis across the production floor; drives, chips, and ancillary items were installed to match customer specifications. As a partly assembled PC arrived at a new workstation, the operator, standing beside a tall steel rack with drawers full of components, was instructed what to do by little red and green lights flashing beside the drawers containing the components the operator needed to install. When the operator was finished, the drawers containing the used components were automatically replenished from the other side, and the PC chassis glided down the line to the next workstation. However, Dell reorganised its plants in 1997, shifting to 'cell manufacturing' techniques whereby a team of workers operating at a group workstation (or cell) assembled an entire PC according to customer specification. The shift to cell manufacturing reduced Dell's assembly times by 75 percent and doubled productivity per square foot of assembly space …

A summary of Dell CC's OIs is now presented so that a whole picture can be obtained:

- build-to-order manufacturing
- partnerships with suppliers
- JIT components inventories
- direct sales to customers
- award-winning customer service and technical support
- pioneering use of the Internet and e-commerce technology
- strong demand forecasting skills
- comparative advertisements
- teamwork at all levels

- avoidance of hierarchical structures in governance.

All these OIs (the TIs were almost entirely absent) made Dell CC the leader in the PC industry, surpassing previous leaders such as IBM and Compaq in a very short time. Now, Dell CC's main competitors are trying to imitate the leader by introducing their own JIT process, their build-to-order manufacturing and to speed new models to market. However, it is hard to duplicate Dell's approach, as previous cases in other industries have shown (see Chapter 7 for more details on the dissemination of the JIT/QC process). Thus, as of mid-1999 Compaq's order-to-delivery time was approximately 12 days versus 3.1 days at Dell (Thomson and Gamble, 2001, p. C-165).

To sum up this fascinating story, M. Dell started his company from scratch in 1984 and today is the world leader followed by Compaq, IBM, Hewlett-Packard, Gateway, Toshiba, about 30 000 resellers of generic or 'house-label' PCs in North America alone and countless thousands more worldwide. All this was achieved by adopting, or introducing, OIs and virtually no major TIs at all.[21] Dell's story in the PC industry is similar to Toyota's story in the car industry, as far as the importance of OIs is concerned.[22]

6.4 CONCLUSIONS

The foregoing discussion, based on some basic quantitative evidence, regarding manufacturing sectoral growth from the 1880s to the 1930s and from the 1960s to the 1990s in the USA and Japan has provided some support to the qualitative and theoretical analysis of the previous four chapters. Briefly, the following major conclusions are worth mentioning.

First, the periods 1899 to 1914 and 1915 to 1929 were the 'golden' periods for many sectors in the USA. Especially during the latter phase of 1915–29 productivity increases were remarkable and were due to the appearance of new OIs (*ceteris paribus*) such as scientific management and Fordism and consolidation of existing OIs such as vertical integration and diversification. The leading sectors in terms of TFP were the automobiles, machinery and rubber industries, but most sectors experienced very high growth rates in TFP and real output. For Japan the golden periods of high growth in real output, mainly in the textiles, metals and machinery sectors took place during World War I and during the 1930s when new leading sectors such as the machinery and metals industries were propelled by the consolidation of focal firms.

Second, during the period 1964 to 1998, the manufacturing sectoral growth in the two countries was quite different; the differences, as detected by graphs and some cointegration tests, were due to differences in the adopted OIs amongst other factors. Thus, during the 1960s up to the mid-1980s, most

Japanese industries exhibited very high growth rates both in real output and TFP, contrary to the American industries. This was primarily due to the high quality and low cost of many Japanese products, and hence it became easy for Japanese firms to penetrate national and foreign markets. This last conclusion is supported by the review of TIs in the two countries which confirms the tendency by Japanese firms to imitate foreign TIs, although in some instances these firms dominated world markets by launching new products (mainly in the consumer electronics sector); thus, the common TIs used in the majority of both American and Japanese firms reinforces the conclusion that it was in the area of differences in OIs that a more comprehensive explanation can be sought for the substantial differences in manufacturing sectoral growth in real output and TFP between these two countries in the period between the 1960s and the mid-1980s. During this period, the most famous Japanese leading firms were in the transport, electrical and electronics sectors.

Third, during the sub-period 1987 to 1997 the situation was reversed between the two countries in several ways. Whereas the machinery, both mechanical and electrical, industries led the American revival of the American economy, in Japan there was no leading sector for the first time in a long time in Japanese economic history. Japanese industries are still in a deep and prolonged recession that started at the beginning of the 1990s; a brief review of the reasons for this downfall seems to strongly indicate that the dual character of the Japanese economy, the non-existence of leading firms and sectors, and a search for new OIs in the management of firms in Japan are the main factors that are contributing to this recession.

In the next chapter a more rigorous empirical analysis will seek to confirm the preliminary conclusions reached in this chapter, especially regarding the American economic revival during the last 20 years or so of the last century.

APPENDIX 6.1: INDICES OF REAL OUTPUT IN THREE-DIGIT SIC SECTORS IN THE USA AND JAPAN, 1963–98

Food

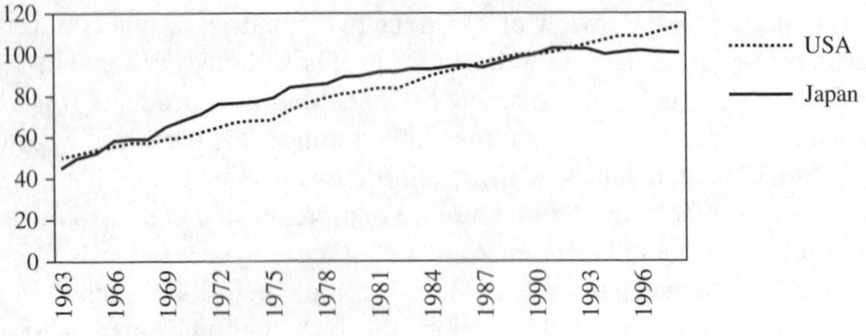

Textiles

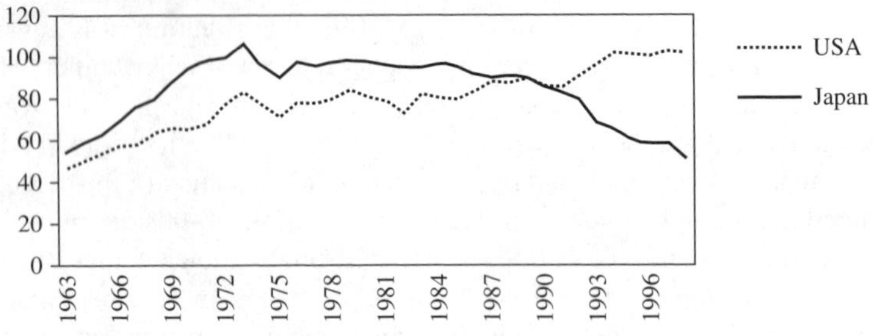

Beverages

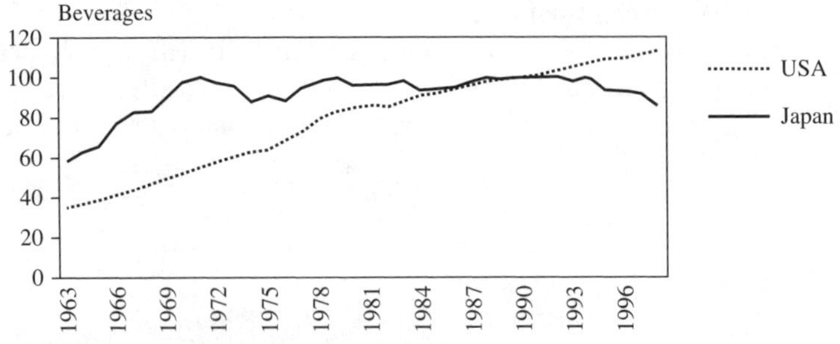

Apparel

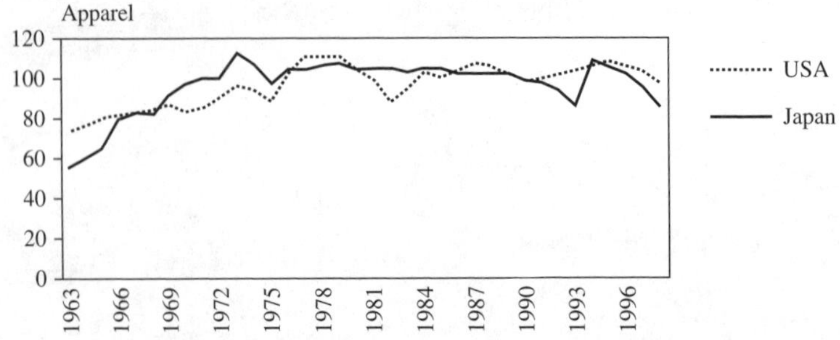

Tobacco

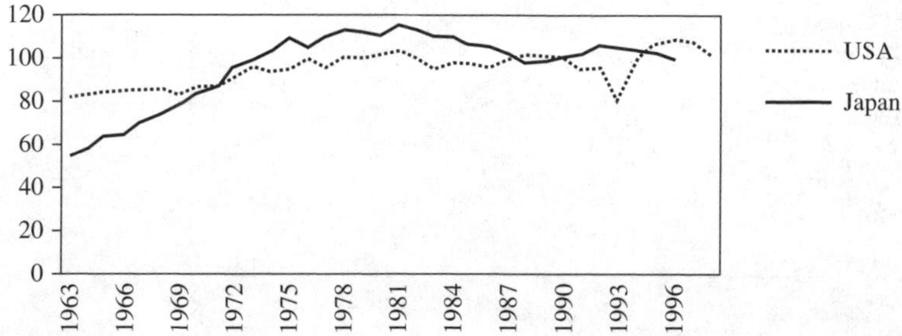

Leather

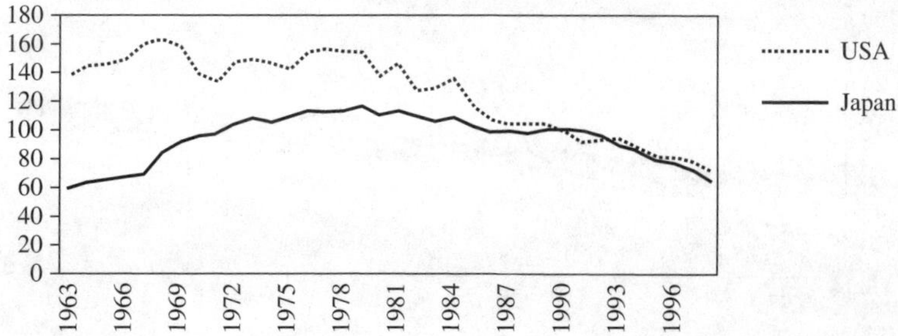

Footwear

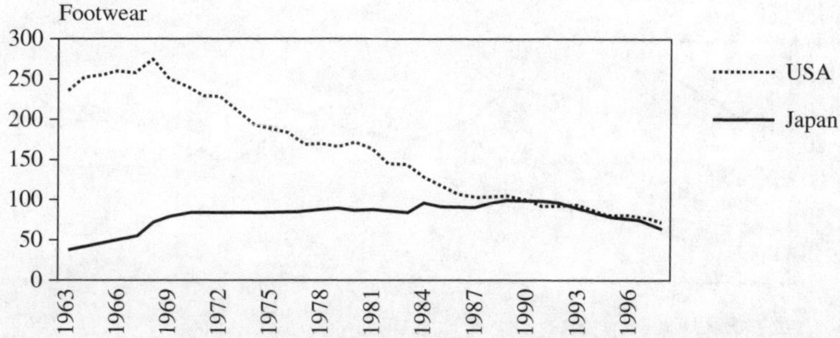

Miscellaneous petroleum and coal

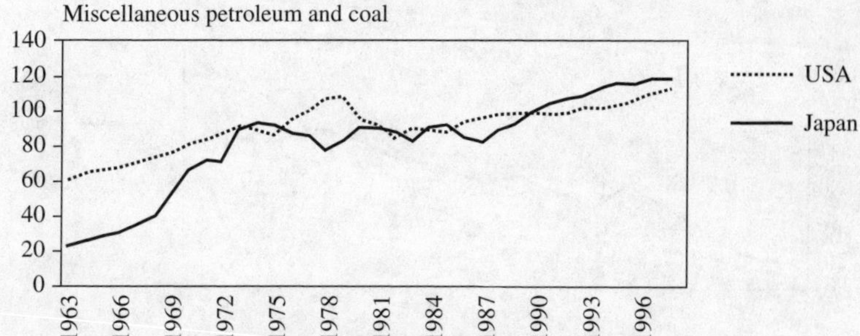

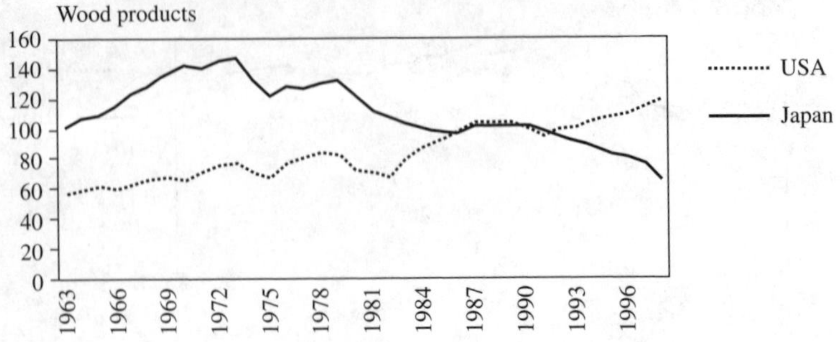
Wood products
160
140
120
100
80
60
40
20
0
1963
1966
1969
1972
1975
1978
1981
1984
1987
1990
1993
1996
USA
Japan

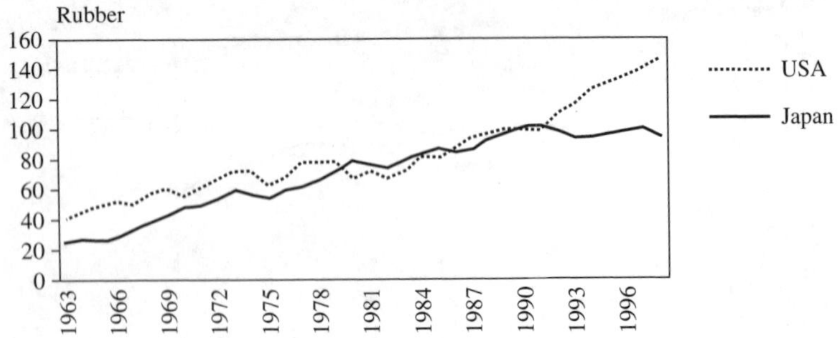
Rubber
160
140
120
100
80
60
40
20
0
1963
1966
1969
1972
1975
1978
1981
1984
1987
1990
1993
1996
USA
Japan

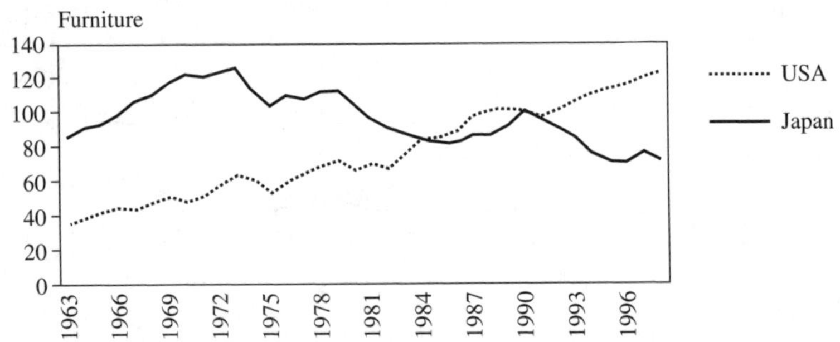
Furniture
140
120
100
80
60
40
20
0
1963
1966
1969
1972
1975
1978
1981
1984
1987
1990
1993
1996
USA
Japan

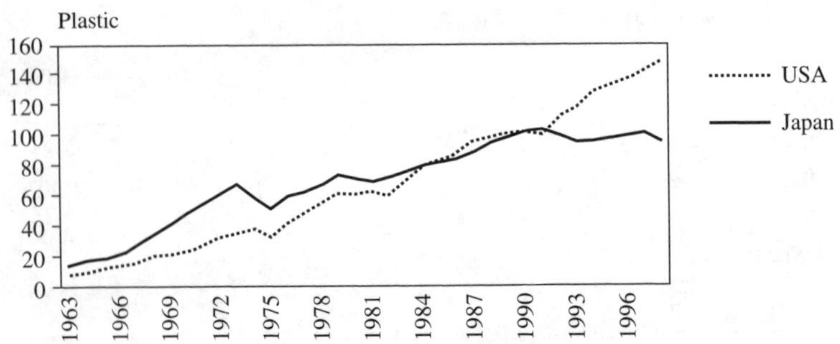
Plastic
160
140
120
100
80
60
40
20
0
1963
1966
1969
1972
1975
1978
1981
1984
1987
1990
1993
1996
USA
Japan

Pottery

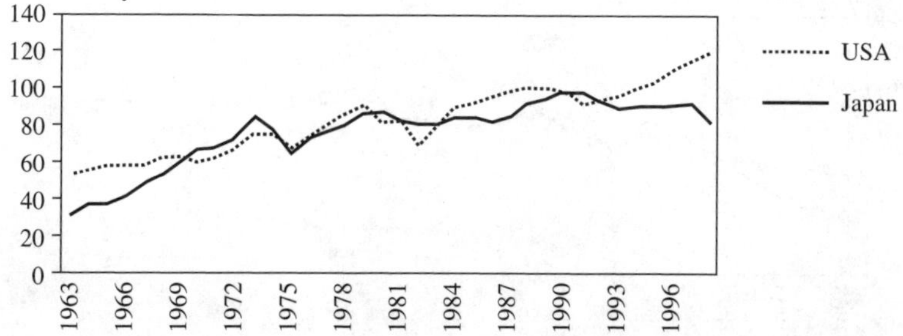

Iron and steel

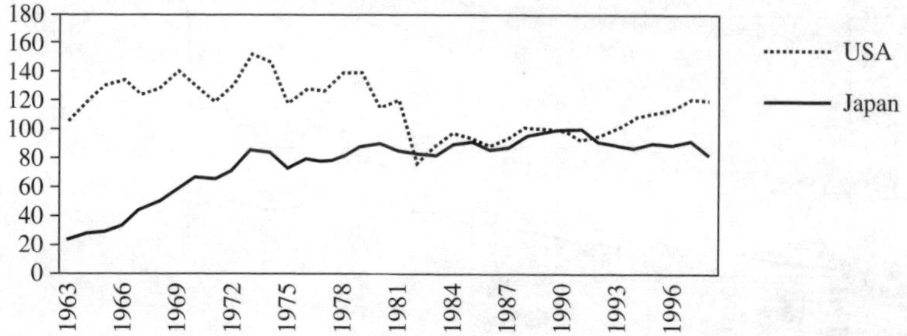

Glass

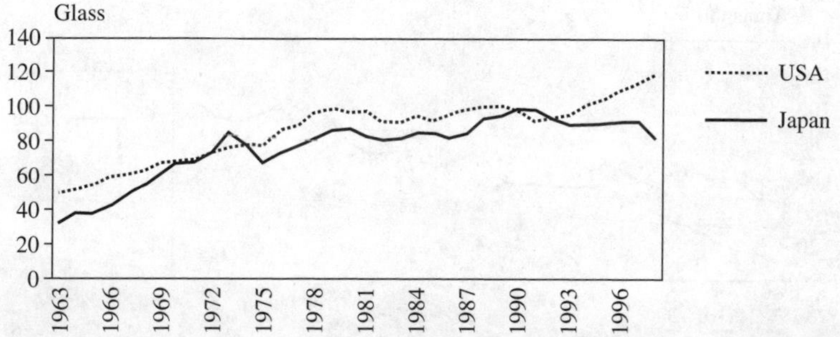

Fabricated metals

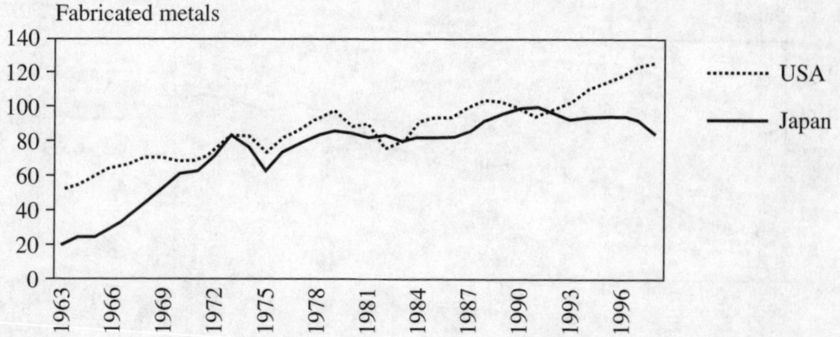

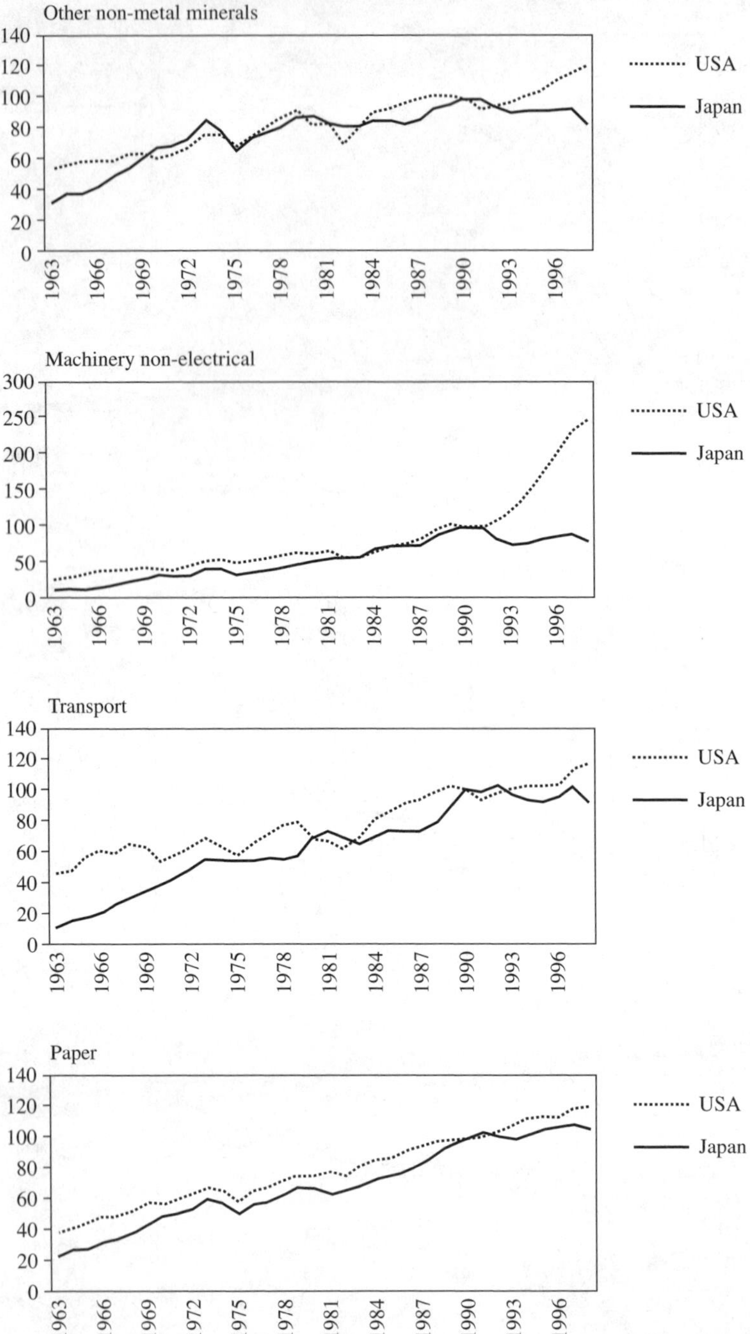
Other non-metal minerals
140
120
100
80
60
40
20
0
1963
1966
1969
1972
1975
1978
1981
1984
1987
1990
1993
1996
USA
Japan
Machinery non-electrical
300
250
200
150
100
50
0
1963
1966
1969
1972
1975
1978
1981
1984
1987
1990
1993
1996
USA
Japan
Transport
140
120
100
80
60
40
20
0
1963
1966
1969
1972
1975
1978
1981
1984
1987
1990
1993
1996
USA
Japan
Paper
140
120
100
80
60
40
20
0
1963
1966
1969
1972
1975
1978
1981
1984
1987
1990
1993
1996
USA
Japan

Paper

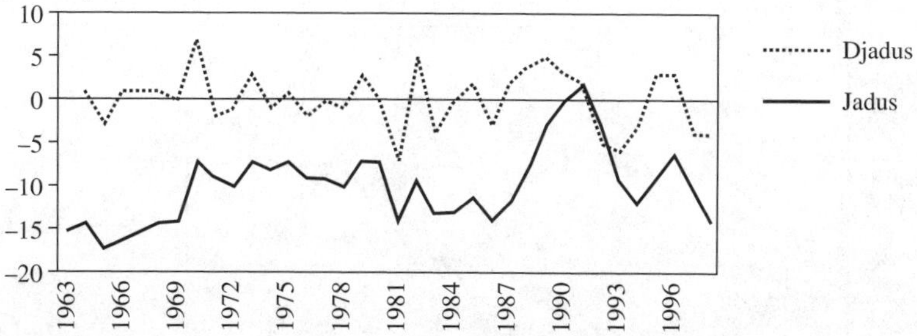

Professional equipment

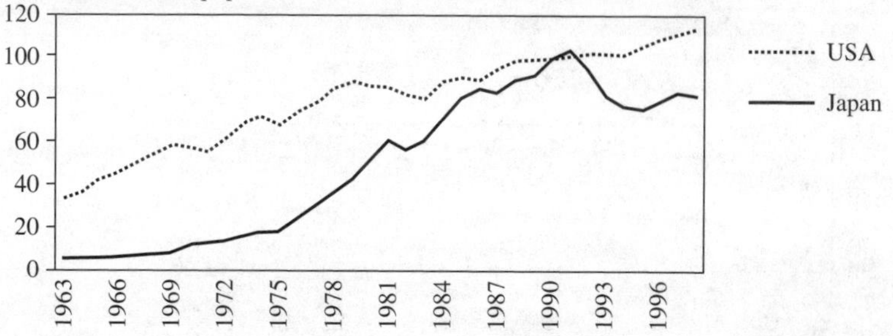

Other manufacturing

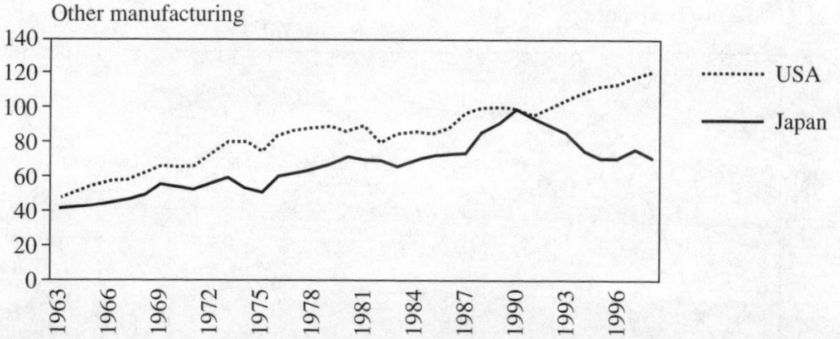

Printing and publishing

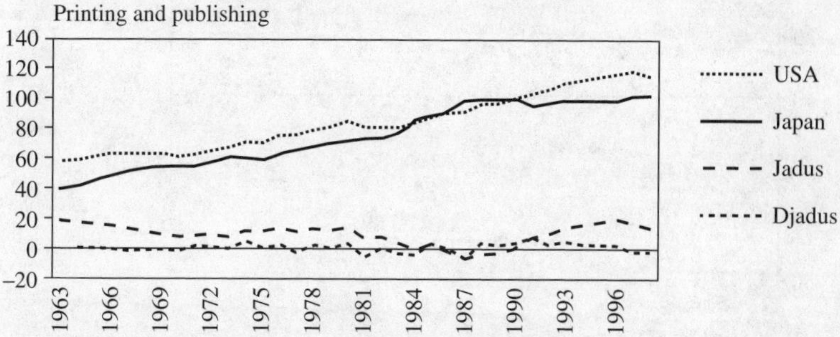

352, Other chemicals

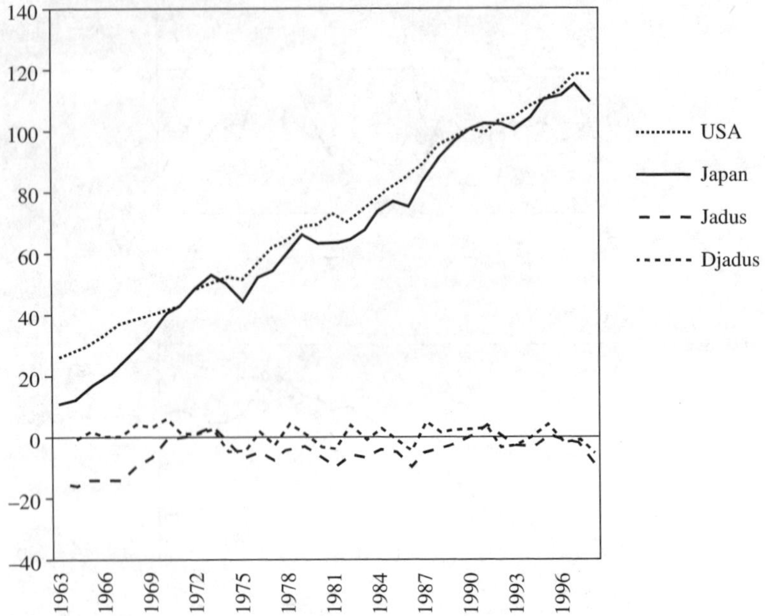

353, Petrol refineries

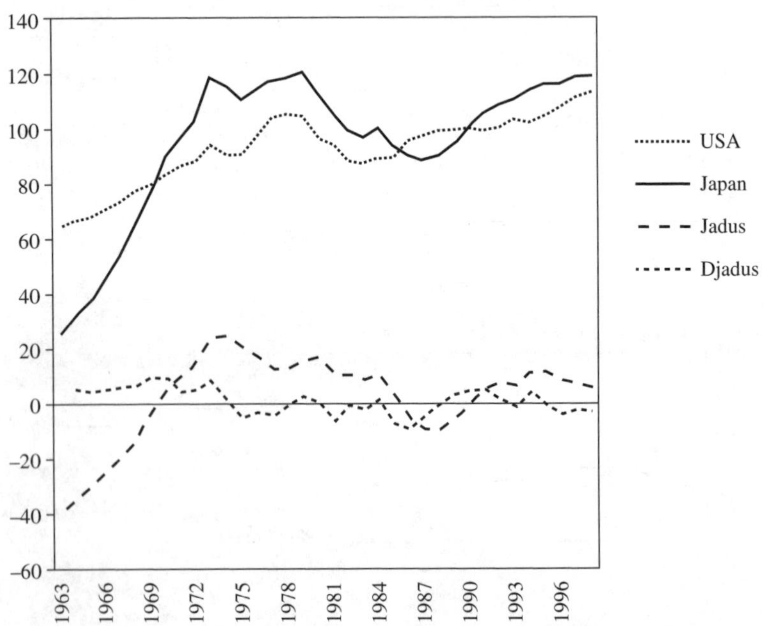

Note: For the significance of 'Jadus' and 'Djadus', see Chapter 6, sub-section 6.3.1.

APPENDIX 6.2: LIST OF MANUFACTURING SECTORS ON A THREE-DIGIT BASIS WITH THEIR CODE NUMBERS

311 Food
313 Beverages
314 Tobacco
321 Textiles
322 Apparel
323 Leather
324 Footwear
331 Wood products
332 Furniture
341 Paper products
342 Printing and publishing
351 Industrial chemicals
352 Other chemicals
353 Petroleum refineries
354 Miscellaneous petroleum and coal
355 Rubber
356 Plastics
361 Pottery and china
362 Glass products
369 Other non-metallic minerals
371 Iron and steel
372 Non-ferrous metals
381 Fabricated metals
382 Machinery (not electric)
383 Electric and electronic machinery
384 Transport equipment
385 Scientific equipment
390 Other

NOTES

1. This discussion is a logical continuation of the last section of the previous chapter.
2. Note that the difference in analysis between the two periods is basically due to the availability and validity of data; thus, a more rigorous analysis will be carried out for the period after World War II.
3. These principles were extensively analysed in Chapter 4.
4. TFP includes OIs, TIs, and other innovations, as explained in the previous chapter.
5. In Chapter 7 more precise relationships between TFP, capital, and other factors are provided.

6. These sectors have played a leading role in American economic history as will be explained below.
7. Thus, for 'industrial chemicals' there is cointegration between the series of the two countries whereas there is no cointegration for 'electrical machinery'. In Appendix 6.1, the graphs clearly show which sectors are cointegrated, that is, they show which sectors had very similar historical evolution in both countries, the USA and Japan. Table 6.3 presents a relevant classification. See also further below for more comments.
8. The machinery sectors, mainly computer related, certainly experienced strong continuous growth in TIs during the last 50 years or so. However, it was only in the third sub-period 1987–98 (see Figure 6.12) that these two sectors really boomed, thus supporting, even more, the assertion made here that OIs was one of the main reasons behind the extra strong growth of output or TFP during 1987–98 for these two sectors. This will be further shown in sub-section 6.3.4, and in Chapter 7. Even the role of a booming demand for computers can be indirectly attributed to the combined effect of, primarily, OIs, and TIs, because these innovations allowed for computer prices to decrease substantially, hence for demand to increase accordingly.
9. This acceleration in growth is due to the introduction of OIs. However, new OIs are needed all the time for the growth to be sustained (*ceteris paribus*).
10. Like in all industries, the TIs in Japan were mostly imported (see previous quote from Arora et al. (1999).
11. Fujimoto (1999, p. 104) considers four dimensions for a product to be competitive: productivity, throughput time, product quality, and flexibility. However, it should be added that all these dimensions – thus including quality – are strongly interrelated when considered in the context of the OI of JIT/QC, as explained elsewhere in this study and by Fujimoto (1999).
12. Even at present, many Japanese industries are still protected in various ways (Porter et al., 2000).
13. Lincoln and Nakata (1997) have similar views on most of these issues.
14. These systems included the autonomous complete assembly line (eventually combining the physical, functional, and organizational elements of the new system), the in-line mechanical automation concept (based on trial and error steps of implementation), the Toyota verification of assembly line (an ergonomics evaluation system, also mentioned in Chapter 5 as an example of measurement of the POM) and the low-cost equipment for better ergonomics (Fujimoto, 1999, ch. 7). As this author mentions, an intraorganizational evolutionary mechanism exists in Toyota which is still based on selection mechanisms generated at the bottom of the management hierarchy, or *genba*, and on a high level of standardization and documentation (ibid., pp. 268–9).
15. For all these concepts see Chapter 5.
16. Some propositions for changes are quite radical. For instance, Tezuka (1997) proposes that the intense competitive nature of many Japanese industries might be an impediment to economic growth. Other propositions are related to macroeconomic policy (see, for example, the liquidity trap argument developed by Krugman, 1998).
17. The sectors examined in this sub-section are: iron and steel, automobile, electrical and electronic. For the chemical industry see Chandler (1977; 1990), and Arora et al. (1999). It is also worth noting Chandler's (1994) American sectoral analysis that combined elements of TIs and OIs. However, this author did not include in his study the new tendencies in OIs that started taking place in the USA in the 1980s, namely the JIT/QC (as will be explained in Chapter 7). In this respect see also Abe (1997) and Lazonick (1992).
18. Toyoda is the original name of Toyota.
19. Langlois and Steinmueller (1999) presented a good analysis of the interplay between OIs and TIs. For instance, many relatively new and smaller specialized manufacturers entered the market when TIs were advancing very fast in the 1970s in the USA (ibid.); this was consistent with the strategy of 'core competences' (Prahalad and Hamel, 1990). See also Malerba (1985) for a good analysis of various aspects of OIs and TIs regarding this industry.

20. This story is based on Thomson and Gamble's (2001, pp. C-132–C-173) analysis of the evolution of this industry in the USA in the most recent period.
21. There are other issues such as the issue of demand that are not included in this brief historical account. A more comprehensive study should include arguments along the lines presented by Bresnahan and Malerba (1999).
22. Allusion is made primarily to the JIT/QC or LPS (lean production system).

7. Empirical evidence on the links between industrial sectoral growth and the lean production system (JIT/QC) in the USA

7.1 INTRODUCTION

In the previous chapter a general discussion was presented about the importance of manufacturing leading firms and leading sectors in Japan and the USA. This discussion was primarily descriptive. In this chapter empirical evidence will be more precise and thorough by considering a primary OI, that of JIT/QC. The latter also represents the main axis of OIs, and is mainly influenced by the POM.[1]

Milgrom and Roberts (1990) examined the complementary relationships between flexible manufacturing technologies, shorter production runs, more frequent product redesigns, lower inventories and other variables of modern American manufacturing, and found that these complementarities are an essential feature of a new era. In particular, lower inventories are an intrinsic part of all these complementarities. They summarized their research as follows (ibid., p. 511): 'Manufacturing is undergoing a revolution. The mass production model is being replaced by a vision of a flexible multiproduct firm that emphasizes quality and speedy response to market conditions while utilizing technologically advanced equipment and new forms of organization … '.

Ebrahimpour and Withers (1993) confirmed that a major shift in manufacturing philosophy from static optimization (initially in the USA) to dynamic evolution (Japanese type) occurred in the USA. In *The Machine that Changed the World*, Womack et al. (1990) describe the lean production system (LPS) that was devised and implemented in Japan in the 1960s and 1970s, and then transplanted in the USA in the late 1980s and 1990s.[2] Whether this system is replacing the mass production system, as the authors strongly suggest, or not, is debatable.[3] However, much research has taken place in the last 20 years to provide evidence of the benefits of the LPS to the Japanese and American economies; for instance, Womack and Jones (1996) provide several real cases, in detail, about the successful implementation and the derived benefits of the lean process. An equivalent system, which initially was extensively analysed by authors such as Schonberger (1982; 1986; 1996), and which is intrinsically related to the LPS is the Just-in-Time cum Quality Control (JIT/QC) system (already discussed in many parts of this study).

Kenney and Florida (1993), in their extensive analysis of the transfer of the Japanese system of production to the USA, summarized the LPS in three elements: efficient use of resources through the elimination of waste, low inventories, and just-in-time production and delivery practices (p. 8). In what follows reference is made primarily to the JIT/QC system, though this can easily be replaced by the LPS or any other equivalent system (in the sense that not all JIT/QC practices are always enforced, or that there are some extensions of the basic system as described elsewhere[4]). The transplanting of the initially implemented LPS in Japan to the American firms has been difficult and slow. It took about 10 years from the late 1970s before the American managers seriously decided to implement the JIT/QC system in a systematic and scientific way, and about ten more years before the actual implementation took place on a substantial scale. Several other significant studies have been produced recently about the whole process of transplanting the JIT/QC system successfully to other countries, for example Liker et al. (1999), Lillrark (1995) and Young (1992). A detailed analysis of the mechanism of this imitation process is of course outside the scope of the present study.

In this chapter the American manufacturing sectors are the focus of empirical analysis (see Sanidas, 2004b, the basis of this chapter). The hypothesis to test is that the LPS or its equivalent JIT/QC has been successfully implemented in the USA in the last 20 years or so.[5] A similar analysis could have been carried out for Japan; however, the relevant quantitative results will provide stronger evidence for this study if the empirical results support the imitator country (the USA) rather than the initiator country (Japan initiated the JIT/QC system). In this study more attention is paid to the intrafirm organizational features of this system, although the interfirm features are also indirectly explored since JIT/QC involves them as well; in addition, the interfirm features in terms of subcontracting relations are as important and more scrutinized in the literature.[6]

Section 7.2 establishes the crucial periods of implementation of JIT cum QC not only in the USA but also in Japan so that a better understanding of the transfer process from Japan to the USA is achieved. The quantitative evidence presented in section 7.3 uses several econometric tools, such as regression and error correction models, and data of varying length for TFP and labour productivity. Section 7.4 provides a detailed account of conclusions.

7.2 IMPLEMENTATION OF THE JIT AND QC SYSTEMS IN THE TWO COUNTRIES

7.2.1 Implementation of the JIT System

Abegglen and Stalk (1985) state that 'Toyota began development of the Just-In-Time system in the late 1930s and made substantial progress in its implementation in the 1950s and 1960s' (p. 93). Abegglen and Stalk (1985) also discuss the crucial relation between factory and suppliers: 'In Japan the pattern has been for the factory to implement JIT first, then for the suppliers to follow. By 1962 Toyota had instituted the JIT system-wide. Only then did it approach its suppliers. Another ten years passed before the JIT system had spread to all of Toyota's suppliers' (p. 115). Overall, it can be suggested that Japanese firms started imitating the pioneer Toyota in the system of JIT from about the early 1970s. What is the penetration of JIT usage in Japanese industries? According to Kim and Takeda (1996), a survey of 81 usable questionnaires revealed that 'consumer electronics' have the highest penetration (about 25–30 per cent), that larger firms are keener to adopt JIT practices and that the non-JIT firms are about 42 per cent. The average length of JIT implementation in Japanese firms is double or more than that of American firms (the average length for US companies was between 1.74 and 3.23 years as Kim and Takeda, 1996, identified). This difference, perhaps, shows that JIT in Japan is part of a more comprehensive and holistic approach to productivity increases than JIT in the USA. In this respect, it is also worth mentioning one of Kim and Takeda's (1996) conclusions: 'When the *rentai* relationships and *wa* culture is truly spread over the entire organization, the implementation of JIT can be successful' (p. 50).[7]

On the other hand, it was only in the 1980s that JIT became known and applicable by American firms (see, for example, the story of the Oregon Cutting Systems company, one of the pioneers of JIT in the USA, as described by Bailes and Kleinsorge, 1992). In his classic book on JIT written in 1982, Schonberger said: 'In one sense it is a bit early to write about just-in-time manufacturing management in the United States. JIT has scarcely been tested here … On the other hand, the Japanese already have at least ten years head start – and much more if one considers Japan's long history of adaptation to its environment of resource scarcities' (p. 83). When Billesbach and Hayen (1994) compared the effect of JIT over the long term for 28 American companies, they examined data between the last three years of the 1970s and the last three years of the 1980s. Overall, it seems that the Americans have been behind the Japanese regarding the implementation of JIT by around ten to 15 years, although this gap is now closing more rapidly.

From a study conducted by Baldwin and Gagnon (1993) some relevant results were obtained from their survey of 200 large Ohio firms (the state of Ohio is part of the Midwest, the industrial heartland of the USA). This sample contained both US and Japanese owned and operated manufacturing companies (for example, Honda of America).[8] These results show that, for example, there were 37 per cent of firms which practised the whole set of JIT principles, 36 per cent practised *ringi*,[9] whereas 81 per cent of surveyed firms used at least one principle of JIT. These figures were confirmed by another study conducted by Stamm and Golhar (1991), according to which 29 out of 77 respondents (that is 37.7 per cent) identified themselves as JIT firms.[10]

Billesbach (1991) analysed the reasons for implementing JIT in 68 American companies. About 50 per cent of respondents considered that JIT was taking place because their competitors were engaging in JIT. This shows that American businesses were seriously recognizing the importance of JIT by 1991. In a more recent study by White et al. (1999), the implementation of JIT practices in a large sample of American firms was much higher, ranging from 50 per cent to 92 per cent depending on the type of practice and the size of firms (p. 7); as these authors said: 'All the manufacturers included in the sample had implemented at least one of the JIT practices, and 96.5 percent had implemented at least three of the JIT practices. About 20 per cent had implemented all 10 JIT practices' (ibid., p. 6). These ten JIT practices included: quality circles, total quality control, focused factory, total productive maintenance, reduced set-up times, group technology, uniform workload, multifunction employees, *kamban* and just-in-time purchasing. It is worth noticing that in the USA the Bose Corporation successfully developed the system JIT II (an extension of JIT practices) during the period 1987 to 1993. Several other large American companies have implemented JIT II in their supply chains, for example, IBM, Intel, AT&T, Honeywell, Roadway Express, Ingersoll-Rand and Westinghouse (Krajewski and Ritzman, 1999, p. 748).

In another study, Daniel and Reitsperger (1991) compared 50 Japanese and 64 American electronics firms. They found the following results. First, in Japan, 45 per cent to 55 per cent of the operating managers received targets for both inventory units and costs, versus 15 per cent to 25 per cent in the USA, thus indicating a Japanese focus on inventory reductions. Second, overall, there was a strong confirmation of the authors' two hypotheses: first, a larger proportion of Japanese than US managers received specific performance goals related to inventory reductions and flexibility; and, second, there is a stronger correlation between the provisions of JIT-related performance goals and feedback in Japan than in the US reflecting the dynamics of working for perfection in production.

Just-in-time is mainly used in the continuous flow industries: iron and steel, metal products, consumer electronics and electrical machinery, automobile, precision instruments and chemical products (see, for instance, Billesbach, 1991; Billesbach and Hayen, 1994; Billesbach and Schniederjans, 1989; Billesbach, et al., 1991; Kim and Takeda, 1996; Meric et al. 1997). To a lesser extent JIT can also be applied in process facilities like those of a textile industry (Billesbach, 1994), or to any other industry which makes customized products such as furniture items (Golhar and Stamm, 1993). Regarding the chemical industry, a recent study by Meric et al. (1997), explains the low inventory level of Japanese chemical firms as the result of an extensive use of the JIT method in the *keiretsu* system. Figure 7.1 shows, in summary, the evolution of JIT implementation in the two countries.

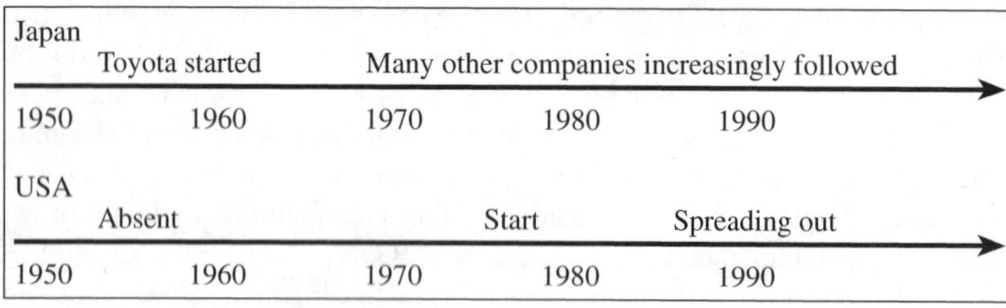

Figure 7.1 Chronology of implementation of JIT in the USA and Japan

7.2.2 Implementation of the QC System

For Japan, quality products are an outcome that took place only after the 1950s (see Chapter 2). Thus, 'by 1960, cameras, sewing machines and transistor radios were obvious success stories. The image of Japanese products as being cheap and shoddy was already being swept away in certain industries' (Nonaka, 1995, p. 542).[11] In Chapter 2 and in some other instances it was briefly described how this reversal of quality manifested itself from the end of World War II. Here, only QC circles will be briefly reviewed, for which data is available. According to Watanabe (1991), 'scientific quality control methods were introduced in pre-war Japan by the National Railway Corporation, navy arsenals and pioneering private sector companies such as Toshiba' (p. 61). In 1958, a study team brought back from the USA Feigenbaum's concept of TQC. In 1962 the JUSE launched its quarterly *QC Circle Magazine* and urged industries to organize workers into small study groups. By 1991, Watanabe remarked 'Japanese workers often discuss subjects related to their quality circle activities during a tea break, at lunch time or at an after-work social gathering' (p. 62). How much did QC circles spread in the Japanese economy after 1962? In 1965 there were 4930 QC circles and 70 920 participants, in 1975 the corresponding figures were 72 475 and 723 201, and in 1990, they were 313 924 and 2 454 635.[12]

Quality control circles are only one side of the quality movement in Japan. Total quality control started earlier and spread earlier. As Goldman (1993) said: 'The transported TQC philosophy radically transformed Japanese organizations (such as Toyota, Nissan, Sanyo, Matsushita, Mitsubishi) during the 1960s, 1970s and 1980s ... In sharp contrast, US and European implementation of TQC did not surface until the late 1970s and early 1980s' (p. 32). To illustrate this contrast, a comprehensive study sponsored by Ernst & Young and the American Quality Foundation in 1991 has confirmed Japanese superiority in terms of quality control vis-à-vis the USA, Germany and Canada (Bowles, 1992; Yearout, 1992).

Senge (1999)[13] has a similar view on the evolution of the quality movement in the USA and Japan. This author distinguishes three stages in quality as a learning process. The primary focus of the first wave has been the front-line workers, and the aim was to champion continual improvement and remove impediments such as quality control experts. This stage has been replaced by the second wave in Japan from the 1980s. The primary focus of the latter wave has been the managers themselves, and the aim was to shift the changes from improving work processes to improving how 'we work'. Quality circles and *kaizen* are an important part of this second stage. The third wave is a blend of the previous two waves. United States industry was operating primarily in the first wave by 1992, and hence the goal of continuous improvement was still an elusive target for most American corporations. To illustrate this elusive target we can use Schroeder and Robinson's (1991, p. 74) remark that, in 1986, Japanese companies received almost 48 million improvement proposals from their employees, while their counterparts in the USA received about 1 million. Kano (1993, p. 29)[14] expressed similar views to those of the previous two authors: 'during the past 10 years, American improvement efforts, including TQM, have gradually led to good results in some industries' (p. 29). For example, in the steel industry, in parallel with new TIs brought into the industry in the late 1980s, TCQ and a better-qualified workforce have been introduced and extensively implemented by most American steel-makers (Dinnen, 1992).

A detailed study of the evolution of the 'quality movement' in the USA in comparison to Japan by Cole (1998), clearly shows how the Americans have only very recently been trying to introduce and implement various quality control systems in the USA. According to Cole, 'quality means maximizing organizational behavior to enhance the satisfaction of present and potential customers' (ibid., p. 43). Furthermore, according to the same author, there are two ideal types or paradigms of quality, as summarized in Table 7.1. In 1980

Table 7.1 The quality paradigms

OLD PARADIGM	NEW PARADIGM
• Internal orientation stressing conformity to requirements • Just one of many functional specialties • Not seen as a competitive element as long as you match your competitors • A specialized function carried out by a small number of experts reporting to operations • Emphasis on downstream fixes • A limited repetitive cycle of detect and repair • A stand-alone effort, with each functional specialty acting to maximize its own goals • Based on whether a product or service is built or delivered according to agreed-upon standards	• Internalize external customer preferences • A common corporate-wide language of problem identification and problem-solving • A strong corporate competitive strategy • All-employee involvement in quality improvements • An upstream prevention focus • A well-defined problem-solving methodology • Training activities tied to continuous quality improvement • Integration of quality into the corporate-wide control system of goals, plans and actions • Emphasis on cross-functional cooperation to achieve quality improvement objectives • Anticipation of customer needs sometimes even before customers are aware of them

Source: Based on Cole (1998), pp. 43–4.

most large US manufacturing companies followed the old quality paradigm, whereas the leading Japanese firms had already begun to assume more of the features of the new paradigm, which was developed in Japan between 1955 and 1980 (ibid., p. 45). The author summarizes the situation in the early 1980s as follows:

> For a great number of large and medium size American manufacturing firms across a broad array of industries (from air-conditioning to autos, from consumer electronics to computers, from copiers to color televisions, from steel to semiconductor equipment, and from metal fabrication to machine tools), especially those subject to international competition, the pressures to find a competitive response

to the Japanese became incredibly intense. In almost all these cases, observers documented a major quality gap between US and Japanese companies in the early 1980s … (ibid., p. 45)

How did the American business world respond? Already, in the 1970s, Japanese competition in sewing machines, cameras, watches and colour televisions had provided the Americans with some good examples of the actual Japanese threat (for instance, the whole American colour television industry was lost to the Japanese during that period). The late 1970s and early 1980s was a period of denial, a slow theoretical and ineffectual debate between the gurus of quality movement (Deming, Duran, Crosby, Feigenbaum, Ishikawa), and a slow trial-and-error process of tactics for quality control (ibid.).

However, from the late 1980s and early 1990s more productive efforts took place, as testified by the proliferation of a large institutional network of users, consultants and other non-market players. Cole analysed seven such organizations: for example, the 1987 established Malcolm Baldridge National Quality Award, the American Supplier Institute whose activities took off in the late 1980s, and the American Society for Quality (ASQ). In particular, the ASQ membership grew fast, as shown in Table 7.2. This can be supplemented by evidence regarding the ASQ's readership (see, for example, ASQ's *Quality Journal*, July 2000). As Cole (1998) testifies, similar trends occurred with the other organizations.

Table 7.2 ASQ's membership

Year	Number of members
1979	32 000
1988	57 000
1997	133 000
2000	187 000

Source: Based on Cole (1998) up to 1997, and on ASQ (2000) for 2000.

At the end of this long process of the American quality movement, what are the results? Let Cole again summarize his findings (ibid., pp. 44, 70):

By the mid-1990s, US firms significantly narrowed (and sometimes closed) the quality performance gap with the Japanese across a broad range of manufacturing industries, such as autos and semiconductors … By the mid- and late-1990s, quality disappeared as a major topic in the media and was less and less a focus of

> top management's attention. This is a natural process manifested in the growing normalization of quality improvement as a management activity …

The conclusion of this brief historical account is that the USA's quality movement has been rapidly catching up with Japan's pioneering efforts to increase the quality of their products after World War II. This catching-up process has been accelerating over the last 20 years or so (in a similar way to JIT).

7.3 QUANTITATIVE EVIDENCE

7.3.1 Preliminary Issues

Based on previous chapters and, in particular, Chapter 4, where a detailed analysis of the JIT/QC system and its benefits was presented, Figure 7.2[15] summarizes some of the most important points. Figure 7.2 needs two extra clarifications. First, the LPS, or its corresponding JIT/QC, system need not be exactly the same as the one pioneered by the Japanese firms, but it can be any system close enough or equivalent to the original one.[16] Thus, it can include extensions, modifications, and so on, providing that the aim of all these alterations is to reduce wastes, increase quality, satisfy customers, reduce costs and reduce inventories. For example, Tomer (2001)[17] said: 'High-performance work systems are organizations that utilize a fundamentally different approach to managing than the traditional hierarchical approach associated with mass production/scientific management. At the heart of this emerging approach is a radically different employer–employee relationship' (p. 63).[18] On the other hand, Im and Lee (1989) mentioned the following 'JIT practices': small lot sizing, *kanban*, flexible workforce, JIT purchasing, dedicated lines, plant compression, quality circle, preventive maintenance, mixed model production, level production, U-shaped layout, cellular manufacturing and automation (all these practices in various ways

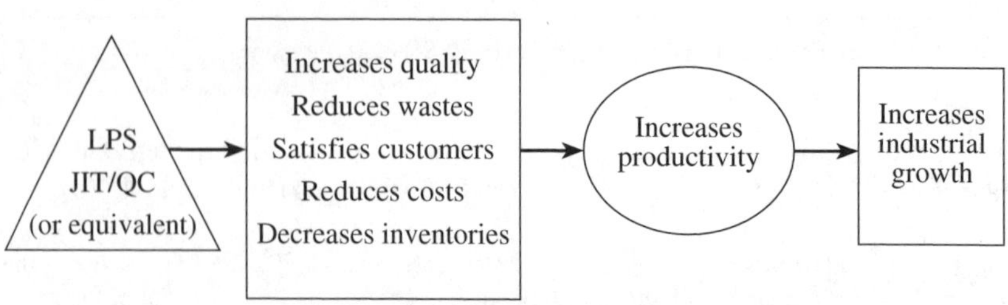

Figure 7.2 From JIT/QC to industrial growth

are, of course, part of the holistic JIT/QC system as explained, mainly, in Chapter 4).

Second, how is the JIT/QC system related to technical innovations such as the adoption and spread out of information technology (IT)? To answer this question reference will be made to an article written by Brynjolfsson and Hitt (2000), which reviewed in some depth the relation between IT, organizational transformation and business performance. These two authors' main conclusion is: 'In particular, both case studies and econometric work point to organizational complements such as new business processes, new skills and new organizational and industry structures as a major driver of the contribution of information technology' (ibid., p. 45). Thus, these writers continue, if computer investments take place without a parallel organizational change, they can create significant productivity losses (ibid., p. 25). For example, Dell Computers and MacroMed successfully combined the principles of the JIT/QC system with the necessary IT investments to, overall, become flourishing in their own business (ibid., pp. 27, 29). In addition, it is not even clear to what extent IT has a positive impact on productivity; in this respect, Solow's famous paradox (that we can see the computer age everywhere except in the productivity statistics[19]) perhaps still holds in many cases. For example, Brynjolfsson and Hitt (2000) wondered whether we should be disappointed by the productivity performance of the downstream firms that have invested in IT. Also, Gordon (2000) arrived at the interesting conclusion that in 88 per cent of the American economy the effects of IT on productivity growth are surprisingly absent. We should also recall that the Japanese firms adopted the LPS without the assistance of personal computers and the Internet in the 1960s and 1970s. Schonberger (1982, p. 15) said in this respect: 'Just-in-time production is simple, requires little use of computers, and in some industries can provide far tighter controls on inventory than are attainable through US computer-based approaches … '.

A crucial point must now be made, as this was identified and stressed in previous chapters (mainly chapters 4 and 5), and as shown in Figure 7.2. Two of the most evident and measurable effects of implementing the JIT/QC system are a corresponding continuous reduction in inventories and a continuous increase in product quality. These reductions are due to many reasons at the same time, because this system affects the whole firm and many other firms via the supplier network. These reasons are the consequences of the holistic JIT/QC process: for example, decreases in the lot size, increases in the repeatability of tasks, flexible automation, reduction in supply uncertainties, improvement in demand forecasts, and so on (cf. also Krajewski and Ritzman, 1999, chs 13 and 16, for a pertinent analysis). Thus, the LPS or JIT/QC is not just one factor, but is a holistic process that encompasses all areas of firm operations (the four PROSIBB of Chapter 5[20]). Hence, unless we suggest that there is another system of firm activities (PROSIBB)[21] that also reduces inventories and increases product quality, the present analysis is valid. At the moment, such a competing system does not exist.

Out of the two main consequences of JIT/QC, only the reduction of inventories, as a ratio to sales (see below for more details), will be used in this study. Data for increases in product quality are not available, especially on a sectoral basis. For an appreciation of how the reduction in inventories is a consequence of the theoretical developments of previous chapters, see Figure 7.8 at the end of this chapter. Furthermore, a significant reduction in inventories in the long-run trend is the important issue here; hence fluctuations due to the business cycle and other factors of a shorter-term length do not affect this trend (see further discussion in sub-section 7.3.2). This longer-term preoccupation links well with significant overall organizational changes such as the adoption of the JIT/QC system.

The empirical relationship between inventories and the JIT/QC system will be thoroughly tested in this section. Only very recently has such a relationship been tested by other researchers in a more systematic way. Thus, Lieberman and Demeester (1999, p. 466) in their abstract said: 'The literature on JIT production suggests a causal link between work-in-process inventory and manufacturing productivity. Such a connection has been described in numerous case studies but never tested statistically. This paper uses historical data for 52 Japanese automotive companies to evaluate the inventory-productivity relationship ... '. These two authors used mainly Granger causality tests to find that a significant elasticity of about –0.1 applies to that relationship. In this section a more general and comprehensive evidence of causality between industrial growth, as represented by total factor productivity (TFP) or labour productivity, and the LPS or its equivalent JIT/QC system, as represented by the inventory to shipments ratio, will be explored.

Another recent study by Lieberman et al. (1999) has attempted to answer the following questions: 'What factors determine the inventories held by manufacturing companies? Does classical inventory theory explain the levels of inventory held in practice? Do Japanese management methods, involving problem solving by shop employees and frequent communication with customers, lead to lower inventories?' (p. 44). The authors' findings are summarized as follows: 'The data show that inventory levels are influenced by set-up and holding costs, production lead times, the extent of customer communication and the involvement of employees in problem solving' (ibid., p. 52), thus providing econometric evidence that the JIT/QC system has considerably reduced inventories in manufacturers of automotive parts in the USA.

It is also worth mentioning an original study undertaken by Baily and Gersbach (1995), who provide quantitative evidence of the importance of work organization in selected manufacturing sub-sectors in the USA, Germany and Japan; the authors referred in particular to the LPS as part of organizational factors that positively impact on labour productivity (for instance see pp. 330–32 of their paper). These authors' findings, in particular the factor entitled 'organization of functions and tasks' (for example, see their graph on ibid., p. 319), which is relevant to OIs,[22] provide evidence of the causes of labour productivity differences in the production process. Some extra relevant references are also worth mentioning here: Callen et al. (2000) in their cross-sectional plant analysis showed that JIT manufacturing is associated with greater plant productivity. Miltenburg (1993) showed in a theoretical model that JIT reduces costs and inventories. Jewkes and Power (1993) in their theoretical microeconomic analysis of investment in JIT manufacturing concluded that JIT firms tend to increase their market share through decreased costs. O'Grady (1988, p. 115) in his case studies concluded: 'The costs associated with each of the implementations are low relative to the returns … '.

Three more remarks will close this introductory sub-section. First, the cross-sectional data (either straight or pooled) analysis circumvents a potential danger: some external factors such as growth in the service sub-sectors are approximately common to all manufacturing sub-sectors and hence do not influence the results. However, in some of the models external or exogenous factors are used, such as the rate of interest, unemployment and so on, in order to complement the analysis.

Second, the dependent variables TFP or labour productivity contain, or imply, not only organizational innovations such as the JIT/QC system but also other types of innovations such as technical or wisdom innovations (see Chapter 5). Thus, an attempt is made to incorporate the impact of technical or embodied innovations on productivity. However, the main purpose in the present exploration is to indicate that OIs such as the JIT/QC holistic system play a (concomitant) role in explaining differences in productivity changes across manufacturing sub-sectors. The proxy used to represent the JIT/QC system, inventories to sales ratio (more on this ratio in the next sub-section), is a direct consequence of this system (as has been shown in several parts of this study). Just-in-time/quality control and the path of this ratio involve all aspects of firm activities (the four PROSIBB of Chapter 5), but in particular they involve the process of movements which influences OIs (Chapter 5 discusses all these points in detail).

Third, the econometric exploration presented in this chapter cannot, in itself, be properly evaluated unless it is supported by the findings of previous sections and chapters. For example, if the JIT/QC system is considered without the theoretical considerations previously developed, and without the evidence already provided by other researchers on this topic, the present quantitative analysis cannot be properly assessed. It is worth summarizing the important findings of all previous sections and chapters that support the econometric analysis carried out in the next sections:[23]

1. The JIT/QC system (a basic description is given in Chapter 4) is a complex, evolving and emergent one that is part of a more complex system involving other dimensions such as sectoral systems and innovations systems (see Chapter 5). It is also part of the main axis of OIs, thus emphasizing internal organizational features that link this axis to the other two axes (Chapters 4 and 5).
2. The JIT/QC system is a major OI, based heavily on the POM. The latter conveys the importance of differentiation and elaboration of internal organization necessary for an open system to become negentropic (see Chapter 5).
3. A major consequence of the negentropic nature of the JIT/QC system is the reduction in inventories and defects, and a parallel increase in productivity.[24] Thus, the JIT/QC system (or LPS) justifies the inclusion of OIs into the process of economic growth (see previous chapters).
4. Examination of some cases of firms and sectors (see Chapter 6) provided evidence that JIT/QC increases productivity and reduces inventories (the major examples used are Toyota motors and Dell computers).
5. Previous econometric studies on a firm basis established that there is a significant elasticity between increases in productivity and reduction in inventories.

In this section, five different approaches, or broad models, will be developed, starting from the simplest (simple ordinary least squares regressions) and finishing with the most complicated (vector error correction model). Thus it becomes possible to cross-check each model's results with as many other different models' results as possible.[25] Consequently, this section becomes a thorough empirical investigation of the hypothesized relation between industrial growth in the USA and JIT/QC. The first and simplest approach will be most scrutinized, as it will lay the foundations for all other approaches. Also, the most complex and last approach will be more scrutinized, as it will thoroughly confirm the overall results. Finally, note that the number of data used in different approaches varies from 12 to 28 to 85 to 420 and, finally, 1008 (the reasons for the choice of the appropriate number of observations will become clear in the sub-sections below). Finally, Appendices 7.1 to 7.3 contain more information on the data used in this chapter. Note that not all of the original data and derived data are included in these appendices, as this would have required additional hundreds of pages.

7.3.2 The Productivity Issue and OIs Proxy Variables

Two databases are used: the UNIDO ISIC three-digit sectoral data from 1963 to 1997 (updated version 2000) and the National Bureau of Economic Research (NBER) four-digit 459 sectoral data from 1958 to 1996 (June 2000). The United Nations Industrial Development Organization (UNIDO) data are supposedly consistent between countries and are recorded according to the International Standard Industrial Classification (ISIC) three-digit classification (28 manufacturing sectors). They include a series for employment (E), value-added (VA), index of industrial production (RO), nominal output (NO), and nominal wages and salaries (W). For the calculation of TFP based on UNIDO (2000) data, the real capital stock (CS) from the NBER publication (2000) was used, and the intermediate inputs or materials (M) were computed as the difference between output and value-added. For the calculation of TFP based directly on NBER data, the initial 459 sector TFP series was transformed into an 85 sector and then into a 28 sector TFP series (to match as much as possible the 28 ISIC sectors) (for more details see also Appendices 7.1 to 7.3).

The TFP (UNIDO based) was calculated for each year from 1964 to 1996 according to the following formulation:

$$TFP' = RO' - w_1 E' - w_2 M' - (1 - w_1 - w_2) CS' \qquad (7.1)$$

The variables, TFP', RO', E', M', CS', … are computed as the first differences in natural logarithms of these variables, thus expressing rates of growth. The weights w_i are the averages of each pair of years for which TFP is computed. For example:

$$w_{t1} = ((W/NO)_t + (W/NO)_{t-1})/2 \qquad (7.2)$$

These formulae are consistent with Jorgenson's work (1990; Jorgenson et al., 1995; Jorgenson and Stiroh 2000); thus, one of the assumptions underlying equations 7.1 and 7.2 is that factors are paid their marginal products, inferring that there are constant returns to scale; otherwise, if there are increasing returns to scale, then equation 7.1 leads to an overestimate of TFP. This is not an undesirable result to obtain in the present analysis as the aim is to pinpoint sectors that experience increasing or decreasing returns to scale. Hence, an overestimated TFP shows an increasing returns to scale sector which grows very fast in terms of productivity.[26] Also, the empirical analysis in this section uses cross sectional data, the assumptions and consequences are the same across all sectoral data, and thus there is no major or significant flaw or bias in the quantitative results. With these formulae, TFP is calculated according to the gross output formula so that it includes the impact of materials and not according to the value-added formula that excludes the impact of materials. However, the results and conclusions presented here are not significantly different from the distribution (across sectors) of TFPs viewpoint, when using either of these formulae.[27]

It is also necessary briefly to interpret the concept of TFP in terms of its relations with capital, production and labour. It can be easily shown that

$$\left(\frac{Y}{L}\right)=\left(\frac{K}{L}\right)+\left(\frac{Y}{K}\right) \tag{7.3}$$

and

$$\mathrm{TFP}=\left(\frac{Y}{L}\right)-\beta\left(\frac{K}{L}\right)=\left(\frac{Y}{K}\right)+(1-\beta)\left(\frac{K}{L}\right) \tag{7.4}$$

where TFP and all terms in brackets are expressed as growth rates (first differences in natural logs). The constant β is capital's (K) share in total revenue, Y is production and L is labour. Equation 7.3 can be interpreted as follows: Capital productivity (*Y*/*K*)[28] expresses (is a proxy of[29]) disembodied technology, or OIs in the present analysis, whereas (*K*/*L*) expresses embodied technology, or TIs in the present analysis (for a good reference regarding this distinction, see Reati, 2001). Consequently, according to equation 7.4 TFP is either the difference between labour productivity (*Y*/*L*) and a fraction of embodied technology (*K*/*L*), or the addition of disembodied technology (*Y*/*K*) and a fraction of embodied technology (*K*/*L*). (An immediate application of these formulae will be shown in sub-section 7.3.3 below.)

Furthermore, labour productivity can be split into two parts (Fujimoto, 1999, p. 331):

(work hours)/(unit) = (work hours)/(value-adding time) × (value-adding time)/(unit) = factor productivity = density of transmission × speed of transmission. Where work hours = value-adding time + non-value-adding time = average cycle time × number of units produced.

Both density and speed of transmission parts of factor productivity apply to both labour and capital productivity. The density of transmission ((work hours)/(value-adding time)) includes the following elements: reducing workers' idle time waiting for the next material to come; walking time for picking up parts and tools; machine breakdown time; tool change time; air-cutting time; and so on. All these elements are primarily related to the process of movements and hence to OIs (as determined in Chapter 5). The speed of transmission ((value-adding time)/(unit)) includes the elements: increasing the worker's standard speed of motions; assembly-line speed; stamping strokes per hour; the cutting and feeding speed with improved tools, materials and machines; and so on. Most[30] of these elements are linked to TIs-related technology (process technology according to Edquist et al.'s, 2001 terminology). One of the main conclusions Fujimoto (1999, p. 111) reached in his Toyota manufacturing system study is the following: 'Traditional American auto factories of the 1970s tended to emphasize the "speedup" (i.e., compression of value-adding time itself) approach and thus the technology-driven approach of mass production. The Toyota-style approach, on the other hand, often brings about significant productivity increases without introducing new production technologies … '.[31] Thus, Japanese makers, who adopted the Toyota production philosophy, emphasized the density approach based on improvements of the POM: 'This latter approach – increasing information transmission density – is the heart of the Toyota-style manufacturing system and the source of its productivity advantage' (ibid., p. 105). This is also the heart of the atomic/systemic model of economic growth presented in Chapter 5, and for which quantitative evidence is provided in this chapter.

Finally, in this introductory sub-section, the proxy of the LPS or its equivalent the JIT/QC system is further explored. As already indicated in previous sub-sections, the reduction in inventories, as one of the two main consequences of the JIT/QC system (the other one being the increase in product quality), will be used in this study to represent the impact of this system on sectoral growth. The variable ratio of inventories to shipments is preferred to that of inventories so that firm size can be taken into account, and thus a bias can be eliminated from the empirical analysis.[32] The use of this proxy follows from the extensive discussion in previous chapters and sections, where it was found that the JIT innovation, and hence reductions in inventories, could only be successful if it is implemented within a more general framework of JIT, TQC, Kaizen and employees' active participation.[33] In other words this proxy is an inherent and integral part of the LPS. In brief, the end-of-the-day aim of JIT management is to reduce inventories as much as possible.

Figure 7.3 exhibits some selected sectors in terms of their inventories to shipments ratio (IRY) from 1958 to 1996. Although not all sectors are shown in Figure 7.3 for clarity reasons, the tendency has been the same for all of them. That is, either a reduction in the ratio of inventories to shipments consistently since the middle of the 1980s or at least a constant ratio through time.[34] In Figure 7.3 one can also see that the impact of the so-called business cycle on the inventories ratio was not as significant in the 1990/91 recession as it was in previous recessions (for example, 1974/75 or 1981/82). In addition, following the early 1980s, and through the most recent recession 1990/91, the inventories ratio has been steadily decreasing.[35] The declining inven-

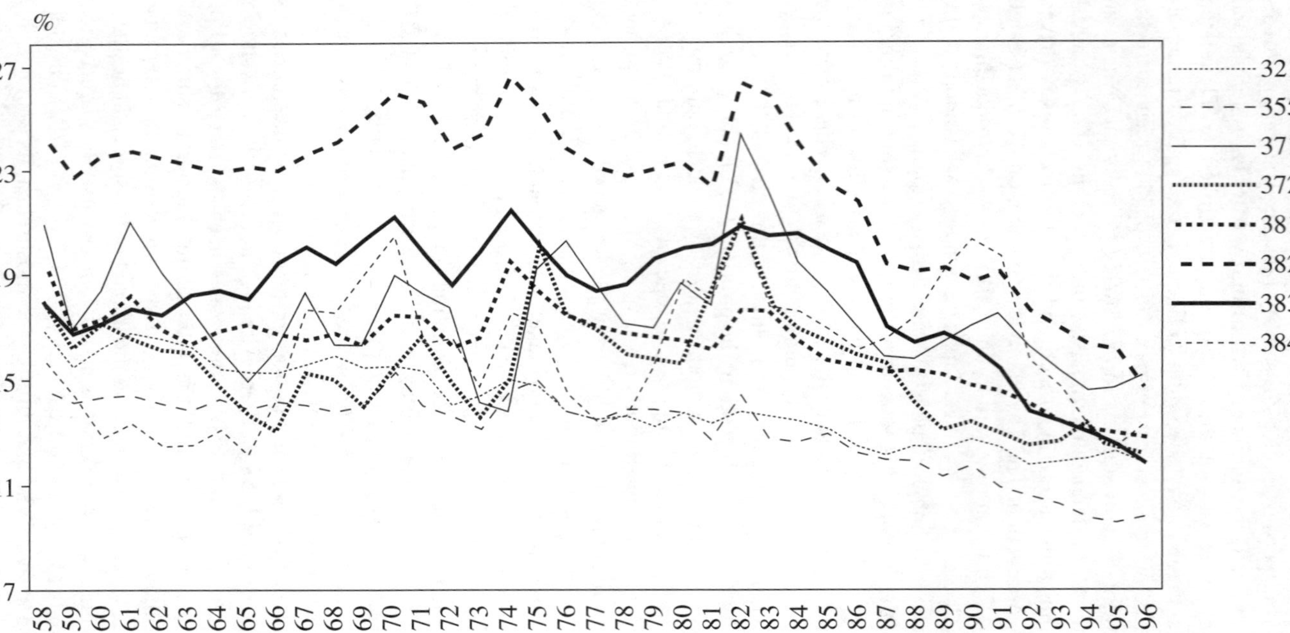

Note: For the corresponding names of sectors see Appendix 6.2.

Source: Author's calculations based on the NBER (2000) database.

Figure 7.3 USA, inventories to shipments ratio, 28 ISIC sectors

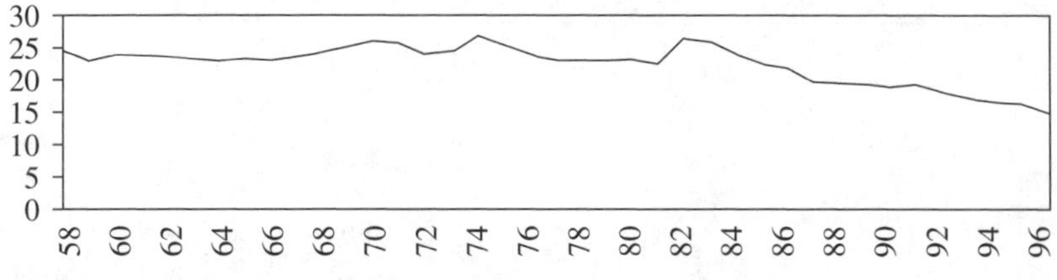

Source: Author's calculations based on the NBER (2000) database.

Figure 7.4a USA, sector 382, inventories to shipments ratio (percentage)

tories ratio from the early 1980s till 1996 can be more clearly seen if we consider the sector 382 (machinery, not electric) in Figure 7.4a. This visual observation can be verified with two simple OLS regressions shown below. In brackets under the coefficients are the t-statistics, which clearly show that the ratio IRY definitely declined through the whole period from the early 1980s till 1996 (the negative and strongly significant coefficient of the variable YEAR indicates this decline):

For the period 1958 to 1984: $IRY = \underset{(9.8)}{21.4} + \underset{(1.2)}{0.04} \times YEAR$

For the period 1985 to 1996: $IRY = \underset{(14.8)}{73.2} - \underset{(-11)}{0.60} \times YEAR$

A similar result is obtained by examining the total manufacturing data. In Figure 7.4b we can see that the continuous decline from the early 1980s is significant and 'unusual'.[36] The thick line is based on consistent data from 1947 to 1997 (see the source and notes to Figure 7.4b). The thin line is based on an indirect way of estimating shipments. The almost parallel thick and thin lines from 1947 to 1970 show that the estimated thin line is valid. Overall, this graph shows that from 1929 to about 1940 the inventories ratio has been influenced by the consequences of the Great Crash and the Keynesian policies of the then American government. The inventories ratio has been strikingly constant (except for short-term two to four-year swings[37]) from the late 1940s to the early 1980s (that is, for more than 30 years).[38] So what happened suddenly from the early 1980s when that ratio started declining in a steady way? It has already been indicated several times in this book that the answer lies in the gradual implementation of the holistic JIT/QC system in the USA from the early 1980s.[39]

In order to see the relationship between productivity and JIT/QC through its proxy, the inventories to shipments ratio, Figure 7.4.c shows three vari-

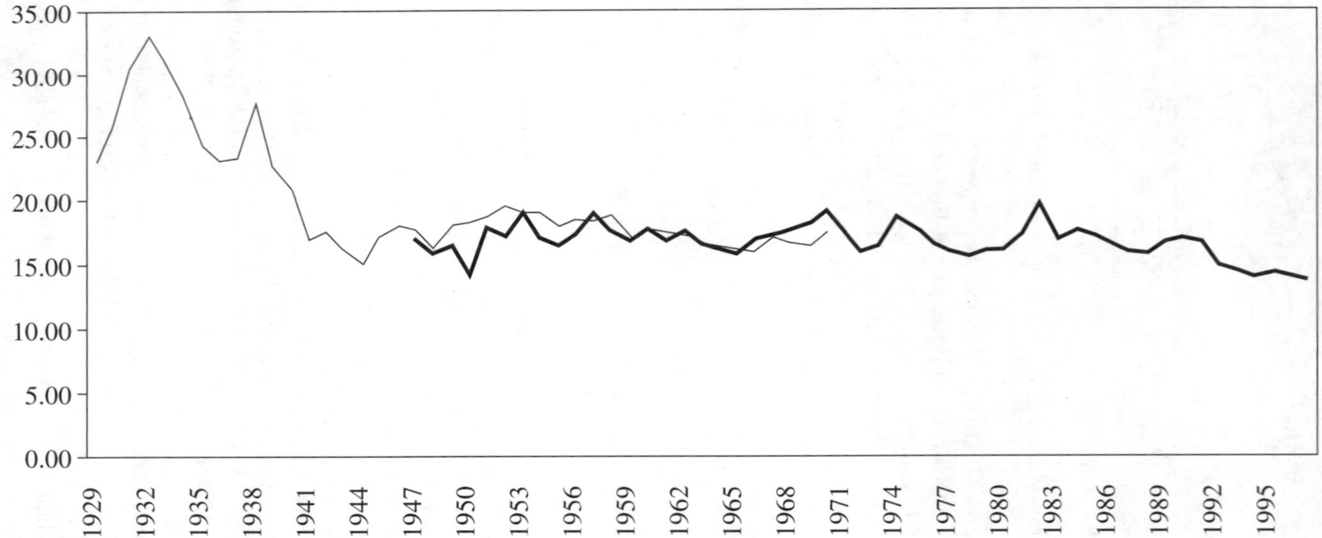

Source and notes: For the period 1929–70, the Tables E-24 (wholesale price index, industrial commodities), P-13 (index of manufacturing production) and T-70 (manufacturers' total book value inventories) of the Bureau of the Census (1975) are used to estimate the ratio of real inventories (T-70 over E-24) to shipments (P-13) (this ratio is multiplied by the constant 20 to be comparable to the other line on the graph). This ratio is shown as a thin line on this figure (stopping at 1970). The thick line shows the ratio of inventories to shipments provided by the Bureau of Census, *Abstract*, various issues, for example, in Table 1308 of the 1991 issue.

Figure 7.4b USA, manufacturing total, inventories to shipments ratio, 1929–97 (percentage)

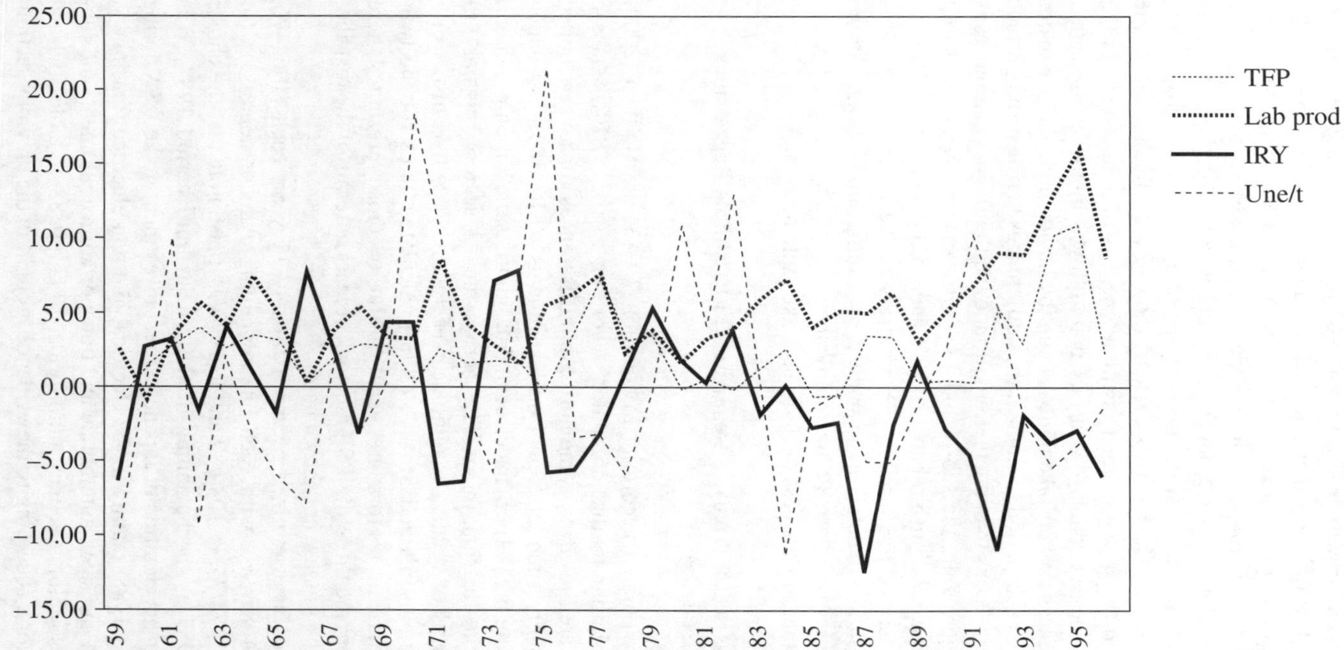

Source: Author's calculations based on the NBER (2000) database; for the unemployment series see Slavin (1999).

Figure 7.4c *USA, sector 383, TFP, labour productivity, inventories to shipments ratio and unemployment (total USA), in percentage changes*

ables as percentage growth changes from 1962 to 1996, namely TFP, labour productivity and IRY, for the dynamic sector of machinery (ISIC 383) in the USA. Also, a good proxy for the business cycle, namely unemployment, is included in the same Figure 7.4.c for comparison purposes.

Once more we can see that from the early 1980s IRY started declining (as a trend) and, at the same time, the two measures of productivity started climbing (as a trend). The business cycle as shown by the fluctuations of the three series, and the proxy of total unemployment is important only in the short term. In the econometric analysis that follows, in the cross-section models the business cycle is taken into account by having included one major recession in each sub-period examined, thus differences between sub-periods are due to other reasons than fluctuations around the trend. For example, as is apparent in Figure 7.4c, the sub-period 1987–96 examined below picks up the strong inverse relationship between IRY and productivity through a significant coefficient of the variable IRY, contrary to the other periods. This will be further shown in detail in the analysis of ensuing sub-sections. For the pooled time series, the variable unemployment is explicitly included in the regressions so that the business cycle is also taken into account.[40]

7.3.3 A Cross-sector Analysis Regarding TFP: Simple Regressions (N = 12, 28, 85)

The TFP, as calculated according to equation 7.1 for the UNIDO (2000) data, is very similar to that computed according to the NBER database (2000). There are, however, some noticeable differences for a very limited number of sectors such as 'scientific instruments' (SIC 385), for which there is a definite break (perhaps error in data) in the series of employment. Nonetheless, the distribution of TFP across the 28 sectors remains unchanged in terms of relative magnitude (see Figure 7.5). It is also worth noting that Jorgenson's and Stiroh's (2000) calculated TFPs for the two-digit SIC American sectors are very similar to those presented here (calculated from the three-digit SIC sector), at least in terms of the distribution of their magnitudes (not shown here).

Three periods are examined of approximately 11 years each (1964–76, 1977–86 and 1987–96), with each one containing a major depression (or recession) (in 1974/75, in 1980/82 and in 1990/91); also note that the choice of the year 1987 as the beginning of the crucial third period 1987–96 is supported by an acceleration of Japanese direct investment in the USA during 1986, 1987 and 1988, mainly in the electrical and non electrical machinery, primary metals and steel, automobile and transportation equipment, and rubber and plastics sectors (Kenney and Florida, 1993, pp. 88–9, tables II.1 and II.2). Furthermore, according to the evidence brought in the previous section,

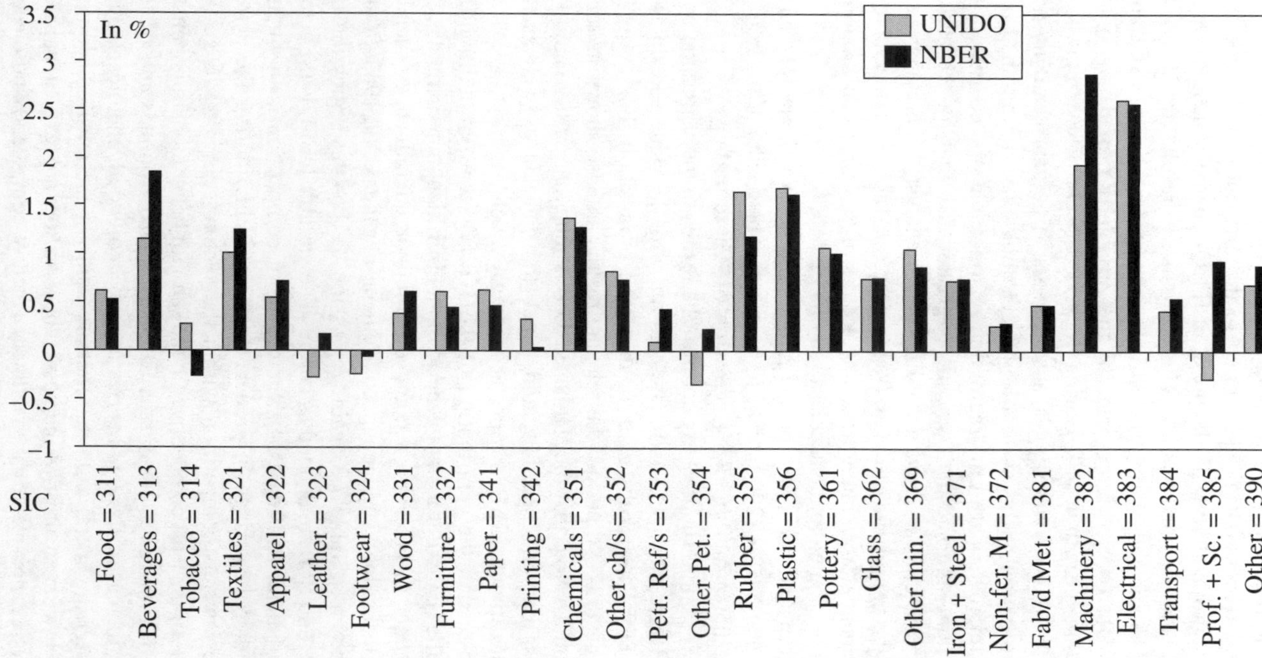

Source: Author's calculations based on the UNIDO (2000) and NBER(2000) databases.

Figure 7.5 USA TFP 28 SIC sectors, 1964–96

an increasing number of American firms started implementing JIT/QC in a more systematic way from the second half of the 1980s.

The simplest regression to be explored in the first place is the one that relates sectoral productivity with the proxy of JIT/QC, namely the inventories to shipments ratio, denoted as IRY. The econometric results are shown in Table 7.3.[41] As expected, during the period 1964–76 (regression 1 of Table 7.3), the LPS (JIT/QC), as represented by the variable IRY, made no contribution at all to the growth in TFP. During the period 1977 to 1986 (regression 2 of Table 7.3) JIT/QC was almost absent in the American economy and hence no contribution of this system to the TFP took place during this time, although the improvement in the regression is noticeable (the coefficient of IRY has the right sign and the t-statistic is greater than that for the period 1964–76). On the contrary, and according to the analysis of previous sections, during the period 1987 to 1996 (regression 3 of Table 7.3), the JIT/QC system implementation in many firms and sectors of the American economy significantly contributed to the acceleration of TFP.

In a similar manner the TFPs for 85 sectors are used, which are calculated from the original NBER (2000) 459 SIC sectors.[42] The results of the regression are also shown in Table 7.3 (regression 6). These findings for a much larger (N = 85) sample significantly confirm the results of the regression with 28 sectors.[43] Regressions 1, 2 and 3 are also run with the value-added based formula for TFP. The results confirm the same proposition. For example, considering the 1987–96 period, the relevant results are also shown in Table 7.3 (regression 5). Note that the VA-based TFP is much larger in magnitude than the gross output based TFP, and hence the constant as well as the standard error estimate (SEE) are also accordingly much larger.

In this sub-section the contribution of the JIT/TQC will, in addition, be examined in the context of a more general model in which more variables are included in order to take into account other important influences such as competition, technology and labour replacement. For the period 1987–96 the results are shown in Table 7.3 again (regression 4). For this regression the variables TFP and IRY are defined as before. The variable DEFL is the actual gross output deflator and represents the degree of competition in the industry.[44] The higher the change in producer prices (DEFL) during the period 1987–96, the lower is competition within the sector (*ceteris paribus*). For instance, some of the least competitive (in the sense of being more oligopolistic in nature or facing fewer substitutes) industries also exhibit some of the highest price increases; these industries are tobacco, wood, chemicals, plastics and non-ferrous metals. On the contrary some of the most competitive sectors show low levels of price increases; these sectors are textiles, apparel, fabricated metals, non-electrical machinery (mainly computers) and electrical machinery (mainly semiconductors).

Table 7.3 Simple cross-sector OLS regression results

Dependent	**TFP 1964–76**	**TFP 1977–86**	**TFP 1987–96**	**TFP 1987–96**	**TFP 1987–96**	**TFP 1987–96**
	(UNIDO derived TFP)				(VA based)	(NBER based)
N (no of data)	28	28	28	28	28	85
Regression no.	1	2	3	4	5	6
Intercept	**0.61**	**0.53**	**0.37**	**1.84**	**1.56**	**0.27**
t-stat	4.4	2.7	1.3	8	4	1.56
IRY	**0.034**	**–0.113**	**–0.47**	**–0.39**	**–0.48**	**–0.497**
t-stat	0.31	–0.84	–3.4	–5.1	–2.4	–6.58
DEFL				**–0.49**		
t-stat				–9		
LAB				**–0.086**		
t-stat				–1.65		
DUM				**2.49**		
t-stat				3.9		
R^2	0.004	0.027	0.3	0.86	0.19	0.34
SE	0.73	1.05	1.16	0.56	1.65	1.37
SD of dep/t	0.72	1.05	1.36	1.36	1.79	1.68
Diagnostic tests (p-values)						
Serial cor/n	0.21	0.54	0.9	0.29	0.76	0.28
Functional	0.18	0.18	0.01	0.16	0.35	0.000
Normality	0.001	0.91	0.43	0.87	0.9	0.000
Heteros/ty	0.48	0.78	0.000	0.46	0.5	0.000

The variable LAB shows the changes in labour input (employment) over the period 1987–96. It represents two tendencies: first, the internal (within the firm) changes in organization of labour and capital (for example, less labour is needed because of the implementation of the JIT system),[45] and second the external (outside the firm but within the sector) changes in organization (for example, more or less labour is needed because of more or fewer firms operating within the sector). A perusal of the relevant data shows that the competitive industries of all types of machinery (several SIC sectors) either increased employment only marginally or even reduced employment (as in the electrical machinery sector).[46]

The variable DUM is a dummy to take into account the errors in the data for the sector 385 (precision instruments) as mentioned above. The exclusion of this sector leaves the other coefficients of regression 4 unchanged, which further confirms the peculiar 'behaviour' of this industry.[47] Also note that both the employment data from the NBER and the UNIDO bases were used in regression 4, in order to check the validity of the results. The latter remained unaltered despite some differences in the two series (again the sector 385 exhibited the greatest difference between the two labour series). Finally, it must be emphasized that the coefficient of the variable IRY is much more significant in this enriched regression 4 than in the previous simple regressions. This, once more, confirms the importance of the proxy for JIT/QC, as it has been analysed in detail so far.

In order to supplement the last model in which some extra variables were introduced, another dimension will now be briefly examined. The number of patents, a good proxy for TIs, by industry in 1990[48] is regressed against the TFP annual growth for 1987–96 (the dependent variable as before), and the IRY variable (as before). Patents are grouped in 12 manufacturing sectors, instead of 28 (due to the restrictions of available data on patents), so the regression also has 12 cross-sectional data (food and beverages, textiles, all chemicals, all petroleum, rubber and plastics, pottery and glass and clay, primary metals, fabricated metals, non-electrical machinery, electrical machinery, transport, and professional and scientific equipment). The results are shown below (T-statistics are under the coefficients):

$$\underset{}{TFP} = \underset{(0.06)}{0.021} - \underset{(-3.91)}{0.68\ IRY} + \underset{(1.87)}{0.793\ PAT} \quad R^2 = 0.81$$

The variable PAT (patents) is only significant at the 9.5 per cent level, whereas the IRY variable remains strongly significant as before.[49] The poor result of the patents variable representing a proxy for technical innovations is not only due to the small sample, and probably collinearity, but also to more substantial reasons: industries like the chemical ones, with a traditionally high rate of technical innovations in new products, have only a relatively low TFP growth, because of their poor organizational innovations in terms of JIT/QC. In the last sub-section below, capital deepening, as represented by the capital to labour ratio (which can partially represent TIs), will be examined, thus the relation between OIs and TIs will be more substantially scrutinized.

A comparison between the three periods in terms of average quantities[50] across the 28 sectors will further clarify the present assertions. Table 7.4 and Figure 7.6 summarize the findings. They reveal some interesting results. The sharp increase in TFP during 1987–96 is accompanied with a sharp increase in JIT/QC (which are represented by a decrease in the inventories to shipments ratio). At the same time competition also increased during 1987–96 as shown by a significantly smaller increase in producers' prices, although there are other factors involved in this increase.

Table 7.4 *Comparison between the three periods, TFP and other variables (in percentage)*

Variable	1964–76	1977–86	1987–96
TFP	0.62	0.54	0.94
IRY	0.21	–0.12	–1.21
Production deflator	4.21	4.70	3.00
Real output	4.21	1.90	2.56
Labour	0.65	–1.15	–0.16
Capital	3.99	2.07	1.23

Note: Figures are average percentage changes.

Source: Author's calculations from the primary NBER (2000) database.

In addition, capital during 1987–96 did not increase as much as in the previous two periods, which suggests that TFP was mainly the outcome of JIT/QC or other similar OIs and competition. Furthermore, labour employment decreased in the same period, which indirectly indicates that again JIT/QC was the central issue at stake during the period 1987–96. The latter point is also confirmed by Brox and Fader (1997), who empirically found that JIT firms are more labour and materials saving than non-JIT firms.[51] In addition, as the use of the Internet became noticeable only from about 1997–98, hence the impact of the Internet on output and TFP growth can only be assessed from that year onwards, the above conclusions are further reinforced, as the tested regressions are based on the period ending in 1996.

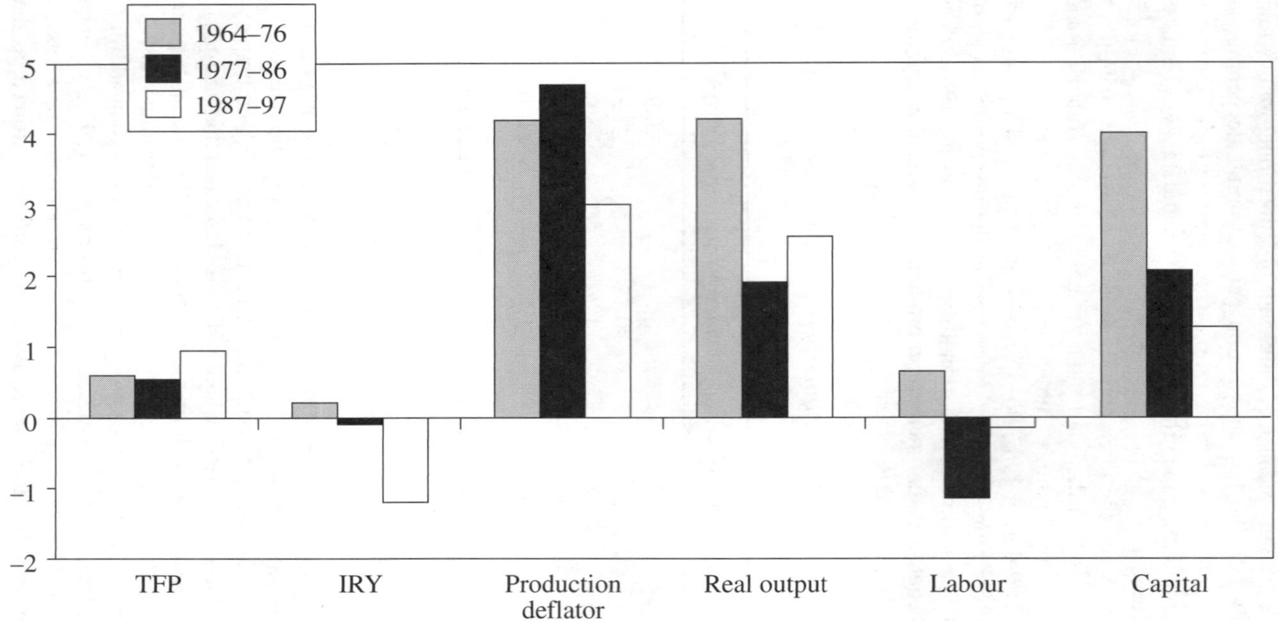

Note: Figures are average percentage changes.

Source: Author's calculations are from the primary NBER (2000) database.

Figure 7.6 USA TFP versus other variables

Table 7.5 Average rates of growth for 85 sectors (percentage)

	IRY	lab pro	cap pro	cap/lab	TFP va
1959–76	–0.15	2.44	–0.41	2.85	0.25
1977–86	–0.58	2.74	–0.52	3.26	0.13
1987–96	–1.35	3.06	1.79	1.27	2.02

Source: Calculations based on the primary NBER (2000) database.

Confirmation of these important findings can be made by examining the summary in Table 7.5, where the following variables are compared for three periods and for 85 sectors: labour productivity ('lab pro'=Y/L), capital productivity ('cap pro'=Y/K), capital to labour ratio ('cap/lab'=K/L),[52] TFA as a value-added formula ('TFP va'), and inventories to shipments ratio (IRY).

Thus, when capital productivity, that is OIs, rose substantially due to the implementation of JIT/QC (hence IRY decreased significantly), only then did TFP move up substantially with a parallel decrease in the capital to labour ratio, that is, the TIs.[53] In addition, if we use equation 7.4 above that links TFP to either capital productivity and the capital to labour ratio, or to labour productivity and the capital to labour ratio, the IRY variable is strongly correlated with capital productivity (OIs) (it increased to a positive 1.79 per cent per annum during 1987–96[54]) but not as strongly to the capital to labour ratio (TIs).

Another confirmation of these important findings can be made by looking into the evolution of TFP across the three periods and across the 28 sectors. Figure 7.7 shows the relevant data. The main increases during the most recent period 1987–96 (in relation to the previous two sub-periods) took place in the SIC categories of 371, 372, 381, 382, 383, 362, 355, 369 and 314 (see Appendix 6.2 for the corresponding names of sectors), thus confirming the leading role most of these sectors have been playing in the recent American economic revival. In particular, the key sectors 371, 372, 381, 382 and 383 experienced most of the applications and implementations of lean production processes such as JIT and QC (as indicated in section 7.2). Furthermore, the economic revival, which took place in the last period from 1987 to 1996 despite a prolonged world recession (especially in Japan), has coincided with a revival in all sectors (apart from some exceptions). In addition, this revival was accompanied by the 'super' leading role the two sectors of 382 (mainly computers) and 383 (mainly semiconductors) played during this latter period.

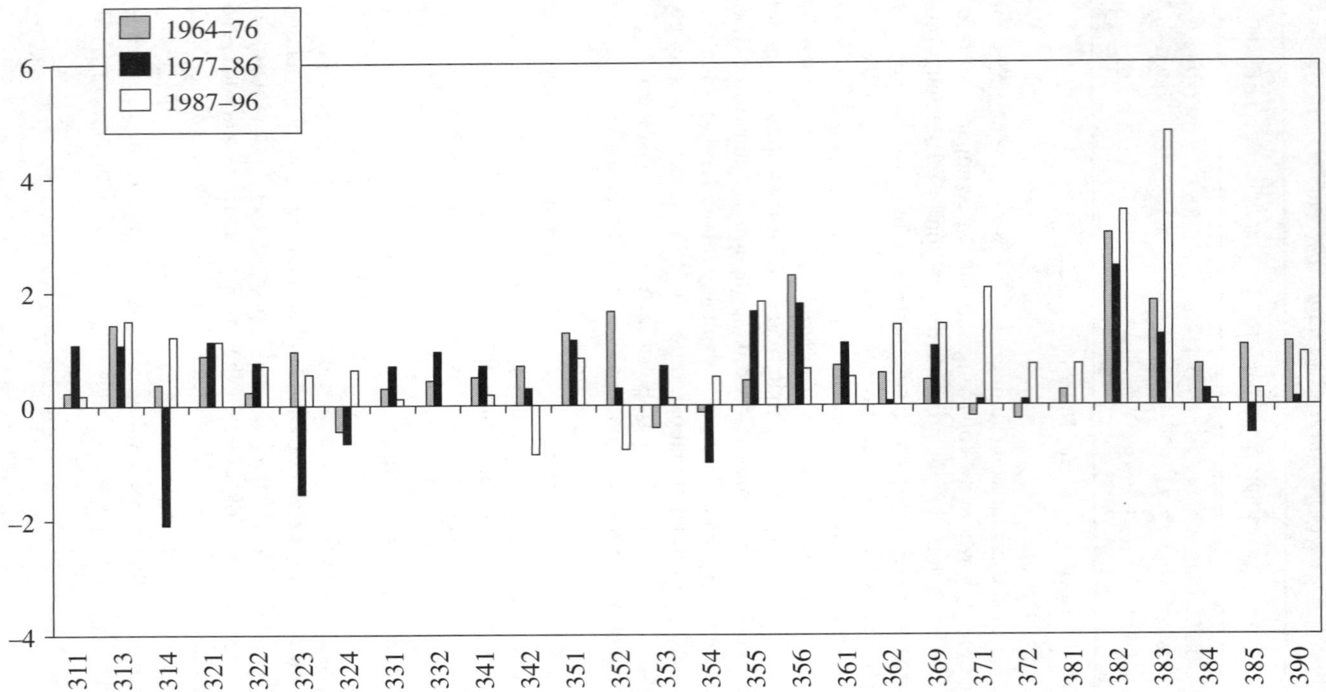

Source: Based on NBER (2000) and UNIDO (2000).

Figure 7.7 USA TFP average NBER and UNIDO for 28 ISIC sectors

7.3.4 A Pooled 28-sector Analysis for TFP and JIT/QC (N = 420)

Another way to bring valid quantitative evidence to the present assertions is to pool all 28 sectors together, and truncate the data to the period 1982 to 1996 during which the American industries have been gradually imitating the Japanese ones in terms of JIT/QC. Another reason for the choice of the year 1982 is that we have detailed census data on three-digit establishments from 1982 to 1996 (every four years). Thus, the number of data is very large (420 observations).

This TFP series will now be regressed against organizational and economic variables of the 28-pooled sectors. Most of these variables are related to the number of establishments for each sector, thus taking into account the effect of small, or big, firms, changes in employment, industrial concentration, and so on (see also Sanidas, 2005a, for the importance of the number of establishments in economic growth). One of these variables is the inventories to shipments ratio as above (IRY), which represents the lean production system. The variable SDLE is the standard deviation of the ratio of employment to the number of establishments during the period 1982 to 1996 for each sector and captures the significance of the variation in the mean size of firms. The variable IMEX is the ratio of imports to exports in 1992 (this year is approximately the middle of the period 1982 to 1996)[55] and captures the competitive strength of each sector. The variable SDMEST is the average ratio of the standard deviation to the mean of establishments from 1982 to 1996. The variable MEANLE is the mean of the ratio of employment to the number of establishments during the period 1982 to 1996 and captures the significance of the mean size of firms. The variable MATSHI is the annual change in the ratio of materials to shipments and captures the degree of vertical integration of each sector. Finally, the variable DUM is a dummy to take into account some outliers in the TFP series.[56] The results of this regression are shown below.

$$TFP = \underset{(6.3)}{0.012} - \underset{(-9.0)}{0.14} * IRY + \underset{(6.4)}{0.0022} * SDLE + \underset{(2.5)}{0.0014} * IMEX - \underset{(-5.3)}{0.0018} * SDMEST -$$

$$-\underset{(-5.6)}{0.0002} * MEANLE - \underset{(-4.3)}{0.19} * MATSHI + \underset{(16.8)}{0.083} * DUM$$

$R^2 = 0.51$ SE = 0.022 SD of TFP = 0.031 DW = 1.84

The diagnostic tests of serial correlation, functional form, normality and heteroscedasticity (not shown here) demonstrate that the model is adequate. The figures in brackets below the coefficients are the T-ratios and show that all variables are significantly different from zero. The coefficients have the right signs. In particular the IRY variable is definitely significant, and has the negative sign as expected. If the variables MATSHI and DUM are dropped from the regression the remaining variables remain significant and their coefficients are unaffected. The aim here is not to analyse the significance and relevance of all variables in the regression but to re-enforce the results obtained in the previous sub-sections regarding the paramount role that the LPS has played in the revival of the American economy in the last 20 years or so. The proxy variable IRY that is inherently linked with the LPS and the JIT/QC is once more undoubtedly significant.

7.3.5 A Pooled 28-sector Analysis of JIT/QC and Various Variables (N = 1008)

This time the whole sample for the 28-pooled sectors will be used, thus forming a very large set of 1008 observations (28 times 36 years, from 1961 to 1996). The aim of this sub-section is to explore the possibility of integrating the proxy variable IRY in the set of other important economic variables such as labour productivity, capital productivity, and so on.[57] Table 7.7 (shown for convenience at the end of section 7.3) contains the results of simple OLS regressions, for the 1008 data, relating each one of the endogenous variables YL, ML, KL, PY and IRY to each other, and to the exogenous variables UN, PKPL, INT[58] (all of them expressed as first differences in natural logs, that is as growth rates except for the rate of interest which is given as a percentage number). Table 7.6 summarizes the definitions of all these variables. YL, ML, KL together constitute the classical determinants of economic growth. The PY represents the output price component and IRY represents the mode of organizational production (the JIT/QC mode). Each one of these five endogenous variables is influenced by all of the others, but not in a simultaneous way (in the next two sub-sections all these variables will be simultaneously determined). They are also influenced by: UN standing for business fluctuations or cycles; PKPL standing for the relative scarcity of capital and labour; and INT standing for the monetary policy of the American government, and the opportunity cost of holding inventories. In addition, each one of the endogenous variables may also represent other issues, for example the KL ratio may embody the impact of TIs and so on.

The variable of immediate interest is the IRY one. Its coefficient is significant for all regressions and has the right sign, at least where YL and PY are the dependent variables. Note that the YL regression is perhaps the most important to observe as it links labour productivity with the LPS proxy IRY. Also the lagged one-year IRY1 and two year IRY2 seem to be significant in some regressions. The elasticities shown are again close to –0.13 (or 0.13

Table 7.6 Definitions of variables[59]

YL	Nominal production per employee (= the classical labour productivity)
ML	Nominal materials per employee (= labour productivity for materials)
KL	Nominal capital per employee (= the classical capital labour ratio)
PY	Deflator of the shipments series
IRY	The ratio of inventories to shipments (= proxy for the JIT/QC system)
UN	Unemployment (= total number of unemployed in the economy)
PKPL	Ratio of capital to labour prices (= investment deflator over the ratio of payments to labour divided by the total number of employed people)
INT	Rate of interest

Source: For UN and INT see Slavin (1999); for all other variables see NBER (2000).

according to the dependent variable) as expected. The importance of the variable IRY can also be examined in the context of the significance of the other dependent variables. In this way, the overall significance of each regression reinforces the significance of each dependent variable, and hence of the IRY variable in particular. More precisely, for the YL regression (that is, when YL is the dependent variable), most dependent variables have significant coefficients (and with the right or expected sign). For example, the higher the growth in ML, KL, KL1, PY and UN, the higher the growth in YL (thus, the higher the unemployment, the higher the ratio YL becomes as the growth in L slows down due to the increase in unemployment). For the IRY regression, the higher the growth rate in YL, IRY1 and IRY2, the lower the growth rate in IRY; the lower the growth rate in KL, ML, PY, UN, INT, the lower rate of growth in IRY. Similar relationships can be found for the other regressions. All these relationships can be further explored and have their own merit in these results. However, a detailed analysis of all these models of the American sectoral economy is outside the scope of this study.

7.3.6 Two-Stage Least Squares Regression Analysis

Still using the 1008-pooled data, this time the aim is to explore the possibility of the integration of IRY into a system of simultaneous equations containing the variables used in the previous sub-section (sub-section 7.3.7 below will explore the simultaneity character of equations in a more rigorous way). These five variables are now simultaneously determined and also influenced by the three endogenous variables UN, PKPL and INT. Since adequate proxies (or instrumental variables) for the exogenous variables are not available, the variables UN, UN1, INT, IRY1, IRY2, PKPL, PKPL1, PKPL2, KL1, YL1 and ML1 (UN1 means UN lagged one year, and so on) are used instead. Despite this inadequacy the method of two-stage least squares (2SLS) can still provide us with some estimates to compare with the simple OLS ones. Table 7.7 shows the results of these two-stage least squares regressions. Again, the variable of immediate interest is IRY. Its coefficient for YL is significant only at the 16 per cent significance level, very significant for PY, and has the right sign at least for YL and PY. Note that the magnitude of the coefficient of IRY in the YL regression is almost the same as that for the simple regression (–0.122).[60]

Also, note that the coefficient of IRY for YL, in both the simple and 2SLS regressions (around –0.13), is lower than that for the 28 data models of the period 1987–96 examined above. This is an expected result as the YL regression covers the whole period of 1961 to 1996. In addition, it is worth emphasizing that the elasticity of IRY (since we have growth rates, the coefficients are also elasticities) in the YL simple and 2SLS regressions is virtually the same as that obtained for the TFP regression of the 420 data model.[61]

7.3.7 Vector Error Correction Model

In this sub-section, the main variables YL, ML, KL, PY, IRY, UN, PKPL and INT (as defined earlier) will be analysed in the context of a general vector error correction model (VECM), which underlies a vector autoregressive (VAR) model and the restrictions imposed upon it via a suitable procedure (see below). The main reasons for using such models for analysing the impact of the JIT/QC system on the American industrial economy can be summarized as follows:

- A VECM can be interpreted as a set of simultaneous equations with endogenous and exogenous variables (see for instance section 5.6 in Maddala and In-Moo Kim, 1998).
- Although the determination of the vectors V (see Appendix 7.4 for a brief derivation of these vectors) of coefficients of error correction models (ECMs) (via the Johansen method for which see Maddala and In-Moo Kim, 1998) is a-theoretical and a priori void of economic interpretation, it is possible to identify the nature of each V as a correspondence between the endogenous variables and these Vs (see below).
- A VECM can provide a Granger type of causality through the error correction terms (see for instance section 5.9.2 in Maddala and In-Moo Kim, 1998).

- It is quite safe to use the VECM and especially Johansen's method if the number of available data is very large.
- Johansen's procedure (see the Appendix 7.4 for a reference to alternatives) is a reliable estimation procedure for determining the coefficients in a system of equations such as the one presented here. This procedure is also related to using the full information maximum likelihood (FIML) modus operandi (Banerjee et al., 1993).
- Johansen's coefficients of the Vs can be used as the restrictions to estimate the ECMs (Enders, 1995; Maddala and In-Moo Kim, 1998).
- A VECM can be considered as a system in the sense of an open complex evolving system (see Chapter 5). In particular, it has a feedback mechanism expressed in terms of the ECMs (see below regarding the results of this mechanism).

In the present analysis 1008 observations of the pooled 28 manufacturing sectors of the American industrial economy are used again. The model of simultaneous equations consists of the five endogenous variables: YL, ML, KL, PY and IRY as well as the three exogenous variables UN, PKPL and INT, as previously defined. The unrestricted VAR model indicated as the optimum order a lag of five if the Akaike information criterion (AIC) is used and a lag of one if the Schwartz Bayesian criterion (SBC) is used.[62] Note that such a discrepancy usually exists and hence one has to experiment with all lags between one and five in order to determine the final model. Thus, for each lag between one and five, the five corresponding Vs were estimated (based on Johansen's method). See Table 7.8 (at the end of section 7.3) for some of these results. The reason for having five Vs and no less for each VAR model, is that Johansen's tests strongly suggest that there are five Vs. This is not surprising, since all the variables used (both endogenous and exogenous) are already log differenced once and hence, as the augmented Dickey-Fuller (ADF) test confirmed, they are all integrated of order 0 (that is, I(0)), which means that they are all stationary.[63]

The next aim is to determine the most representative set of five Vs[64] and accordingly to estimate the ECMs, which will ultimately provide more evidence to this study. In order to determine the most representative set of five Vs, three sets of criteria will be used. First, we already have a relatively good idea as to what each one of the Vs looks like from the previous econometric analysis shown in the previous sub-sections. Second, a priori economic reasoning can identify the expected sign of each long-term coefficient of a given V. Finally, the signs, significance and magnitude of the relevant ECMs will confirm the first two criteria.

Out of all the five sets of Vs the Vs of VAR (1) and VAR (2) were chosen as they satisfied, to a large extent, all three criteria. For instance, virtually all coefficients have the right sign; they also have approximately the right magnitude and to some degree they are similar to the regression results shown in previous sections. It would be possible to test and slightly change some of the coefficients according to the log-likelihood ratio, however this is not the purpose of this analysis. A more detailed examination of the sets of coefficients of the selected Vs is outside the scope of this study since the main purpose is to show the integrated impact of the JIT/QC variable on the economy via the ECMs.

The results regarding the ECMs are now shown in Table 7.9 (at the end of section 7.3).[65] Three sets of ECMs were estimated: the first is based on the VAR (1) model, the second on the VAR (2), and the third on the VAR (5) but using the restrictions (coefficients) of the VAR (1) model (noted as VAR(5)* in Table 7.9). The VAR (1), as expected, still has some serial correlation problems, whereas the VAR (5) does not. All three models have apparently a significant heteroscedasticity, but this is as expected because the 1008 data are pooled data from 28 sectors and hence they have some pattern in terms of variance across these sectors. In addition, a few extreme values (outliers) certainly adversely affected this test. To verify the validity of these explanations the covariance matrix was re-estimated using the White and Newey-West methods (see Pesaran and Pesaran, 1997; also Greene, 1993). The new T-ratios obtained according to the new covariance matrices were smaller, as expected, but were still high enough and close to those shown in Table 7.9. Although the normality test (Bera-Jarque) rejects the assumption that the residuals are normally distributed, the very large sample used in these estimations and the plots of the histograms of residuals (which approximately show a normal distribution) suggest that the normality assumption should not be a problem. Again, outliers have a negative impact on most of these diagnostic tests.[66]

The R^2 is quite high in all regressions, which is encouraging given the nature and stationarity of data. The T-ratios of the ECM coefficients are generally high suggesting the significant impact of the dependent variables underlying the ECMs. In particular, the coefficient of the ECM corresponding to the inventories variable IRY[67] has a considerable impact in determining the other endogenous variables (YL, ML, KL and PY) in a causal way: for instance, it takes the values –0.136, –0.14 and –0.20 for the VAR (1), VAR (5)* and VAR (2) respectively regarding the dYL dependent variable. The negative coefficient of IRY in the ECM, combined with the original negative coefficient of IRY in the V regarding the dependent variable YL, shows that shocks produced by IRY have a positive impact on the mean value of YL.[68] This error correction seems to be cumulative, showing the complexity of the situation, and, in particular, the structural shift in the behaviour of firms regarding overall organization (and inventories) as Figures 7.3, 7.4a, 7.4b and

7.4c have shown. Regarding these figures it was indicated that the inventories to shipments ratio has started to continually decline since the early 1980s. Thus, it is possible that this positive impact of IRY on the long run trend of growth rates of productivities signifies that over a long period of time (even longer than the 15 years of the period 1980s to 1996) the impact of the LPS becomes more permanent.[69] In any case, all this further validates the conclusions of the significant impact of JIT/QC on economic growth. Furthermore, when IRY becomes the dependent variable, the coefficients of the ECMs of all five endogenous variables are also significant, thus re-enforcing the two-way causal and simultaneous interactions of these variables with IRY. Some more remarks will justify the conclusions so far.

First, the choice of the relevant underlying dependent variables of the ECMs[70] is strongly confirmed by the right sign and significance of the appropriate ECM in each regression. For example, for the IRY dependent variable the ECM1 (–1) corresponding to the IRY variable has a positive sign (hence negative when combined with the negative sign of the ECM equation containing IRY), is very significant (a T-statistic ranging from 16.5 to 32.6), and its magnitude (for example, 0.43 for the VAR (1) model; see the bottom left segment of Table 7.9) shows that its impact 'corrects' the long-term path of IRY without too much of a delay. Second, the coefficients of each ECM regression are relatively stable for the three VAR models used, have the right sign and agree with the previous econometric analysis carried out in the subsections above. Third, in the VAR (5)* model the exogenous variables play a significant role (for instance the variable PKPL has a strong impact on KL). This active role of the exogenous variables makes the relationships between the endogenous variables more valid and significant. Fourth, although the results shown and analysed so far are based on the version of 'restricted intercepts and no trends in the VAR', other versions such as the 'unrestricted intercepts and restricted trends in the VAR' produced very similar results.[71] Fifth, further examination of the ECM regressions, involving plots of actual and fitted values, plots of residuals, Wald tests of restrictions imposed on parameters, autocorrelation and spectral functions of residuals, unit root tests for residuals, CUSUM tests and so on, all confirmed the validity of the present analysis.

Sixth, the estimated elasticity between the IRY and productivities is not influenced by any common time trends, since the variables are all stationary. Also, this elasticity takes into account all endogenous variables (as well as exogenous) such as the capital-labour ratio (hence the impact of TIs) and so on in a simultaneous manner. Furthermore, this elasticity is estimated either through the coefficients of the vectors V directly (a rather longer term evaluation) or through the ECMs (a rather shorter term evaluation): for instance, for the VAR (1) and VAR (2) models the V vectors indicate that the IRY elasticity

Table 7.7 Results of the simple OLS and 2SLS regressions

	Simple OLS regressions					Two-SLS regressions				
Dependent	YL	ML	KL	PY	IRY	YL	ML	KL	PY	IRY
Intercept	**0.011**	**–0.0015**	**0.0083**	**–0.029**	**0.004**	**0.0057**	**0.002**	**–0.006**	**–0.029**	**0.03**
	6.3	0.61	2.2	9.1	0.66	1.5	0.41	0.41	4.4	2.8
YL		**1.053**	**0.187**	**0.61**	**–1.304**		**0.884**	**1.86**	**0.429**	**–0.988**
		43.5	2.8	11.3	14.6		4.5	2.6	0.62	1.2
ML	**0.62**		**–0.14**	**0.173**	**0.613**	**0.782**		**–1.76**	**0.8**	**–0.424**
	43.5		2.7	3.9	8.4	5.7		2.9	1.3	0.52
KL	**0.042**	**–0.054**		**–0.033**	**0.309**	**0.17**	**–0.217**		**0.109**	**–0.124**
	2.8	2.7		1.2	6.7	2.7	2.9		0.53	0.48
PY	**0.187**	**0.089**	**–0.045**		**0.372**	**0.078**	**0.165**	**–0.091**		**0.946**
	11.3	3.9	1.2		7	0.71	1.2	0.23		4
IRY	**–0.135**	**0.107**	**0.141**	**0.126**		**–0.122**	**–0.059**	**–0.251**	**0.64**	
	14.6	8.4	6.7	7		1.4	0.49	0.84	4.1	
UN	**0.016**	**–0.027**	**0.142**	**0.017**	**0.049**	**–0.0006**	**0.016**	**0.145**	**–0.043**	**0.08**
	3.3	4.3	15.2	1.9	3.2	0.04	0.97	5	1.3	2.2
PKPL	**–0.221**	**0.078**	**–0.129**	**0.493**	**0.042**	**–0.139**	**0.045**	**0.354**	**0.39**	**–0.309**
	8.8	2.3	2.4	11.1	0.51	2.8	0.55	1.5	3.3	1.7
INT	**0.00009**	**–0.0002**	**0.0009**	**0.0017**	**0.002**	**–9E–05**	**0.0005**	**0.003**	**–0.0007**	**0.002**
	0.47	0.96	2.4	5.1	3.6	0.25	0.98	2.8	0.71	2.1
IRY1	**–0.02**	**0.02**	**0.086**	**0.015**	**–0.102**					
	2	1.5	4.1	0.85	3.3					

IRY2	**–0.008**	**0.004**	**0.024**	**0.001**	**–0.078**					
	–0.88	0.38	1.3	0.06	2.8					
PKPL1	**0.051**	**–0.018**	**–0.066**	**–0.006**	**–0.079**			**–0.183**		
	2	0.54	1.2	0.12	1			1.8		
PKPL2	**–0.006**	**0.009**	**–0.137**	**0.0004**	**0.075**			**–0.122**		
	0.26	0.29	2.9	0.01	1.1			1.5		
UN1	**–0.0063**	**0.019**	**–0.042**	**–0.023**	**–0.013**					
	1.26	3	4	2.6	0.81					
KL1	**0.067**	**–0.057**	**0.196**	**–0.066**	**–0.028**					
	4.5	2.9	6.3	2.4	0.6					
YL1	**0.00003**	**–0.035**	**–0.0094**	**0.258**	**–0.147**					
	0	0.87	0.14	4.7	1.5					
ML1	**–0.0025**	**0.041**	**0.006**	**–0.098**	**0.078**					
	–0.1	1.3	0.1	2.2	1					
R^2 bar	0.88	0.86	0.36	0.68	0.34	0.55	0.43	0.15	0.23	0.13
SE	0.017	0.022	0.037	0.031	0.054	0.019	0.026	0.063	0.052	0.065
SD Dep/t	0.051	0.06	0.046	0.056	0.067					
DW-stat	2.02	2	1.93	1.6	1.98					
Diagnostic tests:										
Ser. Cor/n	0.45	0.93	0.005	0.000	0.401					
Functional	0.4	0.1	0.132	0.011	0.000					
Normality	0.000	0.000	0.000	0.000	0.000					
Heteros/ty	0.000	0.000	0.001	0.000	0.000					

Notes:
1. The figures under the coefficients are T-statistics.
2. The figures of the diagnostic tests are the p-values.

Table 7.8 The vectors V of selected VAR models

VAR (1)	Part I		Initial Vs			Part II		Regression coefficients		
Dependent	IRY	YL	ML	KL	PY	IRY	YL	ML	KL	PY
YL	**–1**	**–1**	**–1**	**–1**	**–1**	**–0.538**	**1.000**	**1.053**	**0.565**	**–0.313**
ML	**1.85**	**0.45**	**0.95**	**2.81**	**0.49**	**0.995**	**0.450**	**1.000**	**–1.588**	**0.154**
KL	**–0.22**	**0.21**	**–0.37**	**1.77**	**–0.77**	**–0.118**	**0.210**	**0.389**	**1.000**	**–0.241**
PY	**–0.75**	**0.32**	**0.04**	**–1.83**	**–3.19**	**–0.403**	**0.320**	**–0.042**	**1.034**	**1.000**
IRY	**–1.86**	**–0.27**	**0.2**	**0.28**	**0.09**	**1.000**	**–0.270**	**–0.211**	**–0.158**	**0.028**
UN	**–0.203**	**0.0206**	**–0.037**	**–0.02**	**–0.162**	**–0.109**	**0.021**	**0.039**	**0.011**	**–0.051**
T-stat	–4.95	3.23	–3.36	–0.47	–1.70	**–0.2239**	**0.0392**	**0.14105**	**0.11254**	**–0.0508**
PKPL	**–0.068**	**–0.008**	**–0.097**	**0.131**	**1.361**	**–0.037**	**–0.008**	**0.102**	**–0.074**	**0.427**
T-stat	–0.38	–0.29	–1.98	0.72	3.25	**–0.4156**	**–0.1455**	**0.546**	**–0.9102**	**0.42677**
INT	**0.0091**	**0.0002**	**0.0023**	**–0.004**	**0.0039**	**0.005**	**0.0002**	**–0.002**	**0.002**	**0.001**
T-stat	6.50	0.72	5.90	–3.00	1.17	**–6E–05**	**0.00049**	**–0.0014**	**0.00028**	**0.00121**
Intercept	**–0.103**	**0.015**	**–0.006**	**–0.055**	**0.1556**	**–0.055**	**0.015**	**0.007**	**0.031**	**0.049**
T-stat	–8.05	7.58	–1.78	–4.26	5.24	**–0.0165**	**0.0103**	**0.00099**	**0.0313**	**0.04878**
VAR (2)	Part I		Initial Vs			Part II		Regression coefficients		
Dependent	IRY	YL	ML	KL	PY	IRY	YL	ML	KL	PY
YL	**–1**	**–1**	**–1**	**–1**	**–1**	**–1.176**	**1.000**	**1.010**	**–1.695**	**0.179**
ML	**0.74**	**0.45**	**0.99**	**–0.13**	**–1.55**	**0.871**	**0.450**	**1.000**	**–0.220**	**0.278**
KL	**0.31**	**0.45**	**–0.14**	**–0.59**	**–2.44**	**0.365**	**0.450**	**0.141**	**1.000**	**0.437**
PY	**0.1**	**0.36**	**–0.06**	**0.33**	**5.58**	**0.118**	**0.360**	**0.061**	**0.559**	**1.000**
IRY	**–0.85**	**–0.11**	**0.054**	**–0.22**	**–0.49**	**1.000**	**–0.110**	**–0.055**	**–0.373**	**0.088**

UN	**0.0008**	**0.0189**	**–0.042**	**0.16**	**0.049**	**0.0009**	**0.0189**	**0.0424**	**0.2712**	**–0.009**
T-stat	0.05	1.77	–3.72	6.84	0.37					
PKPL	**–0.088**	**0.1508**	**–0.193**	**–0.371**	**–1.781**	**–0.1029**	**0.1508**	**0.1949**	**–0.629**	**0.319**
T-stat	–1.15	3.03	–3.67	–3.41	–2.87					
INT	**0.0008**	**–3E–04**	**0.0016**	**0.0005**	**0.0079**	**0.0009**	**–3E–04**	**–0.002**	**0.0009**	**–0.001**
T-stat	1.63	–0.82	4.85	0.74	1.98					
Intercept	**–0.007**	**0.0135**	**–0.007**	**0.0628**	**–0.066**	**–0.0085**	**0.0135**	**0.0073**	**0.1064**	**0.012**
T-stat	–1.71	4.82	–2.45	10.30	–1.88					

VAR (3)	Part I		Initial Vs			Part II		Regression coefficients		
Dependent	IRY	YL	ML	KL	PY	IRY	YL	ML	KL	PY
YL	**–1**	**–1**	**–1**	**–1**	**–1**	**–0.013**	**1.000**	**0.901**	**–1.898**	**0.145**
ML	**3.46**	**0.497**	**1.11**	**0.211**	**–1.568**	**0.045**	**0.497**	**1.000**	**0.400**	**0.227**
KL	**7.92**	**0.437**	**–0.004**	**–0.527**	**–1.167**	**0.104**	**0.437**	**0.004**	**1.000**	**0.169**
PY	**–10.24**	**0.272**	**–0.093**	**0.335**	**6.92**	**–0.135**	**0.272**	**0.084**	**0.636**	**1.000**
IRY	**–76.11**	**–0.261**	**–0.053**	**–0.129**	**–0.913**	**1.000**	**–0.261**	**0.048**	**–0.245**	**0.132**
UN	**1.04**	**0.026**	**–0.094**	**0.118**	**–0.383**	**0.014**	**0.026**	**0.085**	**0.224**	**0.055**
PKPL	**–8.21**	**0.159**	**–0.156**	**–0.513**	**–2.015**	**–0.108**	**0.159**	**0.141**	**–0.973**	**0.291**
INT	**–0.018**	**0.0001**	**0.0016**	**0.001**	**0.0092**	**0.000**	**0.000**	**–0.001**	**0.002**	**–0.001**
Intercept	**–0.403**	**0.0098**	**–0.017**	**0.0355**	**–0.155**	**–0.005**	**0.010**	**0.015**	**0.067**	**0.022**

Note: In Part II of VAR(1), the underlined coefficients are from the VAR(5) model under the restrictions of the endogenous variables' coefficients of the VAR (1) model. The VAR (3) model was discarded for the final ECM estimations.

Table 7.9 ECMs of the three VAR models

Dependent	VAR (1) **dYL**		VAR (5)* **dYL**		VAR (2) **dYL**		VAR (1) **dML**		VAR (5)* **dML**		VAR (2) **dML**		Sign of var/e in the ECM VAR(1)
	Coeff.	T-ratio	Coeff.	T-ratio	Coeff.	T-ratio	Coeff.	T-ratio	Coeff.	T-ratio	Coeff.	T-ratio	
ECM1(–1)(IRY)	–0.136	13.6	–0.14	6	–0.2	4.5	–0.15	12.5	–0.13	4.8	–0.26	4.9	–
ECM2(–1)(YL)	0.896	13	1.05	6.4	0.64	9.1	0.42	5.2	0.52	2.7	0.26	3.1	–
ECM3(–1)(ML)	–0.145	3.5	–0.42	4.7	–0.31	3.9	–0.65	13.1	–0.92	8.9	–1.1	11.7	+
ECM4(–1)(KL)	–0.04	3.1	0.004	0.2	–0.54	13.2	–0.13	8.6	–0.12	4	–0.6	12.5	+
ECM5(–1)(PY)	0.131	16.8	0.15	12	0.085	10.7	0.14	15.5	0.18	12.2	0.085	9.1	–
Bar –R^2	0.39		0.49		0.41		0.4		0.51		0.42		
SE	0.048		0.044		0.048		0.057		0.052		0.055		
SD of dep/t	0.062		0.062		0.062		0.073		0.073		0.073		
DW-stat	2.05		2.02		2.04		2.07		2.02		2.06		
Ser. Corr. Test	0.012		0.107		0.001		0.001		0.128		0.000		
Funct. Form	0.000		0.09		0.001		0.037		0.313		0.019		
Normality	0.000		0.000		0.000		0.000		0.000		0.000		
Heterosc/ty	0.000		0.002		0.000		0.000		0.057		0.000		

Dependent	VAR (1) **dKL**		VAR (5)* **dKL**		VAR (2) **dKL**		VAR (1) **dPY**		VAR (5)* **dPY**		VAR (2) **dPY**		
	Coeff.	T-ratio	Coeff.	T-ratio	Coeff.	T-ratio	Coeff.	T-ratio	Coeff.	T-ratio	Coeff.	T-ratio	
ECM1(–1)(IRY)	0.067	7.8	0.039	1.9	–0.13	3.6	0.044	4.3	0.055	2.2	0.04	0.9	–
ECM2(–1)(YL)	–0.42	7.1	–0.43	3.1	–0.73	12.8	–0.41	5.8	–0.19	1.1	–0.49	6.9	–
ECM3(–1)(ML)	0.76	21.2	0.6	7.8	0.59	9.2	–0.23	5.5	–0.47	5.1	–0.21	2.7	+
ECM4(–1)(KL)	–0.21	18.5	–0.21	9.9	–0.41	12.2	0.044	3.3	–0.004	0.1	–0.3	7.2	+
ECM5(–1)(PY)	0.015	2.3	0.038	3.6	–0.049	7.5	0.15	19.3	0.16	12	0.13	16	–

Bar –R^2	0.47	0.57	0.54	0.31	0.41	0.34
SE	0.042	0.038	0.039	0.049	0.046	0.048
SD of dep/t	0.058	0.058	0.058	0.059	0.059	0.059
DW-stat	1.91	2.05	2.03	2	2.04	2
Ser. Corr. Test	0.001	0.000	0.049	0.985	0.003	0.469
Funct. Form	0.002	0.000	0.018	0.975	0.519	1
Normality	0.000	0.000	0.000	0.000	0.000	0.000
Heterosc/ty	0.074	0.011	0.123	0.000	0.000	0.000

Dependent	VAR (1) **dIRY** Coeff.	T-ratio	VAR (5)* **dIRY** Coeff.	T-ratio	VAR (2) **dIRY** Coeff.	T-ratio	Significant lagged variables of VAR (5)* (at less than 5%, unless otherwise indicated)			
ECM1(–1)(IRY)	0.43	32.6	0.54	16.5	1.31	22.2	**dYL2**	(dKL)	**dINT1**	(dYL,dML,dKL,dPY,dIRY)
ECM2(–1)(YL)	0.31	3.4	0.51	2.3	–0.92	10.1	**dYL3**	(dKL)	**dINT2**	(dYL,dKL,dIRY)
ECM3(–1)(ML)	–0.38	6.9	–0.5	4.2	–0.32	3.2	**dYL4**	(dYL,dML,dPY)	**dINT3**	(dYL,dML,dPY,dIRY)
ECM4(–1)(KL)	–0.17	9.6	–0.24	7.2	–0.15	2.7	**dML1**	(dKL)	**dINT4**	(dML(9.7),dKL,dPY(10.3),dIRY)
ECM5(–1)(PY)	–0.16	1.6	0.004	0.2	–0.048	4.6	**dML2**	(dKL)	**dUN1**	(dYL,dML,dPY)
							dML3	(dKL)	**dUN2**	(dYL,dML,dPY)
Bar –R^2	0.55		0.61		0.58		**dKL1**	(dYL)	**dUN3**	(dYL,dML,dPY)
SE	0.064		0.06		0.062		**dKL2**	(dYL,dKL)	**dUN4**	(dYL,dML,dPY,dIRY)
SD of dep/t	0.095		0.095		0.095		**dKL3**	(dKL)	**dPKPL1**	(dYL(8.8),dKL,dIRY)
DW-stat	2		1.99		2.05		**dKL4**	(dKL)	**dPKPL2**	(dYL,dML,dKL,dPY(8.1))
Ser. Corr. Test	0.697		0.297		0.001		**dPY1**	(dYL,dML,dPY(6.9),dIRY(5.2))		
Funct. Form	0.03		0.308		0.104		**dPY2**	(dPY)	**dPKPL3**	(dYL,dML,dPY)
Normality	0.000		0.000		0.000		**dPY3**	(dPY(9.8))	**dPKPL4**	(dYL(9.7),dKL)
Heterosc/ty	0.107		0.004		0.006		**dPY4**	(dYL,dML)		
							dIRY1	(dPY(6),dIRY)		
							dIRY2	(dPY(10.7),dIRY)		
							dIRY4	(dPY)		

Note: the corresponding dependent variables are in brackets.

with respect to YL is –0.27 and –0.11 respectively; whereas it is –0.136, –0.14 and –0.2 for the VAR (1), VAR (5)* and VAR (2) models respectively regarding the dYL dependent variable of ECM estimation.

Seventh, via the VEC model, the separate impact of OIs and the TIs on sectoral growth has become possible. According to equation 7.4 TFP is a function of either labour productivity (*Y*/*L*) and the capital to labour ratio (*K*/*L*) (all expressed as growth rates), or, a function of capital productivity (*Y*/*K*) and the capital to labour ratio (*K*/*L*) again. Also we know from a previous discussion that (*Y*/*K*) and (*K*/*L*) can represent OIs and TIs respectively. Hence, in the present VEC model, the relationships between the variable KL, which is (*K*/*L*), and the other variables shows the impact of TIs, whereas the impact of OIs (in this case JIT/QC) on the remaining variables is captured by the proxy IRY (the inventories to shipments ratio).

Eighth, the VEC model used in this sub-section is, overall, a model of estimating the impact of productivities of the American industrial sectors through time, and its usefulness can go beyond the estimation of the influence of the JIT/QC system via its proxy of the inventories to shipments ratio. The results of this model seen through the prism of a sectoral growth model of productivities, are significant and rather robust in their own right. If, for example, we consider the relationship between capital deepening as represented by the capital to labour ratio and the labour productivity as represented by the output to labour ratio, then from Tables 7.8 and 7.9 we can see that the corresponding coefficients and their signs are the desirable ones. Thus, the higher capital deepening is the higher labour productivity is, *ceteris paribus*, and so on. Also, the ECM variables corresponding to the same dependent variables (for example, ECM(–1) (YL) of the dependent YL), all have the right sign (negative if the sign of the dependent variable is positive and vice versa). This means that any 'error' of the variation of the dependent variable is brought back to zero (rapidly for YL and ML, slowly for KL and PY, average pace for IRY). In addition, often the exogenous variables are significant, thus providing more robustness to the regressions. Overall, the system of simultaneous equations of the five dependent variables YL, ML, KL, PY and IRY seem to provide a good explanatory tool for the functioning of the economy from the manufacturing sub-sectors point of view. This means that organizational structures such as the JIT/QC, as represented by IRY, are an integrated part of this economy.

7.4 CONCLUSIONS

In previous Chapters (2 to 5) the theoretical foundations for the importance of OIs in economic, and, in particular, industrial growth were laid. More especially one of the major theoretical preoccupations was to define, describe and evaluate the main axis of OIs, namely the JIT/QC cum Fordism axis, which constitutes the intrafirm organization of labour and capital. The value of JIT/QC in the process of economic growth could also be appreciated from the development of the concepts of the four PROSIBB, and, in particular, from the process of movements. In Chapter 6, a qualitative and quantitative analysis of industrial sectoral growth in Japan and the USA provided some evidence, consistent with this study, that the sectors which experienced higher growth rates were those that introduced significant OIs, especially in relation to the main axis of OIs. In this chapter a more detailed, and thorough, quantitative analysis of American manufacturing sectoral growth has provided more evidence as to the confirmation of the hypothesis that the OI of the LPS, or its equivalent the JIT/QC system, has played a significant role in the revival and continuous growth of the American manufacturing sectors.

However, it is necessary to stress that the contribution of the findings of this chapter cannot be properly evaluated unless this contribution is seen in the light of previous chapters. In this study the extensive references to many researchers' work on the substantial role of the LPS or JIT/QC system on industrial growth, including their relevant empirical evidence, have shown that OIs (or disembodied technology) such as JIT/QC are a necessary component of industrial growth. This extensive reference supports the quantitative analysis presented in this chapter and vice versa. As an extra example of this mutual backup between the evidence of previous chapters and the evidence provided in this chapter, consider the article by Brox and Fader (1997) entitled 'Assessing the impact of JIT using economic theory'. These two authors say:

> Our statistical findings support the claims made by JIT proponents that the Japanese JIT manufacturing philosophy is being successfully implemented in North America. More importantly, we find statistical support for the assertions from JIT users that the technology not only enhances cost efficiency, but that the way that productivity is enhanced is through added worker and flexibility and quality improvements … [and] the difference between the elasticities of factor productivity, with respect to output changes, shows JIT firms as being more labour-and materials-saving than the non-JIT firms … (ibid., p. 382)

The results presented in this chapter, based on an econometric study of American manufacturing sectoral growth, and not just on a sample of firms, provide a more comprehensive and robust support for the conclusions reached by the above two authors (and many others cited in previous chapters and sections).

The implementation of the JIT/QC system in the USA (as an extension of that in Japan) took place during the last 20 years or so, and especially from the mid-1980s to the end of the 1990s. Information relating to this provided the necessary background for the subsequent quantitative analysis. The latter was carried out in a robust and comprehensive manner starting from simple models and ending with the usage of much more complicated ones. In brief the results obtained are the following:

1. There is a strong relationship between the proxy for JIT/QC (the ratio of inventories to shipments) and productivity (the lower the ratio, the higher the productivity): this link becomes stronger in the most recent periods examined (1982–96, and especially 1987–96). The elasticity of productivity in relation to the LPS was found to be between –0.14 and –0.65 depending on the variable that the proxy is linked with, the model, and the examined period. These figures are higher or much higher than estimates provided by Lieberman and Demeester (1999), that is, –0.1. This difference is due to three reasons: first, productivity increases are a consequence not only of OIs (the JIT/QC system in this case) but also of other types of innovations such as TIs and strategic innovations.[72] The model that shows more clearly the net impact of JIT/QC is the VECM which provides an elasticity of –0.14 between labour productivity (YL) and the inventories ratio (IRY), since in this model at least the TIs have been considered separately through the proxy of capital to labour ratio (KL). Given that the VEC model was applied to the whole period 1958 to 1996, a higher elasticity is expected for the most recent period of 1987–96. Second, the sectoral models take into account other effects such as the externalities effect (for example, demonstration effects) and hence the elasticities are higher than those provided by Lieberman and Demeester (1999) which are based on a sample of firms. Third, the higher elasticity estimates are more related to the most recent period of 1987–96 for which the JIT/QC implementation has been more spread out and more rigorous. In addition, these elasticity estimates are the direct effect of the JIT/QC impact, as indirectly (through the multiplier mechanism on all sectors including services) the impact of the adoption of JIT/QC on the economy should be much larger.
2. The relationship between the impact of the LPS and the economy is simultaneous and directed in both ways, that is, the JIT/QC influences the economy and vice versa.
3. The leading sectors in the USA during the last 20–25 years or so, mainly machinery (including computers), electronics and plastics experienced the largest reduction in the ratio of inventories to shipments; these sectors were the most efficacious in implementing the LPS (evidence for this last statement was provided in Chapter 6 regarding some of the leading sectors).

4. Other factors have been taken into account in the econometric exploration, such as the number of establishments, number of patents, rate of interest, price levels and so on (within the limits of data availability); the inclusion of these factors enhanced the validity of the results.
5. The impact of TIs, such as patents and IT, was also briefly examined. This impact was also taken more explicitly into account in the econometric analysis by considering the role of the proxy for TIs, namely the K/L ratio.

These conclusions were based on using, in the empirical analysis, the proxy of the ratio of inventories to shipments for JIT/QC. The use of such a proxy (several others could be used instead, or concurrently, such as the number of defects if relevant data were available) was fully justified by the extensive analysis carried out in previous chapters. There is strong evidence that no other production system, such as the neo-Fordist system including variations based on flexible automation, can achieve such substantial reductions in the ratio of inventories to shipments as the LPS can. The empirical analysis carried out in this chapter is strongly supported by the 'theoretical' analysis of Chapters 3, 4, 5 and the general discussion of Chapter 6. Figure 7.8 at the end of this chapter shows the links between the 'theoretical' chapters and the 'quantitative' ones.

These results have far-reaching implications for the theory of industrial and economic growth and policy. First, models of economic growth, whether endogenous or not, should include a new variable, namely, the organizational relation between the traditional inputs of capital and labour. Second, growth policies should encourage the development of quality control, outsourcing of supplies, the lean production process in general, and SMEs (for the link between SMEs and economic growth in this respect see Sanidas, 2005a). Third, the American experience shows that it is possible to transplant with success Japanese-type organizational innovations in other countries. Adjacent to this last point it is worth noting that it is rather important to consider the specific country in which such holistic systems (JIT/QC) are transferred, and thus isolated firms attempting to implement these systems in countries not yet ready to adopt them will most likely fail. In other words it is countries like Japan and the USA, and some other developed nations (perhaps to a lesser extent), which are the friendliest environments for the adoption of the LP or JIT/QC process.[73]

Overall, the main message of this chapter is that the quantitative evidence provided confirms the theoretical propositions made in previous chapters.

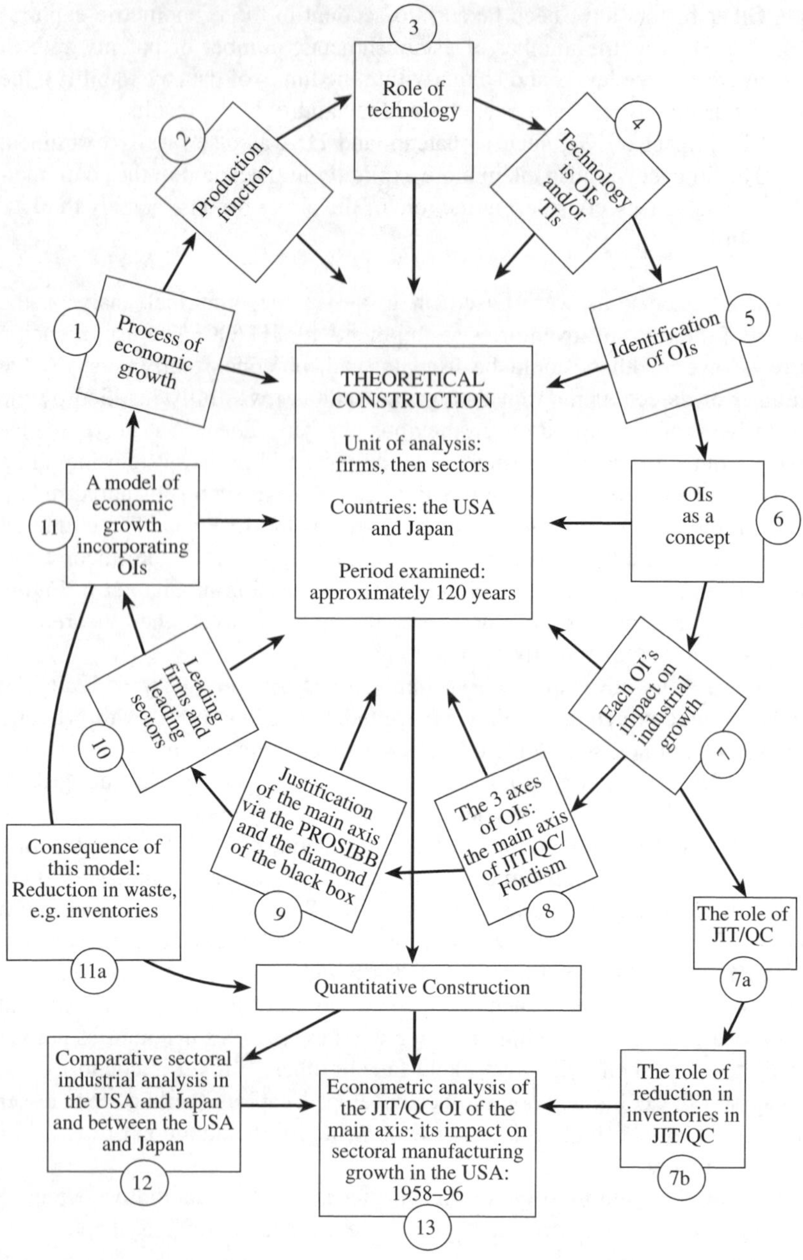

Figure 7.8 The links between theoretical concepts and quantitative analysis

APPENDIX 7.1: TRANSFORMATION OF THE 459-SECTOR DATA TO 85-SECTOR AND 28-SECTOR DATA

For all non-index variables (for example, employment) of the original 459-sector NBER database, the transformation to the two more aggregate tables was made by simple addition of the relevant sectors.

For the index variables of the original 459-sector NBER database (that is, those concerning the four deflator series), the transformation to the two more aggregate tables was made by weighted-averaging the relevant sectors, using as weights the value in dollars of the related series (for example, for the deflator index of the shipment series, the value in dollars of the shipment series of the relevant sectors was used as relative weights).

As an example of the aggregation procedure, the SIC industry no. 201, 'Meat products' (one of the 85 sectors used in the analysis) is aggregated by considering the following three sub-sectors (in the original 459-sector base):

2011: 'Meat packing plants'
2013: 'Sausages and other prepared meals'
2015: 'Poultry slaughtering and processing'

Note that in most cases, the 85 sectors were determined by simply aggregating the four-digit data into the three-digit data according to the 1987 SIC list of industries, as the above example shows. However, at times, the writer decided about the nature of the 85-sector industry according to the principle of homogeneity of products; for instance, the SIC sectors 384 ('Medical instruments and supplies') and 385 ('Ophthalmic goods') were aggregated into a new sector (for the list of the 85 sectors see Appendix 7.3).

The source of the 1987 SIC Industry (459 sectors) list used by NBER and by the writer is available in the NBER Internet site (see references).

The passage from the 85-sector base to the 28-sector base was made by considering the following points:

1. The 28-sector base was determined by simply accepting the UNIDO database classification, which uses 28 sectors (Appendix 6.2 shows the names and codes of these 28 sectors).
2. The correspondence between the 85 aggregated NBER sectors and the UNIDO 28 sectors was not sometimes straightforward, but nonetheless satisfactory for the purpose of the present analysis.

All the above-mentioned tables and calculations (corresponding to hundreds of pages of printing space) are available on request.

APPENDIX 7.2: BASIC DATA FOR THE 28 ISIC SECTORS IN THE USA, EXPRESSED AS PERCENTAGE

	For the period 1964–76					For the period 1977–86					For the period 1987–96				
ISIC	tfp	tfp	x1	x2	x3	tfp	tfp	x1	x2	x3	tfp	tfp	x1	x2	x3
sector	unido	nber	IRY	defl/r	empl	unido	nber	IRY	defl/r	empl	unido	nber	IRY	defl/r	empl
311	0.171	0.31	–0.34	4.75	–0.6	1.28	0.91	–1.4	2.56	–0.712	0.345	0.06	0.01	2.626	1.063
313	0.871	1.97	–2.31	3.47	0.01	1.23	0.98	–2.3	4.89	–1.689	1.298	1.71	0.12	2.394	–1.79
314	0.111	0.68	–1.83	3.84	–1.32	–1.1	–3	–2.3	8.67	–3.288	1.639	0.85	–4.82	5.077	–3.76
321	0.954	0.84	–1.22	2.72	0.21	0.9	1.39	–1	4.23	–2.984	1.069	1.28	–0.49	2.313	–1.05
322	0.41	0.14	0.458	2.52	–0.42	0.66	0.96	0.14	4.82	–2.153	0.509	0.94	–0.41	2.259	–1.53
323	0.281	1.65	–0.66	4.8	–0.77	–0.9	–2.3	3.05	4.59	–5.539	–0.5	1.61	–2.43	7.134	–2.54
324	–0.57	–0.3	1.132	5.49	–2.66	–1.1	–0.2	2.33	6.4	–6.42	0.807	0.47	1.75	2.259	–6.73
331	0.106	0.57	–1.05	6.17	0.44	0.91	0.47	–0.6	3.72	0.032	0.102	0.2	–1.24	4.84	1.901
332	0.423	0.51	1.192	3.68	1.64	1.27	0.68	–0.9	4.42	1.008	0.179	–0.11	–2.25	2.845	1.052
341	0.439	0.64	0.124	4.73	0.43	0.95	0.43	–0.3	3.9	–0.095	0.501	–0.14	–0.54	3.536	0.485
342	0.939	0.5	0.567	4.03	1.37	0.57	0.08	–1.3	6.62	2.415	–0.66	–1.14	–1.07	4.08	1.006
351	1.344	1.29	0.126	3.16	0.74	1.31	0.98	0.74	1.61	–1.85	1.373	0.36	–1.17	3.84	–0.71
352	1.877	1.52	1.17	1.16	1.81	0.91	–0.2	–1.1	4.45	0.742	–0.57	–0.88	–0.49	3.906	0.941
353	0.009	–0.8	–3.71	10.7	–1.12	–0.4	1.81	3.96	6.62	–1.734	0.601	–0.27	–0.42	1.821	–1.03
354	–0.31	0.07	0.561	6.16	1.81	–1.1	–0.9	–0	9.58	–0.516	0.293	0.8	0.944	2.185	–0.43
355	0.69	0.28	0.101	2.67	0.5	2.31	1.04	–1	2.18	–2.096	2.102	1.47	–2.16	0.603	1.325
356	2.969	1.61	0.463	0.56	6.25	1.8	1.75	–0.7	3.15	4.023	–0.03	1.35	–0.56	3.956	3.421
361	0.924	0.51	0.003	4.66	0.66	2.69	–0.4	2.18	2.52	–3.11	–0.3	1.37	–1.72	5.791	3.361
362	0.713	0.5	0.635	4.03	1.27	0.52	–0.4	0.36	4.83	–2.199	0.94	1.89	–1.94	2.918	–0.11
369	0.436	0.52	0.359	4.41	–0.24	1.49	0.52	–0.9	4.07	–0.962	1.308	1.55	–0.82	2.766	0.815
371	–0.14	–0.2	2.031	6.21	–0.06	0.22	–0	–0.8	3.88	–6.291	2.152	1.9	–1.09	2.542	–0.66

372	–0.17	–0.3	1.527	5.46	1.18	–0.1	0.28	–0.5	4.13	–1.134	1.096	0.38	–2.57	3.991	0.745
381	0.389	0.13	0.641	4.18	1.5	0.15	–0.2	–1	4.67	–0.28	0.839	0.65	–1.89	2.052	0.475
382	0.86	5.09	0.304	3.85	2.29	0.68	4.22	–0.7	3.35	–0.402	4.366	2.38	–3.8	–2.11	0.688
383	1.469	2.11	0.465	1.97	1.16	0.7	1.78	0.27	4.8	1.513	5.588	4.05	–4.75	–3.19	0.154
384	0.313	1.1	1.95	5.56	0.08	0.77	–0.3	1.74	4.5	0.724	0.239	–0.09	–1.46	3.284	–1.82
385	0.863	1.16	1.874	3.91	1.58	–1.6	0.51	–1.1	7.47	2.978	0.34	0.15	1.85	7.021	–1.82
390	0.951	1.24	1.181	2.95	0.48	0.29	–0.2	–0.3	4.92	–2.179	0.757	1.1	–0.56	3.352	2.12

Notes:
1. The 'unido tfp' is based on UNIDO data except the capital stock series which is borrowed from the NBER source.
2. The employment ('empl') series is based on the NBER source, whereas the output deflator ('defl/r') series is based on the UNIDO source (estimated as the ratio of nominal output to real output).
3. The 'IRY' series is the ratio of inventories to shipments (derived from the NBER database).
4. All series are expressed as annual growth rates.
5. For the names of the ISIC sectors see Appendix 6.2.
6. The TFP series (both UNIDO and NBER) are estimated by the writer, based on the formula explained in Chapter 7. It includes the inputs capital, labour, and material but it excludes energy. The reason for excluding energy from all calculations of TFP is that it is desirable to have consistency throughout the quantitative work since some data used in other places do not include energy (e.g in UNIDO data). Note that TFP as estimated by the writer (thus excluding energy) and TFP as estimated by NBER are virtually the same since the energy component is usually very small in the production process.

APPENDIX 7.3: THE 85-SECTOR DATABASE, USA, FROM 1987 TO 1996 ANNUAL AVERAGE GROWTH RATE, AS PERCENTAGE

	SIC	names	TFP	IRY
1	201	Meat products	0.360	2.0
2	202	Dairy products	–0.507	3.0
3	203	Preserved fruits and vegetables	–0.109	–1.0
4	204	Grain mill products	–0.289	–0.6
5	205	Bakery products	–1.219	1.0
6	206	Sugar and confectionery products	0.819	–1.1
7	207	Fats and oils	0.195	2.5
8	208	Beverages	1.706	0.1
9	209	Miscellaneous food and kindred products	1.070	–0.7
10	21	Tobacco products	0.853	–4.8
11	221–224	Broadwoven fabric mills, cotton, manmade fibre and silk, wool, narrow fabric mills	1.310	0.9
12	225	Knitting mills	1.876	–2.0
13	226	Textile finishing, except wool	0.834	–0.3
14	227	Carpets and rugs	1.398	0.0
15	228	Yarn and thread mills	1.012	–1.4
16	229	Miscellaneous textile goods	0.321	–0.2
17	231+232	Men's and boys' suits and coats, and furnishings	0.397	–1.2
18	233+234	Women's and misses' outerwear, Women's and children's undergarments	1.581	1.0
19	235–237	Hats, caps, and millinery, girls' and children's outerwear, fur goods	1.524	–1.9
20	238	Miscellaneous apparel and accessories	1.689	0.6
21	239	Miscellaneous fabricated textile products	0.721	–0.6
22	241–243	Logging, sawmills and planning mills, millwork, plywood, and structural members	0.079	–1.0
23	244+245+249	Wood containers, wood buildings and mobile homes, miscellaneous wood products	0.503	–1.9
24	251+252	Household furniture, office furniture	–0.105	–1.5
25	253+254+259	Public building and related furniture, positions and fixtures, miscellaneous furniture and fixtures	–0.200	–3.5
26	261–263+265	Pulp, paper, and paperboard mills, paperboard containers and boxes	–0.236	0.0
27	267	Miscellaneous converted paper products	–0.043	–1.4

28	271–275	Newspapers, Periodicals, Books, Miscellaneous publishing, Commercial printing	–1.192	–0.8
29	276–279	Manifold business forms, Greeting cards, Blank books and bookbinding, Printing trade services	–0.759	–1.9
30	281	Industrial inorganic chemicals	0.819	–0.5
31	282	Plastics materials and synthetics	0.464	–1.0
32	283	Drugs	–2.167	0.1
33	284+285	Soaps, cleaners, and toilet goods, Paints and allied products	0.445	–1.9
34	286+287	Industrial organic chemicals, Agricultural chemicals	0.184	–1.4
35	289	Miscellaneous chemical products	–0.045	–0.1
36	291	Petroleum refining	–0.274	–0.4
37	295+299	Asphalt paving and roofing materials, Miscellaneous petroleum and coal products	0.799	0.9
38	301–2+305–6	Tires and inner tubes, Rubber and plastics footwear, Hose etc., Fabricated rubber products	1.469	–2.2
39	308	Miscellaneous plastics products, n.e.c	1.349	–0.6
40	313+314	Footwear cut stock, Footwear except rubber	0.465	1.7
41	311+315–7+319	Leather tanning and finishing, Leather gloves, Luggage, Handbags etc., Leather goods, n.e.c.	1.608	–2.4
42	321+322+323	Flat glass, Glass and glassware, Products of purchased glass	1.890	–1.9
43	324+325	Cement, Structural clay products	1.939	–1.2
44	326	Pottery and related products	1.370	–1.7
45	327+328	Concrete, gypsum, and plaster products, Cut stone and stone products	1.532	0.2
46	329	Miscellaneous non-metallic mineral products	1.158	–0.2
47	331+332	Blast furnace and basic steel products, Iron and steel foundries	1.897	–1.1
48	333+334+335	Primary and Secondary nonferrous metals, Nonferrous rolling and drawing	0.180	–2.8
49	336+339	Nonferrous foundries, Miscellaneous primary metal products	1.195	–0.6
50	341+342+343	Metal cans and shipping containers, Cutlery, hand tools, and hardware, Plumbing etc.	0.800	–1.4
51	344	Fabricated structural metal products	0.695	–2.0
52	345+346	Screw machine products, bolts, etc., Metal forgings and stampings	1.094	–1.6
53	347+348	Metal services, n.e.c., Ordnance and accessories, n.e.c.	–0.104	–1.7
54	349	Miscellaneous fabricated metal products	0.606	–2.6
55	351	Engines and turbines	0.192	–1.0
56	352	Farm and garden machinery	1.400	–3.1
57	353	Construction and related machinery	0.859	–3.8
58	354	Metalworking machinery	0.720	–1.2
59	355	Special industry machinery	1.778	–1.8
60	356	General industrial machinery	0.424	–2.3

APPENDIX 7.3: CONTINUED

	SIC	names	TFP	IRY
61	357	Computer and office equipment	9.624	–7.6
62	358	Refrigeration and service machinery	0.878	–2.0
63	359	Industrial machinery, n.e.c.	1.666	–1.1
64	361+362	Electric distribution equipment, Electrical industrial apparatus	1.024	–3.4
65	363	Household appliances	1.112	–3.5
66	364	Electric lighting and wiring equipment	0.369	–2.5
67	365+366	Household audio and video equipment, Communications equipment	2.422	–3.3
68	367	Electronic components and accessories	9.912	–7.3
69	369	Miscellaneous electrical equipment and supplies	0.683	–3.9
70	371	Motor vehicles and equipment	0.385	–1.6
71	372	Aircraft and parts	–1.133	1.2
72	373	Ship and boat building and repairing	–0.308	2.7
73	374	Railroad equipment	1.624	–2.1
74	375	Motorcycles, bicycles, and parts	1.914	–1.4
75	376	Guided missiles, space vehicles, parts	–2.734	4.0
76	379	Miscellaneous transportation equipment	0.752	–6.0
77	381+382	Search and navigation equipment, Measuring and controlling devices	–0.120	–1.0
78	384+385	Medical instruments and supplies, Ophthalmic goods	0.439	–2.4
79	386	Photographic equipment and supplies	0.225	–1.5
80	387	Watches, clocks, watchcases, and parts	–1.253	1.7
81	391	Jewellery, silverware, and plated ware	0.699	–1.8
82	393	Musical instruments	–0.136	1.5
83	394	Toys and sporting goods	1.564	–0.1
84	395	Pens, pencils, office, and art supplies	1.783	1.2
85	396+399	Costume jewellery and notions, Miscellaneous manufactures	0.922	0.0

Source: The calculations of TFP and IRY (ratio of inventories to shipments) are based on the initial NBER data.

APPENDIX 7.4: STATIONARY LONG-TERM COEFFICIENTS AND VECTORS V

Consider the system of equations (the variables are expressed as vectors, or matrices):

$$\begin{aligned} x_t &= Ax_{t-1} + \varepsilon_t \\ \Rightarrow \Delta x_t &= -(\mathrm{I} - \mathrm{A})x_{t-1} + \varepsilon_t \\ \Rightarrow \Delta x_t &= \pi x_{t-1} + \varepsilon_t \end{aligned} \qquad \text{(I)}$$

Now, suppose that matrix π has a full rank. Then, the solution to equation I is:

$$\begin{aligned} \pi_{11}x_{1t} + \pi_{12}x_{2t} + \ldots + \pi_{1n}x_{nt} &= 0 \\ \pi_{21}x_{1t} + \pi_{22}x_{2t} + \ldots + \pi_{2n}x_{nt} &= 0 \end{aligned} \qquad \text{(II)}$$

This solution is the 'long-term' solution. This means that the vector x_t must be stationary.

Now, (II) can be transformed into an error correction form. As an illustration, one single equation will only be considered:

$$\Delta x_{1t} = \pi_{11}x_{1t-1} + \pi_{12}x_{2t-1} + \ldots + \pi_{1n}x_{nt-1} + \varepsilon_{1t} \qquad (1)$$

which after normalization becomes:

$$\Delta x_{1t} = s(-x_{1t-1} + s_{12}x_{2t-1} + \ldots + s_{1n}x_{nt-1}) + \varepsilon_{1t} \qquad (2)$$

in which s is the speed of adjustment and the terms inside the brackets constitute the so-called ECM. Equation 2 is derived from a VAR (1) model (only one lag is considered).

The coefficients of this ECM constitute a vector V:

$$V = (-1, s_{12}, s_{13}, \ldots s_{1n}) \qquad (3)$$

This vector V or a matrix of such vectors if we consider many equations from II needs to be determined. There are several methods to determine such vectors V (Maddala and In-Moo Kim, 1998, from p. 165 onwards):

- Johansen's method, based on the partial canonical analysis (significant eigenvalues)
- Box-Tiao method, based on the levels canonical correlation analysis
- Ahn and Reisnel method, based on an algorithm of iterated least squares.

Note that the Johansen's and Ahn and Reisnel's methods yield the same final results (Maddala and In-Moo Kim, 1998, p. 168). In addition, Ahn and Reisnel applied their method on both stationary and non-stationary series in 1988 and in 1990 (ibid.).

Once these vectors V are estimated, the coefficient 's' of the ECM can also be estimated via OLS.

Finally, note that if the matrix π has not a full rank, then the vectors Vs are obviously the well-known cointegrating vectors.

In relation to Chapter 7 results, equation 3 above is the vector V for each one of the endogenous variables in Table 7.8. For example, for the VAR (2) model, the V representing YL is: (–1, 0.45, 0.45, 0.36, –0.11, 0.0189, 0.1508, –0.0003, 0.0135). Although the FIML method used to estimate this V vector (according to Johansen's procedure) does not specify which V corresponds to which endogenous variable, some criteria are used (see sub-section 7.3.7) to determine this correspondence. One of these criteria is the set of corresponding coefficients in the estimation of OLS regressions (see Table 7.7). Thus, the OLS regression for YL provides the coefficients (from Table 7.7) shown in the following table (which also has the corresponding V vector for comparison):

Coefficients	YL	ML	KL	PY	IRY	UN	PKPL	INT	Constant
Of the OLS regression	–1	0.62	0.042	0.187	–0.135	0.016	–0.221	0.00009	0.011
Of the vector V	–1	0.45	0.45	0.36	–0.11	0.0189	0.1508	–0.0003	0.0135

Once these vectors V are determined, as briefly just described, they are inserted in equations 2 above (hence the dependent variable in equation 2 is dYL in this example) in order to estimate through OLS the coefficient 's' (the speed of adjustment) for the ECM representing YL (this 's' has the value 0.64 in Table 7.9, third column, top-left segment); the speed of adjustment of the other ECMs, and the coefficients of lagged variables dYL1=YL(–1)–YL(–2) (–2 means two lags), dML1, dKL1, dPY1, dIRY1, dUN1, dPKPL1, and dINT1. All these results for dYL in VAR (2), in other VAR models and for the other dependent variables are shown in Table 7.9 of Chapter 7. The computer outputs using Microfit 4.0 (Pesaran and Pesaran, 1997) are available on request.

NOTES

1. These concepts were analysed in Chapters 4 and 5.
2. This system was alluded to by Milgrom and Roberts (1990) in the above quote.
3. In this study, it is suggested that there are many common elements between mass production/Fordism and lean production systems, which are both represented by the main axis (see Chapter 4).
4. For example, *cell manufacturing* is closely related to the JIT/QC system (Chan et al., 1999; Magjuka and Schmenner, 1992; Steele et al., 1995). In addition, the most comprehensive JIT/QC system is the one developed and used by Toyota. Fujimoto (1999, app. B) describes 30 interrelated practices and techniques. Other firms in Japan and other countries have adjusted the Toyota system to their own needs. Another example is the 'agile' system, as an extension of the LPS, as described by Rycroft and Kash (1999).
5. For a complex evolving and emergent system as the LPS introduced by Toyota (Fujimoto, 1999), it is not surprising that it takes many years to transfer even a small part of it to a different country. This also agrees with the remark often made in this book that countries and firms develop their own OIs depending on contingency aspects.
6. For example, cf. Taylor and Wiggins (1997) for a relevant comparison between the American and Japanese subcontracting systems.
7. *Rentai* means a joint responsibility between management and workers, whereas *wa* means harmony.
8. One hundred firms were labelled purchasers and were drawn from some end product industries. The vendors were 100 firms drawn from the raw materials and related industries.
9. *Ringi* means management for emergent or unusual situations.
10. Stamm and Golhar (1991) provide evidence that small firms can adopt JIT/QC practices.
11. In this respect, see also Abernathy et al. (1983, ch. 5).
12. However, since the registry's coverage is not complete, Watanabe estimates that in 1989 the total numbers of practising establishments and circles were 65 800 and 743 540 respectively. Also, the total number of participants was estimated to be about 5.5 million workers. Furthermore, QC circles are more concentrated in larger companies (ibid.).
13. Part of this article had appeared previously in the same journal in 1992.
14. Kano, a student of the guru of QC Ishikawa (1985) and an expert in quality management, had been visiting the USA since 1977.
15. The reverse arrows in this figure (not shown) would suggest that the JIT/QC is reinforced and becomes stronger with increases in productivity. This two-way causation will be econometrically explored in a sub-section below when the VEC model is examined. In addition, 'all causative analysis, as I have repeatedly emphasized, depends on theory' (Hicks, 1979, p. 67). Thus, the results of this chapter cannot be properly evaluated unless they are supported by the theoretical outcomes of all previous chapters. In particular, the propositions of Chapter 5 regarding the atomic/systemic model are relevant here as the concept of system implies feedback and reciprocity.
16. This point is also emphasized in previous sub-sections of this chapter.
17. This author introduced the concept of organizational capital in his previous articles, thus he recognizes the importance of OIs.
18. See note 3.
19. Cf. the special issue of the *Canadian Journal of Economics*, 1999, no. 2, for several articles related to Solow's paradox, for example, Triplett (1999).
20. However, as indicated in Chapter 5, JIT/QC is primarily influenced by the POM.
21. Whenever necessary, the corresponding TIOP should also be included.
22. Baily and Gersbach (1995) can be considered pioneers in this respect and certainly provide more credibility to the suggestions and propositions of this study.
23. Again, Figure 7.8 is useful in this respect.
24. Figure 7.8 helps to appreciate this point.
25. Each model has advantages and disadvantages. Thus, it becomes more rigorous to apply

as many models as possible. However, the list of models used here is not exhaustive but rather selective.

26. See also appendix I of Matthews et al. (1982), for a general discussion of the assumptions and measurement of TFP.
27. The value-added formula ($TFP' = VA' - wE' - (1 - w)\ CS'$) usually gives TFPs which are often about double of those TFPs when using the gross output formula. Some regressions in the analysis below use this formula for comparison.
28. This ratio and the others which are part of equations 7.3 and 7.4 are expressed as growth rates in this section.
29. These ratios of (Y/K) and (K/L) are only proxies of OIs and TIs respectively, since they are ratios of quantities of output, capital and labour. Thus by definition, and according to the analysis presented in Chapters 3 and 5, they do not include organizational or technical features *per se*.
30. Not all of these elements are TIs related. For example, performing each individual value-adding task more quickly (a worker's intensification of efforts) is part of the POM.
31. In page 104, Fujimoto (1999) adds that this happened also in the early 1980s. A similar conclusion has already been reached in the present book as will also become more apparent below in this chapter.
32. See also Flood and Lowe (1995) for their use of the stocks to sales ratio in their econometric study of inventories in Australia.
33. Recently, a few articles have established a quantitative relationship between inventories reduction and increases in productivity on a firm basis (for example, Lieberman and Asaba, 1997; Lieberman and Demeester, 1999). For Japan, Fujimoto (1999) mentions that during the late 1960s and early 1970s many of Toyota's first-tier suppliers dramatically decreased their inventory levels, thus contributing to a more efficient network of car production.
34. Runkle (2000) provides some examples of firms from diverse industries that managed to considerably reduce their inventories recently, including Dell Computers, compared with other firms. Chapter 6 has explained that Dell has been a champion in adopting the JIT/QC system.
35. A similar phenomenon took place in Australia, where Flood and Lowe (1995) showed that a significant change in the behaviour of inventories during the period of the early 1980s to the early 1990s occurred that was partly due to an increasing use of sophisticated inventory management techniques and greater production flexibility.
36. A similar regression analysis to the one for the category 382 (see above) was conducted (not shown here) that confirms this assertion.
37. It is important to emphasize that the observed decline of the inventories ratio during the last 20 years in the American economy cannot be part of any long-term economic cycle such as the Kondratieff cycle, because, as can be seen in Figure 7.4b, this ratio was constant for about 30 years prior to the early 1980s.
38. A regression over this period, not shown here, confirms this conclusion.
39. It was extensively shown in Chapter 4 that inventories decrease dramatically when JIT/QC is implemented. Also, in Chapter 5 it was shown that the decease in waste, hence inventories, is a consequence of the negentropic trait of OIs such as the JIT/QC system.
40. For the sector 383, the following results show that even though unemployment (as a proxy of the business cycle) is included in the regressions, the variable IRY is still significant and has the right sign and magnitude: TFP = 2.6 – 0.14 IRY – 0.08 UNE (the t-statistics are 6.2, –1.7 and –2.7 for the constant, IRY, and UNE respectively; $R^2 = 0.25$). Also labour productivity (LAPRO)= 5.3 –0.35 IRY –0.03 UNE (the t-statistics are 11.6, –3.9 and –1.1 for the constant, IRY and UNE respectively; $R^2 = 0.34$). UNE stands for unemployment. The period examined is from 1962 to 1996 and variables are expressed as growth rates. Note that the coefficient of IRY in the TFP regression is less significant (90.5 per cent level) than in the LAPRO regression (99.95 per cent). These two regressions are 'pilot' regressions of the more detailed analysis that follows.
41. Appendix 7.2 shows the relevant data for the three periods. Overall, the sectors experiencing high growth rates in TFP are those experiencing high decline rates of IRY.

42. See Appendix 7.3.
43. Furthermore, similar results (to those above) were obtained when the NBER 28 ISIC 28-sector based estimated TFPs were used instead of the UNIDO-based TFPs (not shown here).
44. This confirms (at this stage) the two-way causality process between productivity and JIT/QC, as competition reinforces the implementation of JIT/QC.
45. The correlation coefficient between IRY and LAB is only 0.13, thus indicating that it is difficult to predict the outcome of employment as a result of JIT/QC. However, in a multiple regression (OLS) in which employment (LAB) is the dependent variable, and the three independent variables are IRY, real output (RO) and the import to export ratio (IMEX) (see sub-section 7.3.4 for its precise definition) for the 28 sectors, and for the period 1987–96 (variables expressed in percentage changes) the outcome is more clear regarding the relationship between employment and JIT/QC (T-statistics under the coefficients):

$$\underset{(0.4)}{LAB = 0.227} \; \underset{(3.7)}{+0.825\ IRY} \; \underset{(3.8)}{+\ 0.634\ RO} \; \underset{(-3.4)}{-0.362\ IMEX} \quad R^2 = 0.61$$

Thus employment and IRY have a significant elasticity of 0.82. The positive sign of 0.82 indicates that the more IRY decreases the lower the employment is, thus confirming Edquist et al. (2001, p. 37) who specifically state that some process organizational innovations related to work organization are labour-saving, providing that the compensating effects are not strong enough to neutralize the primary effects of labour-saving. In this respect it is worth noting that during 1987–96 employment for the whole manufacturing sector remained virtually constant (see also Tables 7.4 and 7.5). This is an important result in itself but its further development is outside the immediate scope of this study. Also note that the variable capital stock (CAP) was included in the above regression but it was found to be insignificant.
46. These data also show that probably an OI such as JIT/QC, a 'capital-saving type' of OI according to evidence provided in Edquist et al. (2001, p. 36), has, overall, neutral or positive effects on employment. However, as mentioned earlier (for example see Chapter 4), there is evidence that JIT/QC is a labour-saving OI (see also previous note), and hence if employment increases after the implementation of JIT/QC this increase must be due to compensation effects such as increases in demand (ibid., p. 84). This can be further verified by the results of the following multiple regression (OLS) (for the definition of variables see previous footnote plus the regressions of Table 7.3):

$$\underset{(3.1)}{RO = 2.1} \; \underset{(3.6)}{+0.48\ LAB} \; \underset{(-3.8)}{-\ 0.65\ IRY} \; \underset{(1.0)}{+0.1\ IMEX} \; \underset{(-2.9)}{-0.32\ DEFL} \; \underset{(2.4)}{+0.35\ CAP} \quad R^2 = 0.70$$

As expected, the variable IRY is highly significant with a very high elasticity of –0.65. The variable DEFL (deflator) is also significant and has an elasticity of –0.32. This latter elasticity can be used as a proxy for the demand elasticity with respect to prices and is definitely below one. This result agrees with Reati's (1998, p. 94) analysis, which indicates that if price elasticity of demand is less than one, then employment declines when a process innovation is implemented (although the income elasticity must be considered as well). This result confirms the conclusion that JIT/QC (a process innovation) increases real product and productivity. All these partial conclusions should be analysed in the light of the effects of various types of elasticities (Edquist et al., 2001, ch. 5). The various models presented in the present chapter contribute in clarifying some of these issues. Another conclusion is that, including the results of Table 7.5 where the capital productivity has substantially increased during the JIT/QC implementation period of 1987–96, JIT/QC is both labour and capital saving. Finally, it is a significant result that in the above regression, the rate of growth of real output (RO) not only depends on capital and labour but also on OIs (the IRY variable), on TIs and other innovations (the constant) and on market conditions (the deflator variable). For the variable CAP see previous note.

47. Thus, the inclusion of the significant dummy for one sector does not influence the other coefficients. This shows that these estimated coefficients are not biased by the presence of the dummy variable.
48. These data are not 'objective' in the sense that individual industries may not add to the total since a patent may be recorded in more than one industry category. The source is table 863 of the *Statistical Abstract* of the US Bureau of the Census (1997, p. 552).
49. The correlation coefficient of TFP and IRY is –0.86, of TFP and PAT is 0.70 and of IRY and PAT is even smaller, at –0.55. It is also worth noticing that some other data was briefly examined in this context and despite the lack of relevant statistics for the period examined here (not all years are covered, and again only 12 sectors were available), the conclusions are similar: TIs as approximated by R&D funds do not and cannot explain the differences in productivity unless OIs are included. This data, more precisely, is 'Total (company, Federal, and other) funds for industrial R&D performance, by industry and by size of company: 1953–98' compiled from data supplied by the National Science Foundation of the USA.
50. Weighted averages (to take into account the weights of sectors) did not significantly alter the results.
51. In this respect cf. also Sayer and Walker (1992), and previous notes in this sub-section.
52. See equations 7.3 and 7.4.
53. These results can be better understood by examining the relationship of equation 7.3, which links the three variables of labour productivity, capital productivity and capital to labour ratio together.
54. Total factor productivity has also grown substantially (see Table 7.5) during 1987–96 because capital productivity increased and the capital to labour ratio decreased.
55. Another year, instead of 1992, does not alter the results.
56. Out of 420 observations, about a dozen outliers were detected. The inclusion of DUM expresses the usual procedure in econometrics to lessen the effect of extreme values.
57. Thus TFP is excluded. This exclusion is due to the fact that TFP is a function of labour productivity and the capital to labour ratio, as shown in equations 7.3 and 7.4.
58. The variables UN, INT are exogenous in the sense that they are common (the same) to all sectors. The variable PKPL showing the relative price of capital to labour, is also a proxy for other exogenous variables such as prices of raw materials which are determined in international markets and so on.
59. All these variables except the rate of interest (which is already given as a percentage) are expressed as first differences in natural logs.
60. The 2SLS regression results are overall the least significant results of all regressions in this chapter. This is expected, because 2SLS heavily depends on the chosen proxies for endogenous variables. The VEC model that follows will better address the simultaneity of equations issue and provide more robust results.
61. This similarity is due to the close relation between TFP and YL, as shown above.
62. The *Microfit* program is used for all the VECM analysis, see Pesaran and Pesaran (1997). The printouts of the tests are not provided here for space reasons unless otherwise indicated.
63. This stationarity is desirable for two reasons: first, because the intention is to examine growth rates, thus being consistent with previous models (with stationary data, OLS estimates are not biased). Second, because the raw pooled 1008 data is also stationary (an examination of the raw data yields similar results for the VEC model).
64. See also Appendix 7.4.
65. See also Appendix 7.4.
66. The elimination of outliers is a major problem in econometric analysis, so it was preferred not to remove them.
67. A similar reasoning process would lead to similar conclusions for the other coefficients (speed of adjustment).
68. If the coefficient of IRY of the ECM were positive then the impact of JIT/QC would have only been temporary (in this case the error correction term draws the impact of the shock back to the trend).

69. This makes sense for JIT/QC because this system is not only about a reduction in the ratio of inventories to shipments, but also about increases in quality of products, and about drastic changes in the PROSIBB.
70. As explained above and in Appendix 7.4, the computations of the Vs does not automatically determine the underlying dependent variables (that is, which V corresponds to which dependent variable).
71. As expected because the data are stationary.
72. However, in notes in sub-section 7.3.3 the estimated regressions suggest that the IRY elasticity could be very high even in the presence of the measurement of other innovations (captured by the constant in that case). More quantitative work is needed to separate more precisely the impact of the various types of innovations on economic growth.
73. This last remark is confirmed by Stern's article (1993, p. 342), although the period examined by this author was the starting period of adoption for many American firms themselves: 'It is perhaps significant that the effect of "just-in-time" inventory policies, which have driven US inventory-sales ratios to historically low levels, is not in evidence in the year-to-year inventory-sales ratios of the foreign affiliates of US parent firms during the 1982–87 span … '. Furthermore, the adoption of an OI such as the JIT/QC system in various countries and firms heavily depends on contingency factors (as repeatedly mentioned in this book).

8. Conclusions, limitations, implications

8.1 INTRODUCTION

The last six chapters have provided evidence of the importance and the role of OIs in shaping industrial and economic growth.[1] In particular, the evidence was directed towards the following points. First, in order to complement the existing theories of economic growth, OIs must be explicitly included in the production process. Second, OIs are distinct from TIs and must be separately analysed when exploring their impact on economic growth. Third, OIs can be grouped into three main interdependent axes with emphasis on the main axis, namely the JIT/QC cum Fordism axis (which is extensively covered in this study). Thus, the meaning of OIs used in the present volume is primarily related to what takes place inside the firm and encompasses not only its organizational form (for example, M-form) but also its organizational system as described by the four processes of firm operations (PROSIBB).

Fourth, the main axis JIT/QC cum Fordism can be further explored by introducing the four fundamental processes of firm operations (PROSIBB) and especially the process of movements and its related kinetic costs; thus the process of growth is ultimately a function of the four PROSIBB and, in particular, the POM. Thus, the 'waste' described by Williamson (1994) and its link with transaction costs is not sufficient to explain why firms differ in terms of growth and performance. Some authors (for instance, Tersine, 2004) have explicitly made the concept of waste the centre of their research. 'Waste' has also been the object of analysis of this volume. Fifth, industrial growth is basically a micro phenomenon; hence emphasis is given to firms and sectors, and especially to leading firms and sectors. Sixth, basic quantitative evidence on a sectoral basis is provided for two countries, the USA and Japan, by examining the links between OIs and productivity (for example, TFP). Finally, econometric evidence on a sectoral basis is provided for the USA, by exploring the impact of the JIT/QC system on sectoral productivity growth (TFP or labour productivity).

To bring adequate evidence for all these points, Chapter 2 established the major OIs by reviewing the organizational features of American and Japanese firms from the last decades of the 19th century till the end of the 20th century. Chapter 3 established the need to separate OIs from TIs and incorporate the

OIs into the production function of firms. Chapter 4 provided evidence for the positive role major OIs played in promoting industrial growth in various periods of the modern economic history of the USA and Japan. Chapter 5 presented an extension of the theory of the firm in order to capture more precisely the importance of the micro foundations of the growth of firms, thus complementing the previous two chapters; thus, the four PROSIBB were introduced. Also in this chapter, the growth on a firm basis was related to the growth on a sectoral basis, in order to establish an atomic/systemic model of economic growth. Chapter 6 provided some basic quantitative evidence for the link between OIs and sectoral industrial growth. Finally, in Chapter 7, rigorous econometric evidence for the positive impact of the JIT/QC system on the revival of the American industrial sectors during the last 20 years of the 20th century was provided. In brief, the qualitative evidence was provided in Chapters 2, 3, 4 and 5, whereas the quantitative evidence was provided in Chapters 6 and 7.

The comparative study between the two economic giants, the USA and Japan, has facilitated the present process of research as theories and empirics are better understood and analysed if this process has the tool of comparisons. In addition, this comparative study was carried out on a historical basis, thus allowing a comparison between various periods, exposing appropriate theories and testing relevant hypotheses.

8.2 SYNTHESIS OF THE MAJOR CONCLUSIONS

Industrial and hence economic growth cannot only be explained by the quantity or quality of capital and labour, or by human capital and TIs. The role of OIs in explaining industrial growth is at least as significant as all these factors. In addition, although OIs are interrelated with all the other inputs of production in an endogenous and simultaneous way, this study has almost exclusively concentrated on how the major OIs influence the process of growth and all the other inputs. This influence was demonstrated in various ways: through review of the relevant literature, contrasting between OIs and TIs, theoretical inclusion of the variable OIs into a general form of a production function, use of pertinent concepts and theories for the inclusion of OIs (for example, transaction costs, core competences and capabilities, and so on), resort to the historical evolution of OIs, an extension to the theory of the firm with the introduction of the four processes of firm operations and hence costs and benefits, resort to the concepts of leading firms and leading sectors of the economy, a quantitative analysis of manufacturing sub-sectors in Japan and the USA, and an econometric analysis of the link between the lean production system and the manufacturing sub-sectors in the USA, plus a few other approaches such as an account of industry cases. All this evidence about the impact of OIs on economic growth is primarily evidence about substantial increases in productivity, and hence growth.

With respect to the major OIs analysed in this study, emphasis was purposely put on only some aspects of OIs, in the belief that these aspects are the moving forces of all OIs. This belief emanated from a detailed analysis of all major OIs in relation to economic growth. These emphasized aspects are the internal organizational features of firms radiating from the main axis of scientific management cum Fordism versus JIT/QC or LPS, and more precisely from the POM. The other two axes, that of vertical integration and that of networks depend on the main axis to a great extent; for example, the recent tendency for disintegration of firms in many industrialized countries (cf. Krajewski and Ritzman, 1999, p. 99; or Grant, 1998, p. 196 or 326) is due to the adoption of the LPS, which is the core trait of the main axis. It is this axis that has generated the American supremacy in the 20th century and later the serious challenge by Japanese leading sectors. The empirical evidence presented in Chapter 7 showed that the American industrial sub-sectors revived and counter-challenged the Japanese ones in the last 20 years because they adopted the LPS (*ceteris paribus*[2]). This main axis can be extended beyond the boundaries of the 20th century; thus, in the 19th century, scientific management was not totally absent but it had a different face based on the effort to rationalize the production processes via various methods; in the 18th century before the First Industrial Revolution in Britain, craftsmen like blacksmiths met individual requirements according to the pull system observed in the modern JIT/QC process (on this point see also Linge, 1991); furthermore, the 21st century might fully 'imitate' the 18th century with a wider adoption of the LPS and the advent of Internet-oriented marketing procedures.

To further explore the main axis and its links with other OIs, a system of the fundamental operations of firms (the four PROSIBB) was presented in a detailed manner in Chapter 5. There, it was argued that we need to take into account not only transaction costs, production costs and competencies in a general and generic way, but also in a more analytical and 'atomic' way so that more light is shed on the nature, functions and usefulness of OIs and other innovations such as strategic ones. Thus, it was suggested that besides transaction costs we also have kinetic costs that are primarily linked to the OIs;[3] we also have wisdom costs, which are related to TIs but not exclusively, and we have strategic costs which are related to the planning process. Such a fundamental exploration of all firm operations and activities (the PROSIBB as shown in detail in Sanidas, 2005b, and as explored in Chapter 5), and hence costs and benefits, show why the main axis is paramount in exploring the impact of OIs on industrial and hence economic growth. With the introduction of the four PROSIBB, their opportunity costs, their unique properties, their strong impact on the tangible inputs of production (TIOP, that is, capital, labour, land and so on), the theory of the firm is extended, and, OIs and other types of innovations become an explicit economic phenomenon of the evolution of firms and the economy. The atomic/systemic model of economic growth, also proposed in Chapter 5, completes the links between the extended theory of the firm, OIs and economic growth.

Moreover, this volume is based on what Tomer (2001) calls the heterodox man (as opposed to the economic man, who maximizes utility) whose many characteristics are included in the PROSIBB. Also, a more general philosophical approach is infused through this study based on philosophers like Plato and Aristotle and the concept of well-being such as that analysed by Tomer (2002). This well-being is partly subjective, hence related to emotional, mental, spiritual and other factors (ibid., pp. 41–2) which establish people's overall level of functioning. Happiness is probably a balance (not an equilibrium[4]) between virtue and utility, between conscious and subconscious, between power and service, or between money and achievements, or between all these elements and others. This happiness is included in the model of the firm proposed in this study through the four PROSIBB and its interactions with the outside world and the outputs produced (see Chapter 5).

In a way, this study suggests that there are no optimal or suboptimal situations from the consumer or worker's viewpoint. An employee's effort depends on the dynamic interplay between the four PROSIBB (and as themselves are influenced by all forces outside the firm). Utility exists (it is part of the POW) but cannot be maximized on its own because there are so many other inputs that constitute the POW. In particular utility is to some extent opposite the input of virtue (see also Tomer, 2002, p. 40, for a similar type of conclusion). On the other hand, production and other tangibles can be optimized. This asymmetry between non-optimization of consumption and optimization of production certainly extracts another evil from Pandora's box that probably invalidates attempts to put the firms, individuals and society under traditional general equilibrium conditions. From the firm's viewpoint there is always the question of power (POW) and survival (POS), which set the motion for the kinesis of the POM and the authority of the POC, and hence the birth, growth, evolution and death of firms is inherently and holistically dependent on the role of the diamond of the black box (see Chapter 5). Amongst all these elements of the four PROSIBB and their interaction with the TIOP, thus including the points on happiness, the role of optimization and so on, OIs play a fundamental role not only in promoting economic growth but also societal growth (the qualitative improvements in the well-being of people (see also Sanidas, 2004a)).

As economic growth is one of the main issues of this study, the very close link between industrial and hence economic growth and the various facets that the main axis can take is now further discussed. From the analysis presented in this study it becomes apparent that it is only when this axis takes the appropriate form that accelerating industrial growth ensues. Thus, it was following the appearance of new TIs and large firms at the end of the 19th century that the main axis had to take the form of scientific management cum Fordism in the USA so that industrial growth could be sustained for many more decades. This form of OIs generated economies of scale and scope given the extensive demand by a growing population with constantly rising real income, given the abundance of resources in that country and given the favourable institutional background. Later, during the 1960s and 1970s, after the challenge of the Japanese leading sectors, which introduced in praxis the JIT/QC system and hence substantially lowered costs, the main axis had to adjust again in the USA, by becoming more flexible, more customer oriented, more quality oriented or, in other words by adopting the JIT/QC philosophy. Thus, industrial growth could be again sustained at relatively high levels despite the already very high income per capita achieved by the USA.

Thus, one of the main conclusions of this study is that the appropriate form of the main axis of scientific management cum Fordism versus JIT/QC or LPS is a necessary condition for sustained industrial growth, but of course is not sufficient as many other factors are part of the complex process of this growth such as TIs, quality of labour and so on. This conclusion is similar to the one reached by Harbison (1956) who emphasized in a rather neglected paper amongst other points: 'Large expenditures for equipment and machinery are likely to be quite unproductive unless there is a corresponding investment in organization … '. In the present study, the importance of the JIT/QC process is emphasized because this process is an overall organizational phenomenon as demonstrated in previous chapters.[5] The contiguous scientific management and its many consequences has been shown to be significantly important for both the USA and Japan, although in different periods and in different ways. Striving for continuous rationalization and improvement has been present one way or another in the USA since the 1910s, whereas for Japan mainly since the end of World War II and mainly via quality control. The different facets of scientific management have been explained in various parts of this volume.

The other two axes of vertical integration and networking are as important to analyse as the main axis as this was shown in Chapter 4.[6] However, from the empirical viewpoint, mostly the main axis was scrutinized in Chapters 6 and 7, where it was found that the manufacturing firms and sectors that led the national economies of the USA and Japan were those that appropriately adopted the characteristics of this axis. Thus, although the main axis was more emphasized throughout this study, the elements of the other two OIs axes were sufficiently explored in order to bring qualitative evidence about their positive impact on industrial growth (the three axes are also intrinsically interrelated). For example, the disappearance of the domestic system and its gradual replacement by the factory system created a substantial increase in economic growth. The reasons for this boost can be summarized in the savings and reduction of transaction costs as well as the appearance of new TIs that in turn shaped the emerging factory system, although, according to the analysis of the four processes of firm operations (PROSIBB) in this study, it becomes clear that all types of costs and benefits ought to be examined in order to understand the passage from one production system to another. Also, the integration of mass consumption and distribution and the parallel increase in vertical big business generated a new era of secular growth for the USA, thus reconfirming Chandler's thesis. In addition, the comparison of the American firm organizational system with that of Japan produced some other significant conclusions.

Thus, one of the positive side effects of this study was the testing of Chandler's thesis regarding the birth and expansion of managerial capitalism in the USA and its contribution to economic growth on a more general basis. One of the major conclusions presented in this study is that, for the USA, both the qualitative and the quantitative analyses provide extra support for Chandler's thesis, at least up to the end of the 1970s.[7] However, the American experience of the 1980s and the 1990s cannot be explained only in typical Chandlerian arguments; as was shown in Chapters 6 and 7, the contribution of the main axis of this analysis based on the importance of intrafirm OIs such as the JIT/QC or lean production system was considerable during the last two decades of the 20th century.[8] In addition, the strong Japanese industrial growth from the very early period of the 1870s to the 1930s, and especially after World War II, cannot be explained strictly with the Chandlerian variables which are mainly based on the exploitation of economies of scale and scope created by big business. Effectively, there is something extra in Japan, expressed in two interrelated ways. First, we have what Fruin (1992) called economies of scope of joint production and cooperation by small focal firms, which gives a different dimension to the explanation of economic growth by small and medium firms; these enterprises are interlinked in a mesh that generates healthy competition and neutralizes the big profits of big monopolistic or oligopolistic corporations, such as those found in the USA; the Japanese model could and has created more value-added than the American model in particular periods of time. Second, we have a reduction in kinetic costs generated by the adoption of the JIT/QC system, as extensively analysed in various parts of this study.

These last points bring forward an important aspect of the assertions seen throughout the whole study: there is no superior model of firm organization and industrial and, hence, economic growth; what is good in one country and in one period is not *necessarily* good in another country or another period. Thus, this conclusion espouses the premises of contingency theories. For instance, the integration of mass production and mass distribution, which took place in the USA at the beginning of the 20th century, took place in Japan only after World War II and to a much lesser extent. This was also due to the rise in the standard of living in Japan during the period 1950 to 1975, which was similar to that of the USA at the beginning of the 20th century; but it was mainly due to the different historical or context contingent evolution of firms during the last 120 years. Therefore, there is no process of determinism that pushes entrepreneurs towards a certain type of organization and management (for a good example of the rather contrary forces of contingency and determinism, cf. Elazar, 1999). The same thing can be said about the adoption of the multidivisional form of big companies, which started propagating in Japan only recently after some other changes took place.

Another way of expressing the contingency approach to industrial development is to examine the conditions under which growth took place in the two countries. Thus, according to Wilken's (1979) terms, the Americans had more resources, more time and more opportunities for growth, but they did not take full advantage of all these extra factors, contrary to the Japanese who, despite much more adverse conditions, exhibited a higher rate of economic growth in various periods of their history. In other words, the opportunity cost in American secular growth was higher than that of Japan, though one has to consider the dual nature of the Japanese economy in order to conclude this with safety. Thus, Japan has a group of modern dynamic sectors (for example, automobiles, electronics) and also a group of much less efficient sectors (for example, food products, banks) that are undergoing some substantial changes at the moment. Porter et al. (2000) emphasize this point well:

> While studying some of Japan's most formidable industries, however, we discovered a Japan that no one was talking about. Alongside the highly competitive industries were other industries that were highly uncompetitive. It was as if there were two Japans. As time passed, the uncompetitive Japan remained obscure and often hidden, and it never showed signs of improvement ... (ibid., p. ix)

On the other hand the American economy has been involved in the rationalization of the quality control process (as part of the holistic JIT/QC process) in a thorough way during the last 15–25 years; hence a higher growth than before has been the relative experience. In Chapter 7 it was shown in a rigorous econometric way that the manufacturing American firms successfully imitated the Japanese lean production system in the last 15–20 years of the 20th century and accordingly grew faster than before, especially in terms of productivity gains. The proxy variable for the LPS used in various cross-section or pooled time series data, namely the inventories to sales ratio, has been found to significantly have an impact on growth (and vice versa). If data were available on other proxies, such as the percentage of defects or the percentage of various types of employment, an equivalent empirical analysis would have been carried out as well. Nonetheless, the proxy of the inventories to sales ratio is *fully supported* by the pertinent qualitative analysis as presented in several parts of this study.

So far the conclusions presented are mainly related to the importance of OIs (on a micro level, that is, firms and industries) in having a positive impact on industrial and hence economic growth. However, there is a link that was established in various parts of this study regarding the way OIs generate growth starting from the firm level and ending at the macro level. This link is the concept of leading firms and sectors that initiate OIs that influence other firms and sectors, and then the whole economy is affected and changes accordingly. This link is also related to the features of systems, such as organization emergency and entropy. Finally, this link is also related to sectoral systems, innovations systems and the evolutionary learning capabilities of firms. In this respect, it is worth consulting again Figures 5.2 and 5.3 in Chapter 5, which summarize the proposed atomic/systemic model of economic growth and the whole study to some extent.

Regarding leading sectors, as it was analysed in Chapter 6, the Second Industrial Revolution in the USA was connected with a strong growth in some managerial and fully integrated industries such as the chemical, oil, transport and machinery ones that heavily invested in OIs and adopted the OIs of scientific management cum Fordism, or, more generally, the process of rationalization. In Japan, the textile industry, also OIs oriented, played a preponderant role for a long time until the transport and machinery industries took over and accelerated the process of growth. More recently, the American economy accelerated its growth because its leading industries – mechanical and electric machinery (mainly computers, semiconductors and telecommunications equipment) have also been heavily investing in OIs, and especially in the JIT/QC system. On the contrary, Japan is still deeply in its 15-year recession because there are no leading sectors (*ceteris paribus*). All this was extensively explored in this study, where it was also emphasized that firms such as Dell Computers or Toyota played an exceptional, and hence a leading, role in the growth of their respective industries.

8.3 LIMITATIONS, OMISSIONS AND FUTURE RESEARCH

This study has several limitations. First, the empirical analysis was limited by the availability of data in the two countries. Thus, regarding the proxy used to empirically test the successful implementation of the LPS in the USA, namely, the inventories to sales ratio, it is not the only one; if relevant data were available on a sub-sectoral basis other proxies could have been used, such as the percentage of defects or changes in employment status, and so on.

Second, the *ceteris paribus* clause has been used in this study in several instances. For example, OIs and industrial growth have not been linked with exports or imports except in some limited cases. This link cannot be neglected in the context of a more comprehensive investigation; for instance, although some of the literature research has revealed that Japanese growth was not export led (for example, Boltho, 1996), thus confirming that other issues have been more prevalent in analysing the Japanese economic miracle, further exploration is needed to examine the impact of the role some leading industrial sub-sectors have played in promoting exports and hence growth. In this respect, the analysis by Campa and Goldberg (1997) has revealed that, as expected, transportation equipment, instruments and electrical machinery have been the main leaders in exports for Japan, thus confirming that there is a strong link between JIT/QC in these key sectors, exports and industrial and hence economic growth. Furthermore, this link could also be reinforced by considering other important empirical results, such as those by Sakakibara and Porter (2001), which show a positive association between domestic rivalry and international trade.

Another example of the *ceteris paribus* clause is the omission of additional evidence to support the tendency for disintegration and for the growth of smaller business in the USA recently; for instance, Stewart (1993, p. 70) shows a graph that compares growth of US GDP versus total sales of the Fortune 500 largest American companies since 1982; this comparison clearly demonstrates the ‘shrinking importance of bigness’ as the author says. The downsizing tendency of big business could bring more evidence to the arguments presented in this study.[9] More specifically, the main axis of JIT/QC has an impact on economic growth (as demonstrated in this book in many ways) and indirectly it has an impact on the vertical integration axis because it encourages outsourcing (hence the evidence just mentioned a few lines above about the ‘shrinking importance of bigness’). Yet another example of the *ceteris paribus* clause is the relationship between OIs and long economic cycles or waves; Freeman and Soete (1997, p. 65, table 3.5), though overwhelmingly preoccupied by TIs, presented some links between Kondratieff cycles and ‘organisation of firms and forms of co-operation and competition’; these links could be further analysed to support the assertions of this study. A final example of the *ceteris paribus* clause is the relationship between business cycles and OIs; thus, the very recent economic recession in the USA (in 2001–02) might be a new beginning for other OIs to emerge in the near future, or it might be the reinforcement of existing OIs, such as the JIT/QC system within *the context of Internet*.

Third, the most recent years, namely the period from 1996/97 to 2002/03 were omitted from the empirical analysis, because the impact of the Internet on the new economy probably started having a significant impact on various industries during that period (thus generating a structural break). This omission, though desirable from the point of view of making the empirical results more valid, must eventually be explored in future research, since the Internet TI will very probably introduce new OIs, so far not easily discernable, that might push productivity gains much higher. Sowinski (2001, p. 54) expresses a similar assertion: 'In recent years, however, use of JIT programs has exploded, and the increased use of the Internet has only pushed the process faster and farther … '. Litan and Rivlin (2001) have summarized the potential of JIT to increase productivity growth in mainly three ways: by reducing transaction costs, by increasing management efficiency (for example, regarding supply chains) and by increasing competition. For the links between the factors of information technology, organization and training and the system of Internet see Manecke and Schoensleben (2004). If we use the tool of PROSIBB developed in this book, it might become apparent that the POW (due to the increased importance of imagination and afflatus, sophistication, tacit knowledge and so on), the POM (due to a higher degree of harmonization of work movements, timing, mechanisms of feedback and so on), and the POS (due to a higher level of readiness and emergence, improvisations and heuristics, momentum and so on) will be more emphasized in the Internet context. On the contrary, the POC will play a lesser role in such a context (due to a more autonomous, less status, more standardized oriented firm, hence due to reduced transaction costs).

Fourth, certainly there is a relation between OIs and TIs, but this link has not been the main topic of this work although it was extensively analysed in many instances. In particular, how is it possible to empirically separate in a more precise way the effects of TIs and OIs on TFP or labour productivity?[10] Although this whole study is an attempt to answer this question, together with others, the future research agenda should include a thorough exploration of the evolution, dependence and links of these two types of technology, and their separate or concomitant impact on industrial and economic growth. As an example of such research the two types of Schumpeterian patterns of innovation (Malerba, 2002; Malerba and Orsenigo, 1995) could be further examined in the light of TIs and OIs (a preliminary relevant discussion took place in this study in previous chapters). In addition, the links between TIs and OIs can be further explored by scrutinising individual economies such as the Japanese one. As already analysed, Japanese firms have mostly imitated foreign technology in terms of TIs; thus Herbig (1995, p. 15) summarizes his work on 'innovation, Japanese style' in this brief statement: 'Adapting, not creating has always been the strong point of the Japanese … '. And yet, Japanese economic growth has mostly been very high. The Japanese case can be contrasted with the Russian case in which TIs have always excelled. And yet, economic growth has mostly been very low. More research of comparative national systems of innovations and sectors (SI in the sense suggested by Edquist, 1997; and SSIP in the sense of Malerba, 2002) is needed to further support the arguments of the present study, namely, that OIs are the neglected force that drives economic growth.

Fifth, the other two sectors of an economy, namely, the primary and services sectors were overall omitted in this study (but often included implicitly and sometimes explicitly). Their consistent inclusion into the present analysis would certainly enrich the findings and conclusions and push the arguments into a higher level of credibility and significance. For example, from a perusal of the already vast relevant literature (for example, see Duclos et al., 1995, or Matsumoto, 2002), it seems that the JIT/QC system is also applicable to the services sector. In addition, despite the fact that recently more than 70 per cent of national income in many developed countries comes from the service sector, it remains true that services and hence the whole economy depend heavily on the development of manufacturing industries (for example, relevant services branching off from the growth of computers). In addition, future research needs to answer, in a methodical way, questions of the type asked in a more comprehensive study such as that by Edquist et al. (2001, p. 42): 'are organisational process innovations more important for productivity and employment than technological ones in service production?' Also, are 'zero' inventories a desirable policy for some distribution sectors such as that of large retail stores?

Sixth, the empirical evidence was limited to only some aspects of OIs, some periods of time, and an emphasis was given to the American economy. This was primarily due to availability of data, space and time limitations. The quantitative analysis presented in this study should be regarded as a modest basis for future more extensive explorations.

Seventh, in this study, the macroeconomy is treated as a derivation of the following sequence: the four PROSIBB, leading firm, then firms, then leading sector, then sectors, then the whole economy. However, it would be possible to start directly with the macroeconomy and examine the impact of OIs within a macroeconomic model, but without denying the micro-foundations of economic growth. For example, starting with the conclusion that 'productivity growth and its determinants are the key to the understanding of the process of economic growth, capital accumulation and savings … (Chaudhri and Wilson, 2000, p. 66), productivity growth can be explored by introducing into Tobin's q (Tobin, 1969) the dimension of OIs. Thus, including an OI such as JIT/QC can enhance the marginal value of a firm, thus altering the value of Tobin's q, which, in turn, has an impact on investment and productivity (ibid., p. 61).[11] Furthermore, if we use a macro model, such as the one explored by Harvie and Kearney (1995), OIs can be incorporated directly into the aggregate supply function of output. In this case, the whole macro model will be affected by the presence of OIs, with significant consequences on employment, wealth and other variables.[12]

Eighth, it is an omission that formal models of economic growth, such as the exogenous and endogenous models, are not included in this study. These models, mainly macroeconomic in nature, could more formally demonstrate the propositions made in this volume, by linking micro and macro considerations.

Ninth, *direct* quantitative evidence of the importance and relevance of the four PROSIBB has not been provided here. However, this is a vast area for future research. For instance, Womack and Jones's (1996) book is a rich source of detailed study cases that describe how firms decided to adopt the JIT/QC system and how they successfully or not adopted this system. Such study cases provide valuable information that can be explored to evaluate the importance of the four PROSIBB. Also, the theoretical links between the PROSIBB, or OIs in general and generic concepts such as the resource-based view of the firm or its core competences have not been sufficiently explored in this study. For example, corporate strategy, governance and leadership can be explained by their embeddedness in the framework of OIs, as suggested in Sanidas (2002).

Tenth, OIs have been explored within the context of the two countries examined, namely, the USA and Japan. However, there are other OIs in other countries that have played a substantial role in their economic development. For example, the importance of China's township and village enterprises (TVEs) can be compared to the Japanese focal firms. Thus, TVEs are small, flexible, market driven, dedicated to human resources, to innovation and quality firms that employ a substantial part of the Chinese labour force and contribute significantly to the process of economic development in China (Harvie, 2000). Moreover, 'the key issue is not whether the TVEs will be able to maintain their industrial momentum in the light of deepening reforms, but rather the organizational and ownership form that will enable them to do so' (ibid., p. 98).[13]

Eleventh, some important elements of the LPS such as the tendency for more systematic outsourcing have not been sufficiently analysed. A more detailed exploration of the links between suppliers and the internal structure of firms due to the implementation of JIT/QC would have further supported the conclusions of this study. For example, the links between the main axis of JIT/QC and the other two axes of OIs, namely, vertical integration and networking, would have become more apparent. In order to carry out such a detailed exploration, Fujimoto's (1999) method and analysis would have been used. As this author says: 'as the case of company B suggests, the process of developing a black box design system requires step-by-step collaboration and tenacious capability building by both the assembler and the supplier ' (ibid., p. 167).[14] As a corollary to this point, the axes of vertical integration and networks need to be further explored in future research.

Twelfth, although this study's object has been the exploration of major OIs such as JIT/QC, minor OIs have a similar impact on economic growth. Minor organizational improvements (for example, re-engineering of a department's tasks) take place frequently in firms especially within the context of JIT/QC, thus the POM is continually altered and generally improved. Also, when an OI is introduced, a priori we do not know whether this OI is minor or major. A good example is the 'total productivity management' movement in Japan as Fruin and Nakamura (1997) explored it. Is this 'movement' (that represents a top-down production management) a minor or major organizational change? The answer is not clear at present. More research is needed to explore minor OIs within the context of major OIs, or without the embeddedness of major OIs. Perhaps, this research needs to be linked with more general families of social change as those suggested by Ven de Ven and Poole (1995), according to whom there are four such ideal families: life cycle, evolution, dialectic and teleology.

Thirteenth, although a substantial econometric analysis was carried out in Chapter 7 in order to provide evidence of the impact of JIT/QC on industrial growth, other methods and more quantification are possible. For example, the seemingly unrelated regressions (SURE) models could be used to evaluate the exact interdependence of manufacturing sectors. Effectively, since there are linkages between groups of sectors according to the input-output interlinkages, a SURE was also used to confirm the results (the relevant tests opted for the use of the SURE method which further strengthened the significance of IRY coefficients). In general, other panel data models (see Greene, 1993) could also be used to further testify the propositions made in this paper (some preliminary analysis – different intercept for each sector, or 'fixed effects' – was conducted that confirmed the present results; however, a detailed analysis is out of the scope of this study). It is also important to note that even 'classical' methods (for example, OLS) could be further explored. Thus, for each one of the individual manufacturing sectors an OLS regression of YL on ML, KL and PY showed a significant coefficient of IRY in most cases as expected (the magnitude of these coefficients varies considerably, for example –0.28 or –0.03).

Fourteenth, in this study emphasis was given to the private sector of the economy. However, the relationship between OIs and the policies of states and governments can also be analysed in the same way, since the non-private sector is an organizational system as well, and is specifically related to the institutional character of such system. Thus, Acharya and Chaudhri (2001, p. 670), in their study of policy strategies in Indian agriculture, say, 'An institutional mechanism needs to be put in place which enables producers or farmers corporate type management to come up for production, marketing and development of waste lands … '. In addition there is interplay between the private and the non-private sectors of the economy. In this respect, Chaudhri and Wilson (2001), in their study of the Indian agricultural growth, employment and poverty, concluded: 'State level factors like institutions, governance, implementation of laws on the statute books that favor the underdogs (for example minimum wage laws implementation in Kerala) and the attitude to a development strategy that pays as much attention to growth as to equity needs serious research' (ibid., p. 642).

Fifteenth, although empirical evidence in this study showed that it is possible to transplant an OI, such as JIT/QC, from the country of origin to another country, more theoretical analysis of such transplant as provided by other researchers like Kenney and Florida (1993), Womack and Jones (1996) and Liker et al. (1999) is needed. The validation of the empirical evidence would be enhanced if these latter references were explored in more detail.

Sixteenth, the actual interactions between the inputs of each PROSIBB and between each PROSIBB, or between the PROSIBB and the outside world have not been properly or fully analysed here. For such interactions see for example, Tomer (1998a), where it can be seen how the individual's behaviour is organizationally responsible. In particular, the process of wisdom is a major area of current research, although expressed in different ways. The interactions between the POW (and its numerous externalities) and the society could be very useful for the present study.[15] Sanidas (2004c) has also linked the PROSIBB to the world 'outside' the firm, by introducing two more levels of PROSIBB, that of the society and that of the individual.

Seventeenth, this study has probably overlooked (due to space and time limitations) some theories regarding the theory of the firm that might be of considerable support to the ideas proposed here. For example, Hannan and Freeman (1989) talk about the 'organistional ecology'[16] aspect of firms according to which – *inter alia* – there is a tendency for structural rigidities amongst established firms. This rigidity might make the imitation process described in Chapter 5 more difficult.[17] Also, the 'regulation theory' was not properly examined in this volume, although a brief review was made of the post-Fordism tendencies after the 1970s in the USA and other countries. This theory is inherently linked with technology and some new ways of organizing firms akin to the JIT/QC system.[18] Furthermore, well-known theories such as the agency theory can be further explored in order to link the role of managers, shareholders, and the PROSIBB (Sanidas, 2005b). In addition, some of the concepts introduced in this book could be compared with some other attempts to introduce similar concepts. For example, Hall (1997) links organizational change with kinetic theory (thus, the POM of the present book is relevant for comparison).

Eighteenth, some aspects of economic growth and fluctuations, such as financial crisis and economic policies have not been linked with technology. However, studies such as that of Moe (2004) can extend and support the conclusions reached in this volume (for example, via innovation systems).

8.4 IMPLICATIONS AND RECOMMENDATIONS

Once the importance of the appropriate OIs to be implemented in a new firm, industry or country in order to achieve higher growth rates in productivity and hence economic growth is *recognized*, there are many implications in terms of economic policy or strategy, or scholarly research. Thus, it is interesting to note that now in the USA, scholars have started talking about the 'end of inventories'; for instance, Oliver (1999) suggested a new acronym to replace JIT, namely JAZ, or just-about-zero. Also, the recent American imitation of the LPS has now produced the increasing awareness that smaller and more focused factories are more efficient (Elmore et al., 1995), thus recognizing the value of the focal Japanese factories, which were extensively analysed in previous chapters (though these authors do not seem to be aware of Fruin's work). The recognition of OIs as an engine of economic growth, not only for firms but also for sectors and whole economies, is well linked with a multidisciplinary approach to the complicated process of economic growth. Thus, as this study has demonstrated, we need economics (which attempt to explain economic growth), as well as management (which attempts to explain organization of firms) and, in particular, operations management (to elucidate the operations part of a firm), accounting (to properly record costs), sociology and biology (to enhance the understanding of systems), physics (to clarify the concept of entropy), operations research for relevant techniques mainly applicable to the POM, and so on.[19]

The importance of OIs (as defined and explored in this study) in respect with economic growth needs to be put into a larger perspective, like the one presented in Chapter 5, in order properly to appreciate it. The core of OIs, the four PROSIBB, and in particular the POM, properly function in the context of sectoral and innovation systems, of system emergence and evolutionary learning capability, and so on. Thus, economic growth is far from being only the quantities of capital, labour and TIs, or knowledge, learning and education, as many books in economics would have suggested. Economic growth is a complex process involving various layers of systemic energy, including OIs, the internal system of firms represented by the fundamental PROSIBB within the context of sectoral and innovations systems, as well as evolutionary learning capability, and so on. This complexity has a significant impact on policy issues. Within the framework of the SI approach, 'innovation policy-making itself should be explicitly viewed as a process of learning' (Edquist et al., 2000, p. 537). Such learning becomes even more critical in the case of new TIs that involve mechanisms of transition: 'Here, the relevant capabilities (which are usually not only technical but also organisational) may lie quite outside the existing structure of firms' competences. What we might call transition problems are thus likely to occur, possibly motivating public policies' (ibid., p. 540).[20] Also, public policy is needed to generate appropriate institutional frameworks, especially in relation to corporate governance so that strategic decision-making processes for the allocation of corporate resources and returns are encouraged (ibid., p. 549). This public policy can be directed towards encouraging all types of organizations to interact with each other in order to adopt specific innovations (TIs, OIs and others).

In addition, the importance of OIs has another dimension. It is implied in this analysis that technology (and its impact on growth) is not just TIs, or TIs and OIs. It is implied that it is possible to consider the OIs as independently affecting industrial growth, and in many cases TIs becoming the handmaiden of OIs. This increased role assigned to OIs should be considered as an important tool of public (government) and private (firms) policy regarding the adoption of new technologies. Should the latter be just the latest state of the art in equipment and machinery? Or, should it be the latest state of the art in new ways of organizing the production process and its relation with other firms? Or, should it be a combination of the two? According to the analysis presented in this study, TIs without the appropriate OIs are doomed to fail, whereas the other way around is not necessarily true. In addition, a growth policy should encourage firms to elaborate on their organizational capabilities, in terms of the elements of the four PROSIBB, and their impact on major OIs such as JIT/QC. Thus, TIs are not necessarily the universal solution to all problems of growth. The government, in particular, should be involved in assisting in the smooth functioning of sectoral and innovation systems as defined in Chapter 5. Furthermore, an eventual adoption of the four PROSIBB could have repercussions on the prevailing accounting rules: costs should measure changes in each of the four PROSIBB so that firms can have a direct knowledge of their organizational operations.[21]

Finally, the evidence provided in the last two chapters has shown that it is possible to implement OIs in a different country from the one in which they originated. The transplantation of OIs might take a long time to be effective, but it is possible. The implication of this conclusion is that governments, managers, entrepreneurs and scholars ought not to waste time in debating whether this transplantation is feasible or not, because there is strong evidence that it is. However, as shown in this study, it takes many years for individual firms to adopt a major system such as JIT/QC and many more years before sectors and the whole economy adopt it. In addition, the tendency for globalization makes it easier to imitate TIs than OIs. As Nelson and Wright (1992, p. 1962) argue: 'the advanced nations of the world have come to share a common technology … '. Consequently, it becomes imperative that policy by firms and government should encourage the generation and introduction of OIs (minor or major) which are more difficult to imitate and yet offer more possibilities for growth, including the development of new TIs.

8.5 EPILOGUE

In most of this study the analysis and synthesis[22] of appropriate topics have aimed at achieving two main targets: first, to increase awareness that a systematic exploration of OIs is necessary in economic thinking in order to dig deeper into the complex process of economic growth. This exploration has methodically identified, defined and categorized OIs so that this process becomes clearer. The distinction between OIs, TIs and other types of innovations is shown to be necessary in economic research in order to explore the well-recognized benefits of technology in promoting and sustaining economic growth. The new concepts introduced, analysed and synthesized in the theoretical parts of this study, namely, the three axes of OIs, the four fundamental firm operations (PROSIBB), in particular the process of movements, and the atomic/systemic model of economic growth have assisted in reaching the target of recognizing the role of OIs in generating economic growth.

Thus, the traditional production function and its extensions, according to which output depends only on physical capital, labour, human capital, knowledge and TIs is not sufficient to explain overall economic growth. We must also include the internal organization of firms (the four PROSIBB, and in particular the POM) and their links with other organizations and agents (the three axes of OIs, sectoral and innovation systems, and overall evolutionary learning capabilities), as demonstrated in this study. In particular, the relationship between OIs, the POM and the traditional production function can be seen in the following overall conclusion. No production is possible unless the POM is *operational*.[23] The other PROSIBB (POS, POW and POC) do not produce anything unless the POM is present. However, the POM on itself is static as it needs the other PROSIBB to become dynamic, changing and evolutionary. The POM is at the core of the main axis, which in turn is at the core of the three axes, which in turn constitute all OIs (see again Figures 5.2 and 5.3).

The second target to reach was to bring some quantitative evidence that OIs generated industrial growth in the two countries explored in this study. Sectoral analysis, used in various ways, demonstrated that OIs play a substantial part in this growth. In particular, the major OI of the JIT/QC system has been a significant force in sustaining industrial growth in the USA during the last 20 years or so,[24] as the thorough econometric study conducted here has demonstrated.

Overall the author of this book hopes that awareness of the important role OIs play in promoting economic (and societal)[25] growth has increased amongst readers, and that further studies will be undertaken to complement and supplement the present effort.[26] He also hopes that the word 'organosis' of the book title is no longer a mystery as it is exclusively related to OIs in their impact on economic growth.

NOTES

1. Economic growth of the macroeconomy depends on the growth of the major sectors of manufacturing industries, services and primary industries and their interlinkages. This study has been mostly concerned with industrial or manufacturing growth, although the results can be easily extended to other major sectors, such as services.
2. See next section for an elaboration of this *ceteris paribus* clause.
3. Organization innovations and the POM (from which the kinetic costs are derived) are intrinsically related, whereas the PROSIBB and their combinations generate the three axes of OIs. An OI such as JIT/QC is a combination of all the PROSIBB but mainly of the POM. Much emphasis in this study is given to the exploration and definition of all kinds of OIs as distinct from TIs and other types of innovations. In addition, all OIs are heavily influenced by the POM.
4. A balance could be a prioritized set of expected outcomes in terms of virtue, utility, money and so on, hence the method of multi-criteria goal programming can be useful in this instance, or the concept of punctuated equilibrium.
5. See also, Sakakibara et al. (1997), who empirically tested JIT, on a firm basis, as an overall organizational phenomenon, and confirmed it as such.
6. These two axes ultimately depend on the main axis.
7. However, even up to the 1970s, this study (also see Kogut, 1996, as previously mentioned) has shown that the rationalization process, which started or accelerated with scientific management cum Fordism, has been a substantial factor in creating and sustaining strong American manufacturing firms and sectors from the early decades of the 20th century.
8. It is easy and frequent to ignore the contribution of OIs even if authors such as Nelson and Wright (1992) are well aware of the role of OIs or at least of the role of some of the OIs in determining the industrial performance of the USA and other countries.
9. For a very recent book on the JIT/QC system and its relationship with strategies, downsizing of business and so on, see Brown (2000).
10. From the empirical point of view some methods were suggested and used in Chapters 5 and 7 to measure the specific impacts of TIs and OIs, but more research is needed in this respect.
11. The author of this study would like to thank Ed Wilson (for example, Chaudhri and Wilson, 2000) for suggesting this idea.
12. The author of this study would like to thank Charles Harvie (for example, Harvie and Kearney, 1995) for suggesting this idea.

13. Harvie (2002) suggests that several OIs are required to take place in TVEs in order for them to maintain their important role in the Chinese economic development (for example, enhancing business alliances, management control and so on).
14. Individual company studies such as the one produced by Fujimoto (1999) are needed to bring more evidence on the issues discussed in this study, especially in relation to the functioning and measurement of the PROSIBB, and in relation to the complexity of an open evolving system of innovations and growth such as that presented in Chapter 5.
15. A good reference regarding these interactions can be found in Lam (2002).
16. An application of this theory is for example the examination of the Danish restaurant sector by Hjalager (2000).
17. The author of this book would like to thank L. Hommen (for example Edquist et al., 2001) for making this remark.
18. Good recent reviews of the regulation theory are found in Friedman (2000) and Jessop (2001).
19. There are other sub-categories of disciplines that can be part of the two main disciplines and that are also involved in this study. For example, the theory of entrepreneurship (part of both economics and management), not directly treated here, but is present in many areas of the analysis. Thus, the evidence provided here about the significant impact of OIs in promoting economic growth can be used as an indirect evidence for the role of entrepreneurs as well. More specifically, the measurement of OIs can be an indirect measurement of entrepreneurship.
20. Edquist et al. (2000) propose several concrete cases where public policy is needed. However, their analysis is predominantly TIs oriented, although they explicitly recognize the implications of OIs.
21. This suggestion hints at the multidisciplinary approach of this study. Accounting should be closely related to what the internal structure of a firm is, which should be the common ground for management and economics as well.
22. Emphasis has been given to 'synthesis' in several parts due to the nature of the topic.
23. If, for example, we put in place five machines, 10 tools and six people – in a building – the output produced will be zero unless these machines, tools and people move around the building, or generally they produce kinetic energy in order to carry out specific tasks according to their organized work. Ultimately, output depends on how all these machines, tools and people are organized and continually reorganized (in terms of layout, timing, hierarchy and so on) according to initiatives, strategies, accumulated knowledge, contracts, psychological factors, rules and so on (all these elements are expressed in the four PROSIBB), and on the links between this factory and the outside world.
24. The transfer of the LPS from Japan to the USA took many years to be successfully implemented, as was shown in this study. This transfer demonstrates that OIs can be imitated but it is a difficult task to achieve, especially on a sectoral or macroeconomic basis. On the contrary, TIs are easier and faster to imitate, especially under the umbrella of increasing globalization.
25. See for example Sanidas (2004a).
26. After all, let us remind ourselves again that the first book in 'economics' by Xenophon (as mentioned in Chapter 1) was related to the importance of organization in promoting growth in firms.

References

Abe, E. (1997), 'The development of modern business in Japan', *Business History Review*, Summer, **71**, 299–308.

Abegglen, J.C. and Stalk, G. Jr (1985), *Kaisha, The Japanese Corporation*, New York: Basic Books.

Abernathy W.J., Clark, K.B. and Kantrow, A.M. (1983), *Industrial Renaissance, Producing a Competitive Future for America*, New York: Basic Books.

Abramovitz, M. (1993), 'The search for the sources of growth: areas of ignorance, old and new', *Journal of Economic History*, June, **53**, 217–43.

Acharya, S.S. and Chaudhri, D.P. (2001), 'Equity-driven trade and marketing policy strategies for Indian agriculture: priorities and agenda', in Acharya, S.S. and Chaudhri, D.P. (eds), *Indian Agricultural Policy at the Crossroads, Priorities and Agenda*, Jaipur: Rawat Publications.

Aglietta, M. (1976), *A Theory of Capitalist Regulation, the US Experience*, translated by D. Fernbach, London: NLB (1979).

Alchian, A.A. and Demsetz, H. (1972), 'Production, information costs and economic organization', *American Economic Review*, **62** (5), 777–95.

Altman, M. (2002), 'Economic theory and the challenge of innovative work practices', *Economic and Industrial Democracy*, **23** (2), 271–90.

American Society for Quality (ASQ) (2000), *Quality Progress*, July.

Andersen, E.S. and Lundvall, B.A. (1997), 'National innovation systems and the dynamics of the division of labor', in Edquist, C. (ed.), *Systems of Innovation, Technologies, Institutions and Organizations*, London: Pinter.

Antonelli, C. (1998), 'Localized technological change and the evolution of standards as economic institutions', in Chandler, A.D., Hagstrom, P. Jr and Solvell, O. (eds), *The Dynamic Firm, The Role of Technology, Strategy, Organization, and Regions*, Oxford: Oxford University Press.

Antonelli, C. (1999), *The Microdynamics of Technological Change*, London: Routledge.

Aoki, M. (1984), *The Co-operative Game Theory of the Firm*, Oxford: Clarendon Press.

Aoki, M. (1990), 'Toward an economic model of the Japanese firm', *Journal of Economic Literature*, March, **28**, 1–27.

Arora, A.A., Landau, R. and Rosenberg, N. (1999), 'Dynamics of comparative advantage in the chemical industry', in Mowery, D.C. and Nelson,

R.R. (eds), *Sources of Industrial Leadership, Studies of Seven Industries*, Cambridge: Cambridge University Press.

Arrow, K.J. (1974), *The Limits of Organization*, New York: W.W. Norton.

Atkinson, W. (1990), 'The customer-responsive manufacturing organization', *Manufacturing Systems*, May, 58–61.

Bailes, J.C. and Kleinsorge, I.K. (1992), 'Cutting waste with JIT', *Management Accounting*, May, 28–32.

Baily, M.N. and Gersbach, H. (1995), 'Efficiency in manufacturing and the need for global competition', *Brookings Papers: Microeconomics*, 307–58.

Baldwin, C.Y. and Clark, K.B. (1994), 'Capital-budgeting systems and capabilities investments in US companies after the Second World War', *Business History Review*, Spring, **68** (1), 73–109.

Baldwin, R. and Gagnon, R.J. (1993), 'The transfer of Japanese manufacturing principles and their impact on purchaser–vendor policies', *Production and Inventory Management Journal*, third quarter, 26–30.

Baldwin, R.E. (1972), *Economic Development and Growth*, New York: John Wiley & Sons.

Banerjee, A., Dolado, J.J., Galbraith, J.W. and Hendry, D.F. (1993), *Co-Integration, Error Correction, and the Econometric Analysis of Non-Stationary Data*, New York: Oxford University Press.

Barro, R.J. and Sala-I-Martin, X. (1995), *Economic Growth*, New York: McGraw-Hill.

Baumol, W.J. (1967), *Business Behavior, Value and Growth*, New York: Harcourt Brace Jovanovich.

Baumol, W.J. (1977), *Economic Theory and Operations Analysis*, Englewood Cliffs, NJ: Prentice-Hall.

Baumol, W.J. (1995), 'Formal entrepreneurship theory in economics: existence and bounds', in Bull, I., Thomas, H. and Willard, G. (eds), *Entrepreneurship, Perspectives on Theory Building*, Oxford: Pergamon.

Beamish, P.W. (1999), 'Sony's Yoshihide Nakamura on structure and decision making', *Academy of Management Executive*, November, **13** (4), 12–16.

Best, M.H. (1990), *The New Competition, Institutions of Industrial Restructuring*, Cambridge: Polity Press.

Bhote, K.R. (1988), 'DOE, the high road to quality', *Management Review*, January, **77** (1), 27–33.

Billesbach, T.J. (1991), 'A study of the implementation of just-in-time in the United States', *Production and Inventory Management Journal*, third quarter, 1–4.

Billesbach, T.J. (1994), 'Applying lean production principles to a process facility', *Production and Inventory Management Journal*, third quarter, 40–44.

Billesbach, T.J. and Hayen, R. (1994), 'Long term impact of just-in-time on inventory performance measures', *Production and Inventory Management Journal*, first quarter, 62–7.

Billesbach, T.J. and Schniederjans, M.J. (1989), 'Applicability of just-in-time techniques in administration', *Production and Inventory Management Journal*, third quarter, 40–45.

Billesbach, T.J., Harrison, A. and Croom-Morgan, S. (1991), 'Supplier performance measures and practices in JIT companies in the US and the UK', *International Journal of Purchasing and Materials Management*, Fall, 24–8.

Blackburn, P., Coombs, R. and Green, K. (1985), *Technology, Economic Growth, and the Labor Process*, New York: St. Martin's Press.

Blaug, M. (1963), 'A survey of the theory of process-innovations', *Economica*, February, 13–32.

Boer, H. (1994), 'Flexible manufacturing systems', in Storey, J. (ed.), *New Wave Manufacturing Strategies: Organizational and Human Resource Management Dimensions*, London: Paul Chapman Publishing.

Boltho, A. (1996), 'Was Japanese growth export-led?', *Oxford Economic Papers*, **48**, 415–32.

Borland, J. and Yang, X. (1992), 'Specialization and a new approach to economic organization and growth', *American Economic Review*, **82** (2), 386–91.

Boulding, K.E. (1981), *Evolutionary Economics*, Beverly Hills, CA, and London: Sage.

Bouttes, J.P. and Hamamdjian, P. (1997), 'Contractual relationships within the firm', in Menard, C. (ed.), *Transaction Costs Economics, Recent Developments*, Cheltenham, UK and Lyme, USA: Edward Elgar.

Bowles, J. (1992), 'Is American management really committed to quality?', *Management Review*, April, 42–6.

Boyce, G. and Ville, S. (2002), *The Development of Modern Business*, Basingstoke: Palgrave.

Breschi, S. and Malerba, F. (1997), 'Sectoral innovation systems: technological regimes, Schumpeterian dynamics and spatial boundaries', in Edquist, C. (ed.), *Systems of Innovation, Technologies, Institutions and Organizations*, London: Pinter.

Bresnahan, T.F. and Malerba, F. (1999), in Mowery, D.C. and Nelson, D.D. (eds), *Sources of Industrial Leadership, Studies of Seven Industries*, Cambridge: Cambridge University Press.

Brown, M. (1968), *On the Theory and Measurement of Technological Change*, Cambridge: Cambridge University Press.

Brown, S. (2000), *Manufacturing the Future, Strategic Resonance for Enlightened Manufacturing*, London: Prentice Hall.

Brox, J.A. and Fader, C. (1997), 'Assessing the impact of JIT using economic theory', *Journal of Operations Management*, November, 371–88.

Brynjolfsson, E. and Hitt, L.M. (2000), 'Beyond computation: information technology, organizational transformation and business performance', *Journal of Economic Perspectives*, **14** (4), 23–48.

Buckley, W. (1967), *Sociology and Modern Systems Theory*, Englewood Cliffs, NJ: Prentice Hall.

Bulmash, S.B. (2002), 'A new paradigm: the wealth effect of the stock market on consumption, in a context of interacting bio-systems', *Journal of Socio-Economics*, **31**, 75–100.

Bureau of the Census (1975), *Historical Statistics of the United States, Colonial Times to 1970*, Washington, DC: US Department of Commerce.

Bureau of the Census (various issues), *Statistical Abstract of the United States, The National Data Book*, Washington, DC: US Department of Commerce, Economics and Statistics Administration.

Cable, J. and Yasuki, H. (1985), 'Internal organization, business groups and corporate performance, an empirical test of the multidivisional hypothesis in Japan', *International Journal of Industrial Organization*, **3**, 401–20.

Callen, J.L., Fader, C. and Krinsky, I. (2000), 'Just-in-time: a cross-sectional plant analysis', *International Journal of Production Economics*, **63**, 277–301.

Campa, J. and Goldberg, L.S. (1997), 'The evolving external orientation of manufacturing: a profile of four countries', *FRBNY Economic Policy Review*, July, **3** (2), 53–81.

Carlton, D.W. and Perloff, J.M. (1990), *Modern Industrial Organization*, New York: HarperCollins.

Caves, R.E. and Uekusa, M. (1976), *Industrial Organisation in Japan*, Washington, DC: The Brookings Institution.

Chan, C.Y., Lam, F.W. and Lee, C. P. (1999), 'Considerations for using cellular manufacturing', *Journal of Materials Processing Technology*, November, **96** (1), 182–7.

Chandler, A.D. Jr (1959), 'The beginnings of "big business" in American industry', in Tedlow, R.S. and John, R.R. Jr (eds), *Managing Big Business, Essays from the Business History Review*, Boston, MA: Harvard Business School Press.

Chandler, A.D. Jr (1962), *Strategy and Structure, Chapters in the History of the Industrial Enterprise*, Garden City, NY: Doubleday.

Chandler, A.D. Jr (1969), 'The structure of American industry in the twentieth century: a historical overview', *Business History Review*, Autumn, **43** (3), 255–98.

Chandler, A.D. Jr (1977), *The Visible Hand, The Managerial Revolution in American Business*, Cambridge: Belknap Press of Harvard University Press.

Chandler, A.D. Jr (1981), 'The American system and modern management', in Mayr, O. and Post, R.C. (eds), *Yankee Enterprise, The Rise of the*

American System of Manufacturers, Washington, DC: Smithsonian Institution Press.

Chandler, A.D. Jr (1990), *Scale and Scope, The Dynamics of Industrial Capitalism*, Cambridge, MA: Belnap Press.

Chandler, A.D. Jr (1992), 'Organizational capabilities and the economic history of the industrial enterprise', *Journal of Economic Perspectives*, Summer, **6** (3), 79–100.

Chandler, A.D. Jr (1994), 'The competitive performance of US industrial enterprise since the Second World War', *Business History Review*, Spring, 1–71.

Chandler, A.D. Jr and Tedlow, R.S. (1985), *The Coming of Managerial Capitalism: a Casebook on the History of American Economic Institutions*, Homewood: R.D. Irwin.

Chaudhri, D.P. and Wilson, E. (2000), 'Savings, investment, productivity and economic growth in Australia 1861–1990: some explorations', *Economic Record*, March, **76** (232), 55–73.

Chaudhri, D.P. and Wilson, E. (2001), 'Agricultural growth, employment and poverty: theoretical and empirical explorations with Indian data (1970–1993)', in Acharya, S.S. and Chaudhri, D.P. (eds), *Indian Agricultural Policy at the Crossroads, Priorities and Agenda*, Jaipur: Rawat Publications.

Cheung, S.N.S. (1998), 'The transaction costs paradigm, 1998 Presidential Address, Western Economic Association', October, *Economic Inquiry*, **36** (4), 514–21.

Clark, N. (1988), 'Some new approaches to evolutionary economics', *Journal of Economic Issues*, **22** (2), June, 511–31.

Clark, P. (2000), *Organizations in Action, Competition between Contexts*, London: Routledge.

Clark, P.A. (1987), *Anglo-American Innovation*, Berlin: Walter de Gruyter.

Coase, R.H. (1937), 'The nature of the firm', *Economica*, new series, November, **4,** 386–405.

Coase, R.H. (1960), 'The problem of social cost', *Journal of Law and Economics*, October, **3,** 1–44.

Coase, R.H. (1992), 'The institutional structure of production', *American Economic Review*, September, **82**, 713–19.

Cole, R.E. (1998), 'Learning from the quality movement: what did and didn't happen and why?', *California Management Review*, Fall, **41** (1), 43–73.

Coriat, B. and Dosi, G. (1998), 'Learning how to govern and learning how to solve problems: on the co-evolution of competences, conflicts and organizational routines', in Chandler, A.D., Hagstrom, P. Jr and Solvell, O. (eds), *The Dynamic Firm, The Role of Technology, Strategy, Organization, and Regions*, New York: Oxford University Press.

Creamer, D., Dobrovolsky, S.P. and Borenstein, I. (1960), *Capital in Manufacturing and Mining, Its Formation and Financing*, Princeton, NJ: A Study by the NBER, Princeton University Press.

Cusumano, M.A. (1994), 'The Limits of "Lean"', *Sloan Management Review*, Summer, **35** (4), 27–32.

Cyert, R.M. and March, J.G. (1963), *A Behavioral Theory of the Firm*, Englewood Cliffs, NJ: Prentice-Hall.

Daniel, S.J. and Reitsperger, W.D. (1991), 'Management control systems for JIT: an empirical comparison of Japan and the USA', *Journal of International Business Studies*, fourth quarter, 603–17.

Dassbach, C.H.A. (1999), 'Lean production in North America: myth and reality', in Bellofiore, R. (ed.), *Global Money, Capital Restructuring and the Changing Patterns of Labour*, Cheltenham, UK and Northampton, MA, USA: Edward Elgar.

Davenport, T.H. (1993), *Process Innovation, Reengineering Work through Information Technology*, Boston, MA: Harvard Business School Press.

David, P.A. (1990), 'The dynamo and the computer: an historical perspective on the modern productivity paradox', *American Economic Review*, **8** (2), 355–61.

Davidson, P. and Griffin, R.W. (2000), *Management, Australia in a Global Context*, Brisbane: John Wiley & Sons.

Dawson, P. (1994), *Organizational Change: A Processual Approach*, London: Paul Chapman Publishing.

Dawson, P. (1997), 'Advanced technology and the development of new forms of work organization', in McLoughlin, I. and Harris, M. (eds), *Innovation, Organizational Change and Technology*, London: International Thomson Business Press.

Deming, W.E. (1986), *Out of the Crisis, Quality, Productivity and Competitive Position*, Cambridge: Cambridge University Press.

Demsetz, H. (1988), 'The theory of the firm revisited', *Journal of the Law, Economics, and Organization*, **4** (1), 141–61.

Demsetz, H. (1997), 'The firm in economic theory: a quiet revolution', *American Economic Review*, Papers and Proceedings, May, **87** (2), 426–9.

Denison, E.F. (1962), *The Sources of Economic Growth in the United States and the Alternatives Before Us*, New York: Committee for Economic Development.

Denny, M., Bernstein, J., Fuss, M., Nakamura, S. and Waverman, L. (1992) 'Productivity in Manufacturing Industries, Canada, Japan and the United States, 1953–1986: Was the "Productivity Slowdown" Reversed?', *Canadian Journal of Economics*, **25** (3), 584–603.

Dicken, P. (1986), *Global Shift: Industrial Change in a Turbulent World*, London: Harper and Row.

Dicken, P. (1998), *Global Shift : Transforming the World Economy*, London: Chapman.

Dietrich, M. (1993a) 'Transaction costs and revenues', in Pitelis, C. (ed.), *Transaction Costs, Markets and Hierarchies*, Oxford: Blackwell.

Dietrich, M. (1993b), 'Total quality control, just-in-time management, and the economics of the firm', *Journal of Economic Studies*, **20**, 17–31.

Dietrich, M. (1994), *Transaction Cost Economics and Beyond, Towards a New Economics of the Firm*, London: Routledge.

Dinnen, S. (1992), 'Can American steel find quality?', *Industry Week*, 20 January, 36–40.

Donchess, C.M. (1990), 'Kanban: just in time for "just-in-time"', *Massachusetts CPA Review*, Spring, **64** (1), 15–20.

Dopfer, K. (1994), 'Kenneth Boulding: A Founder of Evolutionary Economics', *The Journal of Economic Issues*, June, **28** (4), 1201–1204.

Dore, R. (1987), *Taking Japan Seriously: A Confucian Perspective on Leading Economic Issues*, London: Athlone Press.

Dosi, G. and Malerba, F. (eds) (1996a), *Organization and Strategy in the Evolution of the Enterprise*, London: Macmillan Press.

Dosi, G. and Malerba, F. (1996b), 'Organizational learning and institutional embeddedness, an introduction to the diverse evolutionary paths of modern corporations', in Dosi, G. and Malerba, F. (eds), *Organization and Strategy in the Evolution of the Enterprise*, London: Macmillan Press.

Dosi, G., Giannetti, R. and Tononelli, P.A. (eds) (1992), *Technology and Enterprise in a Historical Perspective*, New York: Oxford University Press.

Drury, H.B. (1922), *Scientific Management, A History and Criticism*, New York: AMS Press, reprinted (1968).

Duclos, L.K., Siha, S.M. and Lummus, R.R. (1995), 'JIT in services: a review of current practices and future direction for research', *International Journal of Service Industry Management*, **6** (5), 36–52.

East Asia Analytical Unit (EAAU) (1997), *A New Japan? Change in Asia's Megamarket*, Canberra: Department of Foreign Affairs and Trade.

Ebers, M. (1997a), 'Explaining inter-organizational network formation', in Ebers, M. (ed.), *The Formation of Inter-Organizational Networks,* New York: Oxford University Press.

Ebers, M. (ed.) (1997b), *The Formation of Inter-Organizational Networks*, New York: Oxford University Press.

Ebrahimpour, M. and Withers, B.E. (1993), 'A comparison of manufacturing management in JIT and non-JIT firms', *International Journal of Production Economics*, **32**, 355–64.

Economic Planning Agency (EPA), (2000), *Gross Capital Stock of Private Enterprises, 1983–1998*, Department of National Accounts, Economic Research Institute, February.

Edquist, C. (1997), 'Systems of innovation approaches – their emergence and characteristics', in Edquist, C. (ed.), *Systems of Innovation, Technologies, Institutions and Organizations*, London: Pinter.

Edquist C., Hommen, L. and McKelvey, M. (2001), *Innovation and Employment*, Cheltenham, UK and Northampton, MA, USA: Edward Elgar.

Edquist, C., Hommen, L., Johnson, B., Lemola, T., Malerba, F., Reiss, T. and Smith, K. (2000), 'The systems of innovation approach and its general policy implications', in Edquist, C. and McKelvey, M. (eds), *Systems of Innovation: Growth, Competitiveness and Employment*, vol. 2, Cheltenham, UK and Northampton, MA, USA: Edward Elgar.

Edwards, C.T. and Samimi, R. (1997), 'Japanese interfirm networks: exploring the seminal sources of their success', *Journal of Management Studies*, July, **34** (4), 489–510.

Ehrnberg, E. and Jakobson, S. (1997), 'Technological discontinuities and incumbents' performance: an analytical framework', in Edquist, C. (ed.), *Systems of Innovation, Technologies, Institutions and Organizations*, London: Pinter.

Elazar, D.J. (1999), 'Tocqueville and the cultural basis of American democracy', *Political Science and Politics*, June, **32** (2), 207–10.

Eliasson, G. (1994), 'The theory of the firm and the theory of economic growth, an essay on the economics of institutions, competition, and the capacity of the political system to cope with unexpected change', in Magnusson, L. (ed.), *Evolutionary and Neo-Schumpeterian Approaches to Economics*, Boston: Kluwer Academic.

Eliasson, G. (1996), *Firm Objectives, Controls and Organization, The Use of Information and the Transfer of Knowledge within the Firm*, Dordrecht: Kluwer Academic.

Ellig, J. (2001), 'Internal markets and the theory of the firm', *Managerial and Decision Economics*, **22**, 227–37.

Elmore, R.C., Natarajan, N. and Rezae, Z. (1995), 'Continuous improvement through the focused factory', *Management Accounting Magazine*, February, **69** (1), 21–4.

Enders, W. (1995), *Applied Econometric Time Series*, New York: John Wiley & Sons.

Evans, J.R. and Lindsay, W.M. (2002), *The Management and Control of Quality*, 5th edn, Cincinnati, OH: South-Western.

Flood, D. and Lowe, P. (1995), 'Inventories and the business cycle', *Economic Record*, March, **71** (122), 27–39.

Florida, R. and Kenney, M. (1992), 'Restructuring in place: Japanese investment, production organization, and the geography of steel', *Economic Geography*, **68** (2), 146–73.

Flynn, B.B, Sakakibara, S. and Schroeder, R.G. (1995), 'Relationship be-

tween JIT and TQM: practices and performance', *Academy of Management Journal*, **38** (5), 1325–60.
Foss, N.J. (1997a), 'The classical theory of production and the capabilities view of the firm', *Journal of Economic Studies*, **24** (5), 307–23.
Foss, N.J. (ed.) (1997b), *Resources, Firms, and Strategies: A Reader in the Resource-based Perspective*, Oxford: Oxford University Press.
Foss, N.J. and Knudsen, C. (eds) (1996), *Towards a Competence Theory of the Firm*, London: Routledge.
Foss, N.J. and Loasby, B.J. (eds) (1998), *Economic Organization, Capabilities and Co-ordination, Essays in Honour of G.B. Richardson*, London: Routledge.
Frantz, R.S. (1997), *X-Efficiency: Theory, Evidence and Applications*, 2nd edn, Boston: Kluwer Academic.
Freeman, C. (ed.) (1984), *Long Waves in the World Economy*, London: Pinter.
Freeman, C. and Soete, L. (1997), *The Economics of Industrial Innovation*, 3rd edn, London: Pinter.
Friedman, A.L. (2000), 'Microregulation and post-Fordism: critique and development of regulation theory', *New Political Economy*, **5** (1), 59–76.
Fruin, M.W. (1992), *The Japanese Enterprise System: Competitive Strategies and Cooperative Structures*, Oxford: Clarendon Press.
Fruin, W.M. and Nakamura, M. (1997), 'Top-down production management: a recent trend in the Japanese productivity-enhancement movement', *Managerial and Decision Economics*, **18**, 131–9.
Fuhrer, J.C. and Little, J.S. (1996), 'Technology and growth: an overview', *New England Economic Review*, November/December, 3–25.
Fujimoto, T. (1998), 'Reinterpreting the resource-capability view of the firm: a case of the development-production systems of the Japanese auto-makers', in Chandler, A.D., Hagstrom, P. Jr and Solvell, O. (eds), *The Dynamic Firm, The Role of Technology, Strategy, Organization, and Regions*, Oxford and New York: Oxford University Press.
Fujimoto, T. (1999), *The Evolution of a Manufacturing System at Toyota*, New York: Oxford University Press.
George, C.S. Jr (1972), *The History of Management Thought*, Englewood Cliffs, NJ: Prentice-Hall.
Georgescu-Roegen, N. (1971), *The Entropy Law and the Economic Process*, Cambridge; MA: Harvard University Press.
Gersbach, H. and Baily, M.N. (1996), 'Explanations of international productivity differences: lessons from manufacturing', in Wagner, K. and Van Ark, B. (eds), *International Productivity Differences, Measurement and Explanation*, Amsterdam: Elsevier Science.
Giffi, C., Roth, A.V. and Seal, G.M. (1990), *Competing in World-Class Manu-*

facturing: America's 21st Century Challenge, Homewood, IL: Business One Irwin.

Goldman, A. (1993), 'Implications of Japanese total quality control for western organisations: dimensions of an intercultural hybrid', *Journal of Business Communication*, **30** (1), 29–47.

Golhar, D.Y. and Stamm, C.L. (1993), 'JIT purchasing practices in manufacturing firms', *Production and Inventory Management Journal*, third quarter, 75–9.

Gordon, R.J. (2000), 'Does the 'New Economy' measure up to the great inventions of the past?, *Journal of Economic Perspectives*, **14** (4), 49–74.

Granovetter, M. (1985), 'Economic action and social structure: the problem of embeddedness', *American Journal of Sociology*, **91** (3), 481–510.

Grant, R.M. (1998), *Contemporary Strategy Analysis*, 3rd edn, Malden, MA: Blackwell.

Greene, W.H. (1993), *Econometric Analysis*, New York: Macmillan.

Gregersen, B. and Johnson, B. (2000), 'How do innovations affect economic growth? Some different approaches in economics', in Edquist, C. and McKelvey, M. (eds) *Systems of Innovation: Growth, Competitiveness and Employment*, vol. 2, Cheltenham, UK and Northampton, MA, USA: Edward Elgar.

Habakkuk, H.J. (1962), *American and British Technology in the Nineteenth Century: The Search for Labour-saving Inventions,* Cambridge: Cambridge University Press.

Hagstrom, P. and Hedlund, G. (1998), 'A three-dimensional model of changing internal structure in the firm', in Chandler, A.D. Jr, Hagstrom, P. and Solvell, O. (eds), *The Dynamic Firm, the Role of Technology, Strategy, Organization, and Regions*, Oxford and New York: Oxford University Press.

Hall, D.J. (1997), 'Organisational change: kinetic theory and organizational resonance', *Technovation*, **17** (1), 11–24.

Hall, P. (1994), *Innovation, Economics, and Evolution, Theoretical Perspectives on Changing Technology in Economic Systems*, New York and London: Harvester and Wheatsheaf.

Hannan, M.T. and Freeman, J. (1989), *Organizational Ecology*, Cambridge, MA: Harvard University Press.

Hansson, P.A. (1991), 'Chaos: implications for forecasting', *Futures*, January/February, 50–8.

Harbison, F. (1956), 'Entrepreneurial organization as a factor in economic development', *Quarterly Journal of Economics*, August, **120**, 364–79.

Harris Corporation (1996), 'Founding dates of the 1994 Fortune 500 US companies', *Business History Review*, Spring, **70** (1), 69–89.

Harrison, A. (1992), *Just-in-Time Manufacturing in Perspective*, Hemel Hempstead: Prentice Hall.

Harrison, A. (1994), 'Just-in-Time manufacturing', in Storey, J. (ed.), *New Wave Manufacturing Strategies: Organizational and Human Resource Management Dimensions*, London: Paul Chapman Publishing.

Harrison, B. (1994), 'The dark side of flexible production', *Technology Review*, May/June, 38–45.

Harvie, C. (2000), 'Business alliances, organizational change and township and village enterprises', in Harvie, C. (ed.), *Contemporary Developments and Issues in China's Economic Transition*, Basingstoke: Macmillan; New York: St. Martin's Press.

Harvie, C. (2002), 'China's SMEs: their evolution and future prospects in an evolving market economy', in Harvie, C. and Lee, B.C. (eds), *The Role of SMEs in National Economies in East Asia*, Cheltenham, UK and Northampton, MA, USA: Edward Elgar.

Harvie, C. and Kearney, C.P. (1995), 'Public infrastructure capital and private investment', *Papers and Proceedings of the Twenty Second Annual Meeting of the Midsouth Academy of Economics and Finance*, Biloxi, Mississippi, February, 15–18.

Hatch, M.J. (1997), *Organization Theory, Modern Symbolic and Postmodern Perspectives*, Oxford: Oxford University Press.

Hattwick, R. (1999), 'The future paradigm for socio-economics: three visions and a call for papers', *Journal of Socio-Economics*, **28**, 511–32.

Herbig, P. (1995), *Innovation Japanese Style, A Cultural and Historical Perspective*, Westport, CT: Quorum Books.

Heydebrand, W.V. (1989), 'New organizational forms', *Work and Occupations*, August, **16** (3), 323–57.

Hicks, J. (1979), *Causality in Economics*, Oxford: Basil Blackwell.

Hill, C.W.L. (1994), 'Diversification and Economic Performance: Bringing Structure and Corporate Management Back into the Picture', in Rumelt, R.P., Schendel, D.E. and Teece, D.J. (eds), *Fundamental Issues in Strategy, a Research Agenda*, Boston, MA: Harvard Business School Press.

Hirschman, A.O. (1958), *The Strategy of Economic Development*, 8th printing, New Haven, CT and London: Yale University Press.

Hirschmeier, J. and Yui, T. (1981), *The Development of Japanese Business 1600–1980*, 2nd edn, London: George Allen and Unwin.

Hjalager, A. (2000), 'Organisational ecology in the Danish restaurant sector', *Tourism Management*, **21**, 271–80.

Hodgkinson, G.P. (2003), 'The interface of cognitive and industrial, work, and organizational psychology', *Journal of Occupational and Organizational Psychology*, **76**, 1–25.

Hodgson, G.M. (1998), 'Competence and contract in the "theory of the firm"', *Journal of Economic Behavior and Organization*, **35**, 179–201.

Hunt, S.D. (1997), 'Evolutionary economics, endogenous growth models, and resource-advantage theory', *Eastern Economic Journal*, **23** (4), 425–39.

Im, J.H. and Lee, S.M. (1989), 'Implementation of just-in-time systems in US manufacturing firms', *International Journal of Operations and Production Management*, **9** (1), 5–14.

Imai, K. (1994), 'Enterprise groups', in Imai, K. and Komiya, R. (eds), *Business Enterprises in Japan: Views of Leading Japanese Economists*, Cambridge, MA: MIT Press.

Imai, M. (1991), *Kaizen, the Key to Japan's Competitive Success*, New York: McGraw-Hill.

Imai, M. (1997), *Gemba Kaizen: A Commonsense, Low-Cost Approach to Management*, New York: McGraw-Hill,

Ishikawa, K. (1985), *What is Total Quality Control: The Japanese Way*, Englewood Cliffs, NJ: Prentice-Hall.

Ito, T. (1996), 'Japan and the Asian economies: a miracle in transition', *The Brookings Papers on Economic Activity*, **2**, 205–72.

Jackson, D. (1998), *Technological Change, the Learning Curve and Profitability*, Cheltenham, UK and Lyme, USA: Edward Elgar.

Jacobson, D. and Andreosso-O'Callaghan, B. (1996), *Industrial Economics and Organization, a European Perspective*, London: McGraw-Hill.

Jaeger, C. and Ernste, H. (1989), 'Ways beyond Fordism?', in Ernste, H. and Jaeger, C. (eds), *Information Society and Spatial Structure*, London: Belhaven Press.

Jessop, B. (ed.) (2001), *Regulation Theory and the Crisis of Capitalism*, Cheltenham, UK and Northampton, MA, USA: Edward Elgar.

Jewkes, E. and Power, M. (1993), 'A microeconomic analysis of investment in just-in-time manufacturing', *International Journal of Production Economics*, **29**, 313–21.

Jones, C.I. (1998), *Introduction to Economic Growth*, New York: W.W. Norton.

Jones, S.R.H. (1999), 'The rise of the factory system in Britain, efficiency or exploitation?', in Robertson, P.L. (ed.), *Authority and Control in Modern Industry, Theoretical and Empirical Perspectives*, London: Routledge.

Jorgenson, D.W. (1990), 'Productivity and economic growth' in Berndt, E.R. and Triplett, J.E. (eds), *Fifty Years of Economic Measurement, NBER, Studies in Income and Wealth*, Chicago: University of Chicago Press.

Jorgenson, D.W. and Stiroh, K.J. (2000), 'US economic growth at the industry level', *American Economic Review*, May, **90**, 161–71.

Jorgenson, D.W., Kuroda, M. and Nishimizu, M. (1995), 'Japan–US industry-level productivity comparisons, 1960–1979', in Jorgenson, D.W., *Productivity: vol. 2, International Comparisons of Economic Growth*, Cambridge, MA: MIT Press.

Juran, J.M. (ed.) (1995), *A History of Managing for Quality, the Evolution,*

Trends, and Future Directions of Managing for Quality, Milwaukee, WI: ASQC Quality Press.

Kaldor, N. (1934), 'The equilibrium of the firm', *Economic Journal*, March, 60–76.

Kano, N. (1993), 'A perspective on quality activities in American firms', *California Management Review*, Spring, 12–31.

Kawahara, A. (1997), *The Origin of Competitive Strength, Fifty Years of the Auto Industry in Japan and the USA*, Tokyo (no publisher).

Kenney, M. and Florida, R. (1993), *Beyond Mass Production, The Japanese System and its Transfer to the US*, New York: Oxford University Press.

Keys, J.B., Denton, L.T. and Miller, T.R. (1994), 'The Japanese management theory jungle – revisited', *Journal of Management*, Summer, **20** (2), 373–402.

Khalil, E.L. (1996), 'After the special nature of the firm: beyond the critics of orthodox neoclassical economics', in Groenewegen, J. (ed.), *Transaction Cost Economics and Beyond*, Boston, MA: Kluwer Academic.

Kijne, H.J. (1996), 'Time and motion study: beyond the Taylor-Gilbreth controversy' in Spender, J.C. and Kijne, H.J. (eds), *Scientific Management, Frederick Winslow Taylor's Gift to the World?* Boston, MA: Kluwer Academic.

Kim, G.C. and Takeda, E. (1996), 'The JIT philosophy is the culture in Japan', *Production and Inventory Management Journal*, first quarter, 47–51.

Kirchhoff, B.A. (1994), *Entrepreneurship and Dynamic Capitalism, the Economics of Business Firm Formation and Growth*, Westport, CT: Praeger.

Klein, L. and Ohkawa, K. (eds) (1968), *Economic Growth; the Japanese Experience since the Meiji Era: Proceedings of the Conference of the Japan Economic Research Center*, Homewood, IL: R.D. Irwin.

Knight, F.H. (1921), *Risk, Uncertainty and Profit*, New York: Houghton Mifflin.

Knoedler, J.T. (1990), 'As much technological as pecuniary: turn-of-the-century combinations in three major industries', *Journal of Economic Issues*, June, **24** (2), 397–404.

Kobu, B. and Greenwood, F. (1991), 'Continuous improvement in a competitive global economy', *Production and Inventory Management Journal*, fourth quarter, 58–63.

Kogut, B. (1996), 'National organizing principles of work and the erstwhile dominance of the American multinational corporation', in Dosi, G. and Malerba, F. (eds), *Organization and Strategy in the Evolution of the Enterprise*, London: Macmillan Press.

Kogut, B. and Zander, U. (1992), 'Knowledge of the firm, combinative capabilities, and the replication of technology', *Organization Science*, **3**, 383–97.

Kono, T. and Clegg, S. (2001), *Trends in Japanese Management, Continuing Strengths, Current Problems and Changing Priorities*, Basingstoke: Palgrave.

Krafcik, J.F. (1989) 'A new diet for US manufacturing', *Technology Review*, January, **92** (1), 28–35.

Krajewski, L.J. and Ritzman, L.P. (1999), *Operations Management, Strategy and Analysis*, 5th edn, Reading, MA: Addison-Wesley.

Krugman, P.R. (1998), 'It's baaack: Japan's slump and the return of the liquidity trap', *Brookings Papers on Economic Activity*, **2**, 137–205.

Kuroda, M., Motohashi, K. and Shimpo, K. (1996), 'Issues on the international comparison of productivity: theory and measurement', in chapter 2 of *Industry Productivity, International Comparison and Measurement Issues*, Paris: OECD.

Kuznets, S. (1966), *Modern Economic Growth, Rate, Structure, and Spread*, New Haven, CT: Yale University Press.

Kuznets, S. (1971), *Economic Growth of Nations, Total Output and Production Structure*, Cambridge, MA: Belknap Press of Harvard University Press.

Lam, A. (2002), 'Alternative society models of learning and innovation in the knowledge economy', *International Social Science Journal*, **54** (1), 67–82.

Langlois, R.N. (1992), 'Transaction-cost economics in real time', *Industrial and Corporate Change*, **1** (1), 99–127.

Langlois, R.N. (1999), 'The coevolution of technology and organization in the transition to the factory system' in Robertson, P.L. (ed.), *Authority and Control in Modern Industry, Theoretical and Empirical Perspectives*, London: Routledge.

Langlois, R.N. and Robertson, P.L. (1989), 'Explaining vertical integration: lessons from the American automobile industry', *Journal of Economic History*, **99** (2), 361–75.

Langlois, R.N. and Robertson, P.L. (1995), *Firms, Markets and Economic Change, a Dynamic Theory of Business Institutions*, London: Routledge.

Langlois, R.N. and Steinmueller, W.E. (1999), 'The evolution of competitive advantage in the worldwide semiconductor industry', in Mowery, D.C. and Nelson, D.D. (eds), *Sources of Industrial Leadership, Studies of Seven Industries*, Cambridge: Cambridge University Press.

Lazonick, W. (1990), *Competitive Advantage on the Shop Floor*, Cambridge, MA: Harvard University Press.

Lazonick, W. (1991), *Business Organization and the Myth of the Market Economy*, New York: Cambridge University Press.

Lazonick, W. (1992), *Organization and Technology in Capitalist Development*, Aldershot, UK and Brookfield, US: Edward Elgar.

Leibenstein, H. (1960), *Economic Theory and Organizational Analysis*, Tokyo: Harper and Row; New York, jointly with John Weatherhill, Inc.

Leibenstein, H. (1966), 'Allocative efficiency vs 'X-efficiency', *American Economic Review*, June, **56**, 392–415.

Leibenstein, H. (1987), *Inside the Firm, the Inefficiencies of Hierarchy*, Cambridge, MA: Harvard University Press.

Lieberman, M.B. and Asaba, S. (1997), 'Inventory reduction and productivity growth: a comparison of Japanese and US automotive sectors', *Managerial and Decision Economics*, **18**, 73–85.

Lieberman, M.B. and Demeester, L. (1999), 'Inventory reduction and productivity growth: linkages in the Japanese automotive industry', *Management Science*, April, **45** (4), 466–85.

Lieberman, M.B., Helper, S. and Demeester, L. (1999), 'Empirical determinants of inventory levels in high-volume manufacturing', *Production and Operations Management*, Spring, **8** (1), 44–55.

Liker, J.K., Fruin, W.M. and Adler, P.S. (1999), *Remade in America, Transplanting and Transforming Japanese Management Systems*, New York: Oxford University Press.

Lillrark, P. (1995), 'The transfer of management innovations from Japan', *Organization Studies*, **16** (6), 971–89.

Lincoln, J.R. and Nakata, Y. (1997), 'The transformation of the Japanese employment system', *Work and Occupations*, **24** (1), 33–55.

Lincoln, J.R., Ahmadjian, C.L. and Mason, E. (1998), 'Organizational learning and purchase–supply relations in Japan: Hitachi, Matsushita, and Toyota compared', *California Management Review*, Spring, **40** (3), 241–64.

Linge, G.J.R. (1991), 'Just-in-time: more or less flexible? (Usage of just-in-time inventory systems in business)', *Economic Geography*, October, **67** (4), 316–32.

Litan, R.E. and Rivlin, A.M. (2001), 'Projecting the economic impact of the Internet', *American Economic Review*, May, **91** (2), 313–17.

Litterer, J.A. (1961), 'Systematic management: the search for order and integration', *Business History Review*, Winter, **35** (4), 461–76.

Litterer, J.A. (1963), 'Systematic management: design for organizational recoupling in American manufacturing firms', *Business History Review*, Winter, **37** (4), 369–91.

Littler, C.R. (1982), *The development of the labor process in capitalist societies: a comparative study of the tranformation of work organization in Britain, Japan, and the USA*, London: Heinemann Educational.

Locke, R.R. (1996), *The Collapse of the American Management Mystique*, Oxford: Oxford University Press.

Lockwood, W.W. (1968), *The Economic Development of Japan, Growth and Structural Change*, expanded edn, Princeton, NJ: Princeton University Press.

Lux, W. (1997), 'Japanese management evolves again', *Management Review*, June, **86** (6), 36–8.

Lydall, H. (1998), *A Critique of Orthodox Economics, an Alternative Model*, Basingstoke: Macmillan Press and St. Martin's Press.

Maddala, G.S. and In-Moo Kim (1998), *Unit Roots, Cointegration, and Structural Change*, Cambridge: Cambridge University Press.

Maddison, A. (1982), *Phases of Capitalist Development*, Oxford and New York: Oxford University Press.

Magjuka, R.J. and Schmenner, R.W. (1992), 'Cellular manufacturing and plant administration: some initial evidence', *Labor Studies Journal*, **17** (2), 44–63.

Malerba, F. (1985), *The Semiconductor Business, the Economics of Rapid Growth and Decline*, London: Pinter.

Malerba, F. (2002), 'Sectoral systems of innovation and production', *Research Policy*, **31**, 247–64.

Malerba, F. and Orsenigo, L. (1995), 'Schumpeterian patterns of innovation', *Cambridge Journal of Economics*, **19** (1), 47–65.

Malerba, F. and Orsenigo, L. (1996), 'Technological regimes and firm behavior', in Dosi, G. and Malerba, F. (eds), *Organization and Strategy in the Evolution of the Enterprise*, London: Macmillan Press.

Manecke, N. and Schoensleben, P. (2004), 'Cost and benefit of Internet-based support of business processes', *International Journal of Production Economics*, **87**, 213–29.

Mansfield, E. (1968), *The Economics of Technological Change*, New York: Norton.

Marglin, S.A (1974), 'What do bosses really do?', *Review of Radical Political Economy*, Summer, **6**, 33–60.

Marrewijk, C. van (1999), 'Capital accumulation, learning, and endogenous growth', *Oxford Economic Papers*, **51**, 453–75.

Marris, R. (1966), *The Economic Theory of Managerial Capitalism*, London: Macmillan.

Marshall, A. (1890), reprint 8th edn (1949), *Principles of Economics*, London: Macmillan.

Matsumoto, J. (2002), 'The interactive kitchen', *Restaurants and Institutions*, April, **112** (8), 99–102.

Matthews, R.C.O., Feinstein, C.H. and Odling-Smee, J.C. (1982), *British Economic Growth 1856–1973*, Oxford: Clarendon Press.

McCloskey, D.N. (2000), 'How to be scientific in economics', *Eastern Economic Journal*, **26** (2), 241–6.

McGee, J. (1995), Comment on Baumol W.J., (1995), 'Formal entrepreneurship theory in economics: existence and bounds', in Bull, I., Thomas, H.

and Willard, G. (eds), *Entrepreneurship, Perspectives on Theory Building*, Oxford: Pergamon.

McGraw, T.K. (ed.) (1997), *Creating Modern Capitalism, How Entrepreneurs, Companies, and Countries Triumphed in Three Industrial Revolutions*, Cambridge, MA: Harvard University Press.

McLoughlin, I. and Clark, J. (1994), *Technological Change at Work*, 2nd edn, Buckingham: Open University Press.

McLoughlin, I. and Harris, M. (eds) (1997), *Innovation, Organizational Change and Technology*, London: International Thomson Business Press.

McMillan, C.J. (1984, 1996), *The Japanese Industrial System*, Berlin and New York: W. de Gruyter.

Mefford, R.N. (1986), 'Introducing management into the production function', *Review of Economics and Statistics*, February, **68** (1), 96–104.

Menard, C. (1996), 'Why organizations matter: a journey away from the fairy tale', *Atlantic Economic Journal*, **24** (4), 281–300.

Meric, I., Ross, L.W., Weidman, S.M. and Meric, G. (1997), 'A comparison of the financial characteristics of US and Japanese chemical firms', *Multinational Business Review*, Fall, 23–7.

Milech, D. and Finucane, M. (1998), 'Decision support and behavioral decision theory', in Kirsner, K., Speelman, C., Maybery, M., O'Brien-Malone, A., Anderson, M. and MacLeod, C., (eds), *Implicit and Explicit Mental Processes*, Manhah, NJ: Lawrence Erlbaum Associates.

Milgrom, P. and Roberts, J. (1990), 'The economics of modern manufacturing: technology, strategy, and organization', *American Economic Review*, **80** (3), 511–28.

Milgrom, P. and Roberts, J. (1992), *Economics, Organization and Management*, Englewood Cliffs, NJ: Prentice-Hall.

Miltenburg, J. (1993), 'A theoretical framework for understanding why JIT reduces cost and cycle time and improves quality', *International Journal of Production Economics*, **30–31**, 195–204.

Minami, R. (1986), *The Economic Development of Japan, a Quantitative Study*, London: Macmillan Press.

Mintzberg, H. and Quinn, J.B. (1991), *The Strategy Process, Concepts, Contexts, Cases*, 2nd edn, Englewood Cliffs, NJ: Prentice-Hall.

Mitchell, B.R. (1998), *International Historical Statistics, 1750–1993*, 3rd and 4th edn, London: Macmillan Reference.

Moe, E. (2004), 'An interpretation of the Asian financial crisis: innovation systems and economic performance in a period of transformation', *Asian Affairs, an American Review*, Winter, **30** (4), 227–48.

Montgomery, D. (1987), *The Fall of the House of Labor: The Workplace, the State, and Labor Activism, 1865–1925*, Cambridge: Cambridge University Press.

Moravcsik, M.J. (1988), 'The limits of science and the scientific method', *Research Policy*, **17**, 293–9.

Morgan, G. (1998), *Images of Organization*, executive edn, San Francisco, CA: Berrett-Koehler.

Morikawa, H. (1992), *Zaibatsu: the Rise and Fall of Family Enterprise Groups in Japan*, Tokyo: University of Tokyo Press.

Mowery, D.C. and Rosenberg, N. (1998), *Paths of Innovation, Technological Change in 20th-Century America*, Cambridge: Cambridge University Press.

Mroczkowski, T., and Hanaoka, M. (1998), 'The end of Japanese management: how soon?', *Human Resource Planning*, **21** (3), 20–30.

Nafziger, E.W. (1995), *Learning from the Japanese, Japan's Pre-war Development and the Third World*, Armonk, NY: M.E. Sharpe.

Nakagawa, S. (1996), 'Scientific management and Japanese management 1910–1945', in Spender, J.C. and Kijne, H.J. (eds), *Scientific Management, Frederick Winslow Taylor's Gift to the World?* Boston, MA: Kluwer Academic.

National Bureau of Economic Research (NBER), (2000), *NBER-CES Manufacturing Industry Database*, E.J. Bartelsman, R.A. Becker, and W.B. Gray, June, NBER Internet site: www.nber.org.

Nelson, D. (1974) 'Scientific management, systematic management, and labor, 1880–1915', *Business History Review*, Winter, **98** (4), 479–500.

Nelson, D. (ed.) (1992), *A Mental Revolution, Paths of Innovation, Technological Change in 20th-Century America*, Columbus, OH: Ohio State University Press.

Nelson, R.R. (1994), 'The role of firm differences in an evolutionary theory of technical advance', in Magnusson, L. (ed.), *Evolutionary and Neo-Schumpeterian Approaches to Economics*, Boston, MA: Kluwer Academic.

Nelson, R.R. (1996), *The Sources of Economic Growth*, Cambridge, MA: Harvard University Press.

Nelson, R.R. (1997), 'How new is New Growth theory?', *Challenge*, September/October, **40** (5), 29–58.

Nelson, R.R. and Winter, S.G. (1982), *An Evolutionary Theory of Economic Change*, Cambridge, MA: Belknap Press.

Nelson, R.R. and Wright, G. (1992), 'The rise and fall of American technological leadership: the postwar era in historical perspective', *Journal of Economic Literature*, December, **30**, 1931–64.

Nicholson, W. (2002), *Microeconomic Theory, Basic Principles and Extensions*, Cincinnati, OH: South-Western/Thomson Learning.

Nonaka, I. (1995), 'The recent history of managing for quality in Japan', in Juran, J.M. (ed.), *A History of Managing for Quality, The Evolution, Trends, and Future Directions of Managing for Quality*, Milwaukee, WI: ASQC Quality Press.

North, D.C. (1990), 'Institutions, transaction costs and economic growth', *Economic Inquiry*, July, **25** (3), 419–428.

North, D.C. (1990a), *Institutions, Institutional Change, and Economic Performance*, Cambridge: Cambridge University Press.

North, D.C. (1992), 'Institutions and economic theory', *American Economist*, Spring, **36** (1), 3–6.

North, D. and Thomas, P. (1973), *The Rise of the Western World*, Cambridge: Cambridge University Press.

North, D.C., Davis, L.E. and Smoroding, C. (1971), *Institutional Change and American Economic Growth*, London: Cambridge University Press.

Nyland, C. (1989), *Reduced Worktime and the Management of Production*, Cambridge: Cambridge University Press.

Nyland, C. (1996), 'Taylorism, John R. Commons, and the Hoxie Report', *Journal of Economic Issues*, December, **30** (4), 985–1016.

O'Grady, P.J. (1988), *Just-in-Time Philosophy into Practice, A Strategy for Production Managers*, London: Kogan Page.

Odagiri, H. (1981), *The Theory of Growth in a Corporate Economy, Management Preference Research and Development and Economic Growth*, Cambridge: Cambridge University Press.

Odagiri, H. and Goto, A. (1996), *Technology and Industrial Development in Japan, Building Capabilities by Learning, Innovations and Public Policy*, Oxford: Oxford University Press.

Ohno, T., with Mito, S. (1988), *Just-In-Time for Today and Tomorrow*, Cambridge: Productivity Press.

Oliver, R.W. (1999), 'The end of inventory', *Journal of Strategy*, January/February, 8–11.

Ouchi, W. (1981), *Theory Z – How American Business Can Meet the Japanese Challenge*, Reading, MA: Addison-Wesley.

Pack, H. (1994), 'Endogenous growth theory: intellectual appeal and empirical shortcomings', *Journal of Economic Perspectives*, Winter, **8** (1), 55–72.

Penrose, E.T. (1959), *The Theory of the Growth of the Firm*, Oxford: Blackwell.

Perroux, F. (1955), 'Note sur la notion de "pole de croissance"', *Economie appliquee*, January–June, 307–20.

Perrow, C. (1986), *Complex Organizations: a Critical Essay*, 3rd edn, New York: McGraw-Hill.

Perry, M. (1999), *Small Firms and Network Economies*, London: Routledge.

Pesaran, M.H. and Pesaran, B. (1997), *Working with Microfit 4.0, Interactive Econometric Analysis*, Oxford: Oxford University Press.

Peters, E.E. (1980), 'Scientific management and technological change', PhD dissertation, Morgantown, West Virginia University.

Pettigrew, A. and Whipp, R. (1991), *Managing Change for Competitive Success*, Oxford: Blackwell Publishers.

Piore, M.J. and Sabel, C.F. (1984), *The Second Industrial Divide: Possibilities for Prosperity*, New York: Basic Books.

Pitelis, C.N. (1998), 'Transaction costs and the historical evolution of the capitalist firm', *Journal of Economic Issues*, **32** (4), 999–1017.

Plato, *The Republic*, 2nd edn, Harmondsworth Penguin Books.

Pollard, S. (1968), *The Genesis of Modern Management*, Harmondsworth: Penguin.

Pollard, S. and Robertson, P. (1979), *The British Shipbuilding Industry, 1870–1914*, Cambridge, MA: Harvard University Press.

Porter, M.E. (1980), *Competitive Strategy: Techniques for Analyzing Industries and Competitors*, New York: Free Press.

Porter, M.E. (1990), *The Competitive Advantage of Nations*, London: Macmillan.

Porter, M.E. (1998), 'The Adam Smith address: location, clusters, and the "new" microeconomics of competition', *Business Economics*, January, **33** (1), 7–13.

Porter, M.E. and Takeuchi, H. (1999), 'Fixing what really ails Japan', *Foreign Affairs*, May/June, **78** (3), 66–81.

Porter, M.E., Takeuchi, H. and Sakakibara, S. (2000), *Can Japan Compete?* London: Macmillan Press.

Powell, W. (1990), 'Neither market nor hierarchy: network forms of organization', in Cummings, L. and Shaw, B. (eds), *Research in Organizational Behavior*, 12, Greenwich, CT: JAI Press.

Prahalad, C.K. and Hamel, G. (1990), 'The core competence of the corporation', *Harvard Business Review*, May–June, **68** (3), 79–91.

Reati, A. (1998), 'The present technological change: growth and employment perspectives', in Michie, J. and Reati, A. (eds), *Employment, Technology and Economic Needs, Theory, Evidence, and Public Policy*, Cheltenham, UK and Lyme, USA: Edward Elgar.

Reati, A. (2001), 'Total factor productivity – a misleading concept', *BNL Quarterly Review*, September (218), 313–32.

Resnick, R., Halliday, D. and Krane, K.S. (1992), *Physics*, New York: John Wiley & Sons.

Ricketts, M. (2002), *The Economics of Business Enterprise, an Introduction to Economic Organization and the Theory of the Firm*, 3rd edn, Cheltenham, UK and Northampton, MA, USA: Edward Elgar.

Robertson, P.L. and Alston, L.J. (1992), 'Technological choice and the organization of work in capitalist firms', *Economic History Review*, **95** (2), 330–49.

Robertson, P.L., Pol, E. and Carroll, P. (2001), 'Technological diffusion through an historical lens', paper presented in Eindhoven Conference on Innovation Studies, 'The Future of Innovation Studies', Eindhoven University of Technology, the Netherlands, September.

Rohwer, J. (2000), 'Japan goes web crazy', *Fortune*, 7 February, **141** (1), 115.

Romanelli, E. and Tushman, M.L. (1994), 'Organizational transformation as punctuated equilibrium: an empirical test', *Academy of Management Journal*, **37** (5), 1141–66.

Romer, P.M. (1990), 'Endogenous technological change', *Journal of Political Economy*, **98** (5, part 2), October, S71–102.

Romer, P.M. (1994), 'The origins of endogenous growth', *Journal of Economic Perspectives*, Winter, **8** (1), 3–22.

Rosenberg, N. (ed.) (1969), *The American System of Manufactures*, Edinburgh: The University Press.

Rosenberg, N. (1972), *Technology and American Economic Growth*, New York: Harper and Row.

Rosenberg, N. (1976), *Perspectives on Technology*, Cambridge: Cambridge University Press.

Rostow, W.W. (1960 [1971]), *The Stages of Economic Growth*, Cambridge: Cambridge University Press.

Rostow, W.W. (1990), *Theorists of Economic Growth from David Hume to the Present, with a Perspective on the Next Century*, New York: Oxford University Press.

Roy, W.G. (1997), *Socializing Capital, the Rise of the Large Industrial Corporation in America*, Princeton, NJ: Princeton University Press.

Rumelt, R.P., Schendel, D.E. and Teece, D.J. (eds) (1994), *Fundamental Issues in Strategy*, Boston, MA: Harvard Business School Press.

Runkle, D.E. (2000), 'Old ideas at work in the new economy', *The Region*, June, 14–17 and 40–2.

Ruttan, V.W. (1998), 'The new growth theory and development economics: a survey' *Journal of Development Studies*, **35** (2), 1–31.

Rycroft, R.W. and Kash, D.E. (1999), *The Complexity Challenge, Technological Innovation for the 21st Century*, London: Pinter.

Sakakibara, M. and Porter, M.E. (2001), 'Competing at home to win abroad: evidence from Japanese industry', *Review of Economics and Statistics*, May, **83** (2), 310–22.

Sakakibara, S., Flynn, B.B., Schroeder, R.G. and Morris, W.T. (1997), 'The impact of just-in-time manufacturing and its infrastructure on manufacturing performance', *Management Science*, September, **43** (9), 1246–57.

Sako, M. (1992), *Prices, Quality and Trust: Inter-Firm Relations in Britain and Japan*, Cambridge: Cambridge University Press.

Sanchez, R. (2000), 'Product and process architectures in the management of knowledge resources', in Foss, N.J. and Robertson, P.L. (eds), *Resources, Technology, and Strategy*, London: Routledge.

Sanidas, E. (2002), 'Leading manufacturing sectors in the USA and Japan during 1899–1937 and organizational innovations: embeddedness for corporate strategy', refereed paper, *Proceedings of the Conference on Corporate Leadership in Australia*, Wollongong, Australia, 5–6 September, CD.

Sanidas, E. (2004a), 'Technology, technical and organizational innovations, economic and societal growth', *Technology in Society*, **26** (1), 67–84. Reprinted with permission from Elsevier.

Sanidas, E. (2004b), 'Impact of the lean production system on economic growth: evidence from US manufacturing industries', *International Journal of Applied Business and Economic Research*, **2** (1), 21–45.

Sanidas, E. (2004c), 'A holistic model for the emergence and success of entrepreneurs with applications to Asian countries', *Journal of International Business and Entrepreneurship*, **10** (2), 113–28.

Sanidas, E. (2005a), 'On the evolution of organization of firms, SMEs, and economic growth in the USA and Japan', in Harvie, C. and Lee, Boon-Chye (eds), *Sustaining Growth and Performance in East Asia: The Role of Small and Medium Sized Enterprises*, Cheltenham, UK and Northampton, MA, USA: Edward Elgar.

Sanidas, E. (2005b), 'The open system of four dynamic bio-socio-economic processes of the firm: the diamond of the black box', *The Journal of Socio-Economics*, forthcoming.

Saviotti, P.P. (1997), 'Innovation systems and evolutionary theories', in Edquist, C. (ed.), *Systems of Innovation, Technologies, Institutions and Organizations*, London: Pinter.

Sawyer, R.K. (1999), 'The emergence of creativity', *Philosophical Psychology*, **12** (4), 447–69.

Sayer, A. and Walker, R. (1992), *The New Social Economy: Reworking the Division of Labor*, Cambridge, MA: Blackwell.

Schleifer, A. and Vishny, R.W. (1994), 'Takeovers in the 1960s and the 1980s: evidence and implications', in Rumelt, R.P., Schendel, D.E. and Teece, D.J. (eds) (1994), *Fundamental Issues in Strategy*, Boston, MA: Harvard Business School Press.

Schonberger, R.J. (1982), *Japanese Manufacturing Techniques, Nine Hidden Lessons in Simplicity*, New York: Free Press.

Schonberger, R.J. (1986), *World Class Manufacturing, the Lessons of Simplicity Applied*, New York: Free Press.

Schonberger, R.J. (1996), *World Class Manufacturing: The Next Decade, Building Power, Strength, and Value*, New York: Free Press.

Schroeder, D.M. and Robinson, A.G. (1991), 'America's most successful export to Japan: continuous improvement programs', *Sloan Management Review*, Spring, 67–81.

Schumpeter, J.A. (1934), *The Theory of Economic Development*, Cambridge MA: Harvard University Press.

Schumpeter, J.A. (1942a), *Capitalism, Socialism and Democracy*, New York: Harper Colophon Books, edition of 1975.

Schumpeter, J.A. (1942b), *Capitalism, Socialism and Democracy*, 2nd edn, New York: Harper Brothers.

Scott, B.R. (1973), 'The industrial state: old myths and new realities', *Harvard Business Review*, March–April, 133–48.

Scranton, P. (1997), *Endless Novelty, Specialty Production and American Industrialization, 1865–1925*, Princeton, NJ: Princeton University Press.

Senge, P. (1999), 'It's learning: the real lesson of the quality movement', *Journal for Quality and Participation*, November/December, 34–40.

Simon, H.A. (1997), *Models of Bounded Rationality, Volume 3, Empirically Grounded Economic Reason*, Cambridge, MA: MIT Press.

Sklar, M.J. (1988), *The Corporate Reconstruction of American Capitalism, 1890–1916 : The Market, the Law, and Politics*, Cambridge: Cambridge University Press.

Slavin, S.L. (1999), *Microeconomics*, 5th edn, Boston, MA: Irwin, McGraw-Hill.

Smith, A. (1776), *An Enquiry into the Nature and Causes of the Wealth of Nations*, Bicentenary edn, Oxford: Clarendon Press.

Solow, R. (1957), 'Technical change and the aggregate production function', *Review of Economics and Statistics*, August, **39**, 312–20.

Solow, R. (1960), 'Investment and technical progress', in Arrow, K.J., Karlin, S. and Suppes, P. (eds), *Mathematical Methods in the Social Sciences*, Stanford, CA: Stanford University Press.

Solow, R.M. (1956), 'A contribution to the theory of economic growth', *Quarterly Journal of Economics*, February, **70** (1), 65–94.

Sowinski, L.L. (2001), 'Move over, JIT, a new delivery model is in the driver's seat' *World Trade*, April, 54–6.

Stamm, C.L. and Golhar, D.Y. (1991), 'Customer and supplier linkages for small JIT manufacturing firms', *Journal of Small Business Management*, July, 43–9.

Steele, D.C., Berry, W.L. and Chapman, S.N. (1995), 'Planning and control in multi-cell manufacturing', *Decision Sciences*, **26** (1), 1–34.

Stern, A. (1993), 'Inventories of US parent companies and their foreign affiliates: an analysis of recent data', *Applied Economics*, **25**, 335–42.

Stern, P.C. (1990), 'The social construction of the economy', *Challenge*, January–February, 38–45.

Stewart, T.A. (1993), 'Welcome to the revolution', *Fortune*, 13 December, **135** (15), 66–72.

Stiles, R.A. and Mick, S.S. (1997), 'Components of the costs of controlling quality: a transaction cost economics approach', *Hospital and Health Services Administration*, Summer, **42** (2), 205–19.

Street, J.H. (1988), 'The contribution of Simon S. Kuznets to institutionalist development theory', *Journal of Economic Issues,* June, **31** (2), 499–505.

Sugiura, K. (2002), 'Japan: the role of SMEs in the mature economy', in Harvie, C. and Lee, Boon-Chye (eds), *The Role of SMEs in National Economies in East Asia*, Cheltenham, UK and Northampton, MA, USA: Edward Elgar.

Taylor, B.W., III (2002), *Introduction to Management Science*, 7th edn, Upper Saddle River, NJ: Prentice-Hall.

Taylor, C.R. and Wiggins, S.N. (1997), 'Competition or compensation: supplier incentives under the American and Japanese subcontracting systems', *American Economic Review*, **87** (4), 598–618.

Taylor, F.W. (1947), *Scientific Management: comprising Shop Management, the Principles of Scientific Management [and] Testimony before the Special House Committee*, Westport, CT: Greenwood Press, 1972 edn.

Teece, D.J. (1998), 'Design issues for innovative firms: bureacracy, incentives and industrial structure', in Chandler, A.D. Jr, Hagstrom, P. and Solvell, O. (eds), *The Dynamic Firm, The Role of Technology, Strategy, Organization, and Regions*, Oxford and New York: Oxford University Press.

Tersine, R.J. (2004), 'The primary drivers for continuous improvement: the reduction of the triad of waste', *Journal of Managerial Issues*, Spring, **16** (1), 15–28.

Tezuka, H. (1997), 'Success as the source of failure? Competition and cooperation in the Japanese economy', *Sloan Management Review*, Winter, 83–93.

Thomas, M.D. (1972), 'Growth pole theory: an examination of some of its basic concepts', in Hansen, N.M. (ed.), *Growth Centers in Regional Economic Development*, New York: Free Press.

Thomson, A.A. and Gamble, J.E. (2001), 'Dell Computer Corporation: strategy and challenges for the 21st century', in Thomson, A.A. Jr and Strickland, A.J. III, *Strategic Management, Concepts and Cases*, 12th edn, Boston, MA: McGraw-Hill/Irwin.

Thomson, A.A. Jr and Strickland, A.J. III (2001), *Strategic Management, Concepts and Cases*, 12th edn, Boston, MA: McGraw-Hill/Irwin.

Thurow, L.C. (1996), *The Future of Capitalism, How Today's Economic Forces Will Shape Tomorrow's World*, St Leonards, NSW: Allen and Unwin.

Tisdell, C. (1996), *Bounded Rationality and Economic Evolution, a Contribution to Decision Making, Economics and Management*, Cheltenham, UK and Brookfield, USA: Edward Elgar.

Tobin, J. (1969), 'A general equilibrium approach to monetary theory', *Journal of Money, Credit and Banking*, **1**, 15–29.

Tomer, J.F. (1987), *Organizational Capital: The Path to Higher Productivity and Well-being*, New York: Praeger.

Tomer, J.F. (1990), 'Developing world class organization: investing in organizational capital', *Technovation*, **10** (4), 253–63.

Tomer, J.F. (1998a), 'Organizational capital and joining-up: linking the individual to the organization and to society', *Human Relations*, **51** (6), 825–46.

Tomer, J.F. (1998b), 'Beyond the machine model of the firm, toward a holistic human model' *Journal of Socio-Economics*, **27** (3), 323–40.

Tomer, J.F. (2001), 'Understanding high-performance work systems: the joint contribution of economics and human resource management', *Journal of Socio-Economics*, **30**, 63–73.

Tomer, J.F. (2002), 'Human well-being: a new approach based on overall and ordinary functionings'. *Review of Social Economy*, March, **60** (1), 23–45.

Triplett, J.E. (1999), 'The Solow productivity paradox: what do computers do to productivity?', *Canadian Journal of Economics*, April, **32** (2), 309–34.

Tsutsui, W.M. (1998), *Manufacturing Ideology, Scientific Management in Twentieth-Century Japan*, Princeton, NJ: Princeton University Press.

Tugwell, R.G. (1927), *Industry's Coming of Age*, New York: Harcourt Brace.

Turner, L. (1994), 'Japan suffers in the slump', *International Management*, March, **49** (2), 30–2.

United Nations Centre on Transnational Corporations (UNCTC), (1985), *Transnational Corporations in World Development, Third Survey*, London: UN and Graham and Trotman.

United Nations Industrial Development Organization (UNIDO) (2000), *Industrial Statistics Database, 3–digit ISIC 2000*, CD.

Urwick, L. and Brech, E.F.L. (1994), *The Making of Scientific Management, Vol. 1, Thirteen Pioneers*, Bristol: Thoemmes Press.

Van De Ven, A.H. (1995), 'The development of an infrastructure for entrepreneurship' in Bull, I., Thomas, H. and Willard, G. (eds), *Entrepreneurship, Perspectives on Theory Building*, Oxford: Pergamon.

Van De Ven, A.H. and Poole, M.S. (1995), 'Explaining development and change in organizations', *Academy of Management Review*, **20** (3), 510–40.

VDA (Association of the German Automotive Industry) (2000), http://www.vda.de/en/aktuell/statistk/auto+wirtschaft/schluesselbranche.html, 22 November, 2 pages.

Von Tunzelmann, G.N. (1995), *Technology and Industrial Progress, The Foundations of Economic Growth*, Aldershot, UK and Brookfield, US: Edward Elgar.

Wallis, J.J. and North, D.C. (1986), 'Measuring the transaction sector in the American Economy, 1870–1970', in Engerman, S.L. and Gallman, R.E. (eds), *Long-term Factors in American Economic Growth*, Chicago: University of Chicago Press.

Watanabe, S. (1991), 'The Japanese quality control circle: why it works', *International Labor Review*, **130** (1), 57–80.

Wheatley, M.J. (1994), *Leadership and the New Science, Learning about Organization from an Orderly Universe*, San Francisco, CA: Berrett-Koehler.

White, R.E., Pearson, J.N. and Wilson, J.R. (1999), 'JIT manufacturing: a survey of implementations in small and large US manufacturers', *Management Science*, January, **45** (1), 1–15.

Whittaker, D.H. (1997), *Small Firms in the Japanese Economy*, Cambridge: Cambridge University Press.

Wilken, P.H. (1979), *Entrepreneurship, a Comparative and Historical Study*, Norwood: Ablex.

Williamson, O.E. (1975), *Markets and Hierarchies*, New York: Free Press/ Macmillan.

Williamson, O.E. (1980), 'The organization of work', *Journal of Economic Behavior and Organization*, **1,** 5–38.

Williamson, O.E. (1985), *The Economic Institutions of Capitalism: Firms, Markets, Relational Contracting*, New York: Free Press.

Williamson, O.E. (1994), 'Strategizing, economizing, and economic organization' in Rumelt, R.P., Schendel, D.E. and Teece, D.J. (eds), *Fundamental Issues in Strategy, a Research Agenda*, Boston, MA: Harvard Business School Press.

Williamson, O.E. (1997), 'Hierarchies, markets and power in the economy: an economic perspective', in Menard, C. (ed.), *Transaction Cost Economics, Recent Developments*, Cheltenham, UK and Lyme, USA: Edward Elgar.

Wilson, J.M. (1996), 'Henry Ford: a just-in-time pioneer', *Production and Inventory Management Journa*l, second quarter, 26–31.

Wollin, A. (1999), 'Punctuated equilibrium: reconciling theory of evolutionary and incremental change', *Systems Research and Behavioral Science*, July, **16** (4), 359–66.

Womack, J.P. and Jones, D.T. (1996), *Lean Thinking, Banish Waste and Create Wealth in Your Corporation*, London: Touchstone Books.

Womack, J.P., Jones, D.T. and Roos, D. (1990), *The Machine that Changed the World*, New York: Rawson Associates.

Wood, S. (1989), 'The Japanese management model, tacit skills in shop floor participation', *Work and Occupations*, **16** (4), 446–60.

World Bank Group (2001), *World Development Indicators*, various issues.

Wren, D.A. (1979), *The Evolution of Management Thought*, New York: John Wiley & Sons.

Xenophon (*circa* 370 BC) (1990), *Oeconomicus* (English title: *The Estate Manager*), Harmondsworth: Penguin Classics.

Yang, B. and Lester, D. (1995), 'New directions for economics', *Journal of Socio-Economics*, **24** (3), 433–46.

Yang, X. and Borland, J. (1991), 'A microeconomic mechanism for economic growth', *Journal of Political Economy*, **99** (3), 460–82.

Yearout, S.L. (1992), 'The international quality study reveals which countries lead the race for total quality', *Journal of European Business*, March/April, 27–30.

Young, A.A. (1928), 'Increasing returns and economic progress', *Economic Journal*, December, **38** (152), 529–41.

Young, S.M. (1992), 'A framework for successful adoption and performance of Japanese manufacturing practices in the United States', *Academy of Management Review*, **17** (4), 677–700.

Yui, T. and Nakagawa, K. (1989), *Japanese Management in Historical Perspective*, Tokyo: University of Tokyo Press.

Zucker, L. (1986), 'Production of trust: institutional sources of economic structure, 1840–1920', *Research in Organizational Behavior*, **8**, 53–111.

Index